Flip Chip
Technologies

Electronic Packaging and Interconnection Series
Charles M. Harper, Series Advisor

ALVINO • *Plastics for Electronics*

CLASSON • *Surface Mount Technology for Concurrent Engineering and Manufacturing*

GINSBERG AND SCHNORR • *Multichip Modules and Related Technologies*

HARPER • *Electronic Packaging and Interconnection Handbook*

HARPER AND MILLER • *Electronic Packaging, Microelectronics, and Interconnection Dictionary*

HARPER AND SAMPSON • *Electronic Materials and Processes Handbook, 2/e*

HWANG • *Modern Solders and Soldering for Competitive Electronics Manufacturing*

LAU • *Ball Grid Array Technology*

LICARI • *Multichip Module Design, Fabrication, and Testing*

SERGENT AND HARPER • *Hybrid Microelectronics Handbook, 2/e*

STEARNS • *Flexible Printed Circuitry*

SOLBERG • *Design Guidelines for Surface Mount Technology, 2/e*

Related Books of Interest

BOSWELL • *Subcontracting Electronics*

BOSWELL AND WICKAM • *Surface Mount Guidelines for Process Control, Quality, and Reliability*

BYERS • *Printed Circuit Board Design with Microcomputers*

CAPILLO • *Surface Mount Technology*

CHEN • *Computer Engineering Handbook*

COOMBS • *Printed Circuits Handbook, 4/e*

DI GIACOMO • *Digital Bus Handbook*

FINK AND CHRISTIANSEN • *Electronics Engineers' Handbook, 3/e*

GINSBERG • *Printed Circuits Design*

JURAN AND GRYNA • *Juran's Quality Control Handbook*

MANKO • *Solders and Soldering, 3/e*

RAO • *Multilevel Interconnect Technology*

SZE • *VLSI Technology*

VAN ZANT • *Microchip Fabrication, 3/e*

To order or receive additional information on these or any other McGraw-Hill titles, in the United States please call 1-800-822-8158. In other countries, contact your local McGraw-Hill representative.   **BC15XXA**

Flip Chip Technologies

John H. Lau
Editor

McGraw-Hill

New York   San Francisco   Washington, D.C.   Auckland   Bogotá
Caracas   Lisbon   London   Madrid   Mexico City   Milan
Montreal   New Delhi   San Juan   Singapore
Sydney   Tokyo   Toronto

Library of Congress Cataloging-in-Publication Data

Flip chip technologies / John H. Lau. editor.
      p.   cm. — (Electronic packaging and interconnection series)
    Includes bibliographic references and index.
    ISBN 0-07-036609-8 (hc)
    1. Microelectronic packaging. 2. Multichip modules
(Microelectronics)—Design and construction. I. Lau, John H.
II. Series.
TK7874.F5897   1995
621.381'046—dc20   95-44649
                                                        CIP

McGraw-Hill

A Division of The McGraw·Hill Companies

ISBN 0-07-036609-8

*The sponsoring editor for this book was Stephen S. Chapman, the
editing supervisor was David E. Fogarty, and the production supervisor
was Donald Schmidt. It was set in Century Schoolbook by Ron Painter
of McGraw-Hill's Professional Book Group composition unit.*

Printed and bound by Quebecor-Book Press

This book is printed on acid-free paper.

Contents

Chapter 8. Anisotropic Conductive Adhesive for Fire-Pitch Flip Chip Interconnections   289

Chapter 9. Flip Chip Interconnection Technology Using Anisotropic Conductive Adhesive Films   301

Chapter 10. Anistropic Conductive Flip Chip-on-Glass Technology   317

Foreword

The electronics industry is one of the most dynamic, fascinating, and important industries. In a relatively short period of time, it has become the largest and most pervasive manufacturing industry in the developed world. It provides many products that affect our daily lives, e.g., telephones, high-definition televisions, electronic organizers, computers, wireless phones, pagers, portable electronics products, video camcorders, CD-ROMs (compact disc–read-only memory), PCMCIA (Personal Computer Memory Card International Association) cards, and multimedia products.

The major trend in the electronics industry today is to make products more personal by making them smarter, lighter, smaller, thinner, shorter, and faster, while at the same time making them more friendly, functional, powerful, reliable, robust, innovative, creative, and less expensive. As the trend toward minature and compact products continues, the introduction of products that are more user-friendly and contain a wider variety of functions will provide growth in the market. One of the key technologies that is helping to make these product design goals possible is electronics packaging and assembly technology.

Electronics packaging is an art based on the science of establishing interconnections ranging from zero-level packages (chip-level connections), first-level packages (either single-chip or multichip modules), second-level packages (e.g., printed circuit boards), and third-level packages (e.g., mother boards). There are at least three popular chip-level connections, namely, face-up wire bonding, face-up tape-automated bonding, and flip chip. Among these three technologies, flip chip provides the highest packaging density and performance and the lowest packaging profile. Unfortunately, for most of the practicing engineers and managers, flip chip is the least understood of the chip connection technologies.

Thus, there was an urgent need, both in industry and in universities, to create a comprehensive book on the current state of knowledge in design, materials, process, manufacturing, and reliability of flip chip technology. This book should be written for everyone who can quickly learn about the technology, understand the tradeoffs, and make system-level decisions.

To meet this need, Dr. John H. Lau has assembled an outstanding team of packaging experts from industry and universities. Together, they have produced *Flip Chip Technologies*, an excellent book for both industry and university use. It is equally appropriate as an introduction to flip chip technologies for those just entering the field and as an up-to-date reference for those already engaged in packaging design and development.

This book deals with the subject of flip chip technology in three categories: the classical solder-bumped flip chip technologies, the next generation flip chip technologies, and known-good-die testing for multichip module applications. I join the authors in hoping that this book will help focus the attention of practicing engineers, managers, and scientists, as well as faculty and students, on the complex flip chip packaging challenges which must be solved. I also hope that a future generation of engineers, scientists, and students will continue to further the science and engineering of flip chip technologies. Armed with their predecessors' knowledge and their own creativity, there is no doubt that these wishes will come true.

Terry T. M. Gou
Chairman and C.E.O.
Hon Hai Precision Industry Co., Ltd.

Preface

Flip chip technology, in this book, is defined as mounting the chip to a substrate with any kind of interconnect materials and methods (e.g., fluxless solder bumps, tape-automated bonding (TAB), wire interconnects, conductive polymers, anisotropic conductive adhesives, metallurgy bumps, compliant bumps, and pressure contacts), as long as the chip surface (active area) is facing to the substrate.

One of the earliest flip chip technologies was introduced by IBM (solder-bumped flip chip technology) in the early 1960s, as a possible replacement for the expensive, unreliable, low-productivity, and manually operated face-up wire-bonding technology. Until recently, however, research and development efforts into flip chip technology have not been particularly aggressive because high-speed automatic wire bonders met the needs of the semiconductor device to the next-level package interconnection.

The past decade witnessed an explosive growth in the research and development efforts devoted to flip chip technology (more than 1000 patents were granted) as a direct result of the higher requirements of package density, performance, and interconnection; the limitation of face-up wire bonding technology; and the growing use of multichip module technology. (There are about 600 flip chip patents listed at the end of Chapter 1.) By comparing with the popular face-up wire-bonding technology and face-up TAB technology, flip chip technology provides shorter possible leads, lower inductance, higher frequency, better noise control, higher density, greater input/output (I/O), smaller device footprints, and lower profile.

Today, the most well-known and successful flip chip technology is IBM's solder-bumped flip chip technology. However, it will not be emphasized in this book because it has been published in many technical journals (see, for example, Refs. 82-113 in Chapter 1) and books, e.g., *Microelectronics Packaging Handbook* edited by Tummala and Rymaszewski, *Principles of Electronic Packaging* edited by Seraphim,

Lasky, and Li, and *Chip On Board Technologies for Multichip Modules* edited by Lau. The focus in this book is to present the next generation of flip chip technologies developed within the past 5 years.

This book is divided into eight parts. The first part (Chapter 1) presents a brief introduction to integrated circuit (IC) chip yield, known good die (KGD), unknown bad die (UBD), multichip module (MCM) yield, and some of the necessary information for flip chip technologies. Also, a couple methods for estimating the thermal fatigue life of flip chip solder joints are provided.

The second part of this book discusses solder-bumped flip chip technology. In most of the classical fine-pitch solder-bumped flip chip technology, flux residues are entrapped underneath the chip and are very difficult to clean. These residues could degrade the flip chip assembly realiability with corrosion effects, gross solder voids, passivation delaminations, etc. Thus, a technique that will eliminate both flux and postsolder cleaning is desperately needed for low-cost and reliable flip chip applications. A new fluxless process called PADS (plasma-assisted dry soldering) was developed by MCNC and is presented by Nick Koopman and Sundeep Nangalia in Chapter 2. PADS is a dry pretreatment that converts the surface oxides to oxyfluorides. This conversion film passivates the solder surface and breaks when the solder melts in inert (nitrogen) oven or even in oxidizing (air) ambients.

Some important aspects of the electrical and thermal performance of solder-bumped flip chip assemblies are discussed by Ravi Sharma and Ravi Subrahmanyan in Chapter 3. Also, based on a damage integral approach, they outline a methodology to develop accelerated test strategies for evaluating new solder-bumped flip chip technologies.

Flip chip, especially for very large and high I/O chip, has been recognized as one of the most important technologies necessary to meet the stringent requirements of workstation computer applications. In Chapter 4, Tom Chung, Tom Dolbear, and Richard Nelson examine the objectives, approaches, key technology elements, and some of the test results of the thermal management, manufacturability, and reliability of very large and high I/O flip chip technology.

The needs for increased interconnection I/O and performance have led AT&T to develop a platform which consists of a composite of substrate, attachment, and packaging that they called "microinterconnect technology." It exploits the unique capabilities of the silicon MCM-D interconnection architectures. In Chapter 5, Yinon Degani, T. Dixon Dudderar, Eobert Frye, King Tai, Maureen Lau, and Byung Joon Han describe the design, development, and optimization of these manufacturing processes for cost-effective large-volume assembly of flip chip silicon-on-silicon MCM-D through the use of a custom solder-bumped flip chip technology and high-speed automated assembly.

The third part of this book presents the conductive polymer flip chip technology. It is known that conductive adhesive polymer materials have been used by the electronics industry for a long time. However, these materials are only now beginning to be recognized for high-volume and commercial flip chip applications. Chapter 6, by Richard Estes and Frank Kulesza, presents an overview of various conductive adhesive polymer materials and their applications to flip chip technology. Some of the design, process, electrical, and mechanical reliability data are also provided.

Some of the drawbacks of conductive adhesive flip chip technology are nonreworkable, and movement in the z-axis is caused by the thermal expansion coefficient and water absorption of the adhesive. A new compliant bump, which are polymer-core bumps with a metal coating, was developed by MCC. In Chapter 7, Mark Breen, Diana Duane, Randy German, Kathryn Keswick, and Rick Nolan describe polymer bump formation and metallization, assembly and rework processes, and fabrication cost pictures. Also, some reliability data are presented.

The fourth part of this book examines the applications of anisotropic conductive materials for flip chip technology. In general, the anisotropic conductive materials come in two different forms, namely, ACF (anisotropic conductive film) and ACA (anisotropic conductive adhesive paste). They are also called z-axis conductive adhesive because they are designed to be conductive between the electrodes of the chip and substrate in the vertical direction and nonconductive (no shorts) in the horizontal direction.

Fujitsu developed a new ACA material for fine pitch interconnection between a flip chip and a printed circuit board (PCB). Their ACA material consists of conductive particles coated with a thin dielectric resin. During bonding, the high pressure between the pad on the chip and the corresponding pad on the PCB destroys the resin layer between the pads. The electrical connection is established through the Ag conductive particles. In Chapter 8, Hiroaki Date, Yuko Hozumi, Hideshi Tokuhira, Makoto Usui, Eiji Horikoshi, and Takehiko Sato present the characteristics of their ACA material and the reliability data of the ACA interconnection between the chip and PCB.

A few ACF materials have been developed by Hitachi for flip chip applications. In Chapter 9, Itsuo Watanabe, Naoyuki Shiozawa, Kenzo Takemura, and Tomohisa Ohta describe the principle of ACF interconnects and the role of conducting particles dispersed in ACF through voltage, current, resistance measurements. Also, they disclose a double-layer ACF material which consists of a nonfilled adhesive layer and a conducting particle-filled adhesive layer.

The applications of ACA and ACF materails to flip chip on glass (COG) technology are presented in Chapter 10. The processes and materials for several flip COG assembly methods using ACA or ACF have been discussed along with each method's technical merits and demerits by Chang Hoon Lee.

The fifth part of this book presents some face-down wired flip chip technologies. It is known that more than 90 percent of the chips used today are face-up wire bonded on the substrate. Thus, making the Au stud with a wire bonder should be very straightforward, and the reliability data are already available. However, as mentioned earlier, face-down chip technology has many advantages over the face-up chip technology. Thus, in this book a few wired flip chip technologies are presented.

Less than five years ago a new assembly process called BIP (bonded interconnect pin) technology was developed by Raychem Corporation. It utilizes thermosonic ball bonding and reflow soldering to create electrical interconnections between the flip chip and the substrate. In Chapter 11, Stacey Baba and William Carlomagno present the design, fabrication, and assembly processes of the BIP technology. Also some reliability data of the BIP assembly under temperature, mechanical, shock and vibration, and stabilization bake conditions are provided.

Fujitsu developed a slightly different wired flip chip technology and put it into production for their low-end computer products. They used a wire bonder to make the stud bumps and conductive paste to attach the stud bumps on the PCB. Reliability of the assembly is enhanced by underfill encapsulant. Kazuhisa Tsunoi, Toshihiro Kusagaya, and Hidehiko Kira (Chapter 12) describe the structure, stud-bump formation and leveling, conductive paste application, assembly process, and rework. Reliability testing results are also presented.

Flip chip technologies using the wire bonder can only apply to peripheral array pads. For area array pads, FCPT (Fujitsu Computer Packaging Technologies) developed a new wire interconnect technology (WIT) for high-density and performance flip chip application. WIT "grows" very small metal posts (usually 10 μm in diameter and 50 μm long) from the chip pad and the other end of the posts are soldered on the substrate. In Chapter 13, Larry Moresco, David Love, William Chou, and Ven Holalkere disclose their WIT fabrication and assembly process, and reliability testing results. The flip chip WIT technical characteristics are also presented through electrical, mechanical stress, heat transfer, α-particle radiation, and routing analyses and measurements.

The sixth part of this book presents an area array TAB type of flip chip technology. By combining most advantages of area array TAB, wire bonding, and flip chip technologies, Tessera develped a small, thin, and high-performance CSP (chip size package) called micro-ball grid array (μBGA). It allows a chip pitch as small as 50 μm (which is competitive with advanced TAB pad pitch) and the flexible gold beam lead is assembled to the chip pad with conventional thermosonic wire-bonding equipment. The leads fan inwardly from the chip pads to a NiAu (or solder) bump array that serves to make connection to the next level package. A high level of reliability can be achieved with the elastomer compliant layer, which also provides the opportunity for full test and burn-in for KGD prior to final assembly. In Chapter 14, Thomas DiStefano and Joseph Fjlstad present the design, material, bump metallurgy, and assembly process of their μBGA. The electrical and thermal performance, reliability testing, and burn-in of the μBGA are also provided.

The seventh part of this book discusses the metallurgy-bumped flip chip technology. As mentioned earlier, a fluxless and no-clean process is desperately needed for reliable flip chip applications. In Chapter 15, Elke Zakel and Herbert Reichl describe their bumping processes and characteristics for the Au (by electroplating), AuSn (by electroplating), and NiAu (by electroless) metallurgies, and Au stud and solder-stud bumps. Also, the assembly processes and test data of flip chip assemblies using Au, AuSn, NiAu, Au-stud, and solder-stud bumps are provided.

The last part of this book (Chapters 16 and 17) presents a very important subject in the application of flip chip technology to multichip module packaging. In Chapter 16, Larry Gilg addresses the KGD definition and standards and the cost for producing KGD. Various methods (such as process control, statistical sampling, test every die, test and burn-in carriers, and wafer-level burn-in) of assuring KGD for flip chip application are also presented.

In Chapter 17, Glenn Rinne explains the principles of burn-in for flip chip ICs. Some key flip chip burn-in considerations such as alignment, contact area, differential expansion, oxidation, solid-state diffusion, bump integrity, and plastic and creep deformation of the bump are also provided.

Some duplication of material between chapters is necessary if each chapter is to offer the reader all information essential for understanding the subject matter. An attempt has been made to provide a degree of uniformity in perspectives, but diverse views on certain aspects of flip chip technology are a reality. I hope that their inclusion here is seen as an unvarnished reflection of the state of the art and a useful feature of the book. However, since the final responsibility for the selection and treatment of materials falls on my shoulders, I will receive the blame for any deficiencies.

For whom is this book intended? Undoubtedly it will be of interest to three groups of specialists: (1) those who are active or intend to become active in research and development of flip chip technologys;

(2) those who have encountered practical flip chip technology problems and wish to understand and learn more methods of solving such problems; and (3) those who have to choose a creative, high-performance, robust, and cost-effective packaging technique for their interconnect system. This book also can be used as a text for college and graduate students who could become our future leaders, scientists, and engineers in electronics industry.

I hope this book will serve as a valuable source of reference to all those faced with the challenging problems created by the ever more expanding use of flip chip technology in electronics packaging and interconnection. I also hope that it will aid in stimulating further research and development on ICs, bare wafer/chip testing and burn-in, solder bumps, metallurgy bumps, compliant bumps, polymer bumps, anisotropic conductive adhesives, face-down wire bonding, face-down TAB, substrates, chip attachment, encapsulant and under-fill epoxy, flux, printing, placement, mass reflow, cleaning, inspection, rework, testing, electrical and thermal managements, reliability, equipment, material, process, design, and more sound use of flip chip technology in either single-chip or multichip packaging applications.

The organizations that learn how to design and manufacture flip chip technologies in their interconnect systems have the potential to make major advances in electronics packaging and to gain great benefits in cost, performance, quality, size, and weight. It is our hope that the information presented in this book may assist in removing road-blocks, avoiding unnecessary false starts, and accelerating design, material, and process development of these technologies. The flip chip technologies are limited only by business constraints, the ingenuity and imagination of engineers, the vision of management, and infrastructures.

John H. Lau, Ph.D., PE
Electronics Packaging Services

Acknowledgments

Development and preparation of *Flip Chip Technologies* was facilitated by the efforts of a number of dedicated people at McGraw-Hill. I would like to thank them all, with special mention to Cathy Hertz for her skillful copyediting, and to Roger Kasunic, David Fogarty, and Donna Namorato for their unswerving support and advocacy. My special thanks to Steve Chapman who made my dream of this book come true by effectively sponsoring the project and solving many problems that arose during the book's preparation. It has been a great pleasure and fruitful experience to work with them.

The material in this book has clearly been derived from many sources including individuals, companies, and organizations, and the various contributing authors have attempted to acknowledge, in the appropriate parts of the book, the assistance that they have been given. It would be quite impossible for them to express their thanks to everyone concerned for their cooperation in producing this book, but on their behalf, I would like to extend due gratitude. Especially, I want to thank several professional societies and publishers for permitting us to reproduce some of their illustrations and information in this book. For examples, the American Society of Mechanical Engineers (ASME) Conferences, Proceedings, and Transactions (*Journal of Electronic Packaging*); the Institute of Electrical and Electronic Engineers (IEEE) Conferences, Proceedings, and Transactions (*Hybrids, Packaging, and Manufacturing Technology*); the International Society of Hybrid Microelectronics (ISHM) and the International Electronic Packaging Society (IEPS) Conferences, Proceedings, and Transactions (*Microcircuits & Electronic Packaging*); American Society of Metals (ASM) Conferences, Proceedings, and books (e.g., *Electronic Materials Handbook*, Volume 1, *Packaging*); the National Electronic Packaging Conferences (NEPCON) and Proceedings; the Surface Mount International (SMI) Conferences and Proceedings; the *IBM Journal of Research and*

Development; Electronic Packaging & Production; Circuits Assembly; Surface Mount Technology; Connection Technology; Solid State Technology; Circuit World; Hybrid Circuits; and *Soldering and Surface Mount Technology.*

I express my deep appreciation to the 47 contributing authors, all experts in their respective fields, for their many helpful suggestions and their cooperation in responding to requests for revisions. Their depth of knowledge, dedication, and patience have been demonstrated throughout the process of preparing this book. Working with them has been an adventure and a privilege, and I learned a lot about flip chip and other electronic packaging technologies from them. Their brief technical biographies are presented at the end of this book.

Lastly, I want to thank my former employer, Hewlett-Packard, and former managers Anita Danford and Helmut Kroener, for providing an excellent environment in which completing this book was possible. I also want to thank my colleagues (both in HP and the industry) for their stimulating discussions on electronic packaging and interconnections. I learned a lot from them. Finally, I want to thank my daughter (Judy, who also helped to design the book cover) and my wife (Teresa) for their love, consideration, and patience by allowing me to work on many weekends for this private project. Their simple belief that I am making my small contribution to the electronics industry was a strong motivation for me, and to them I have dedicated my efforts on this book.

John H. Lau, IEEE Fellow
Palo Alto, California

A Brief Introduction to Flip Chip Technologies for Multichip Module Applications

John H. Lau

1.1 Introduction

Figure 1.1 is a schematic representation of electronic package hierarchy.[1,2] Although the wafer is not typically included in the packaging hierarchy, it is included in Fig. 1.1 to show where the integrated cir-

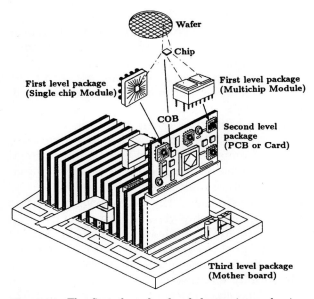

Figure 1.1 The first three levels of electronics packaging. Although not a standard part of the packaging hierarchy, the wafer shows the origin of the IC chip.

cuit (IC) chip originates. Packaging focuses on how a chip (or many chips) is (or are) packaged efficiently and reliably.[1-172]

The chip is not an isolated entity. It communicates with other chips in the circuit through an input/output (I/O) system of interconnects, and the delicate chip and its embedded circuitry are dependent on the package for support and protection. Consequently, the major functions of the electronic package are to (1) provide a path for the electrical current that powers the circuits on the chip, (2) distribute the signals onto and off of the silicon chip, (3) or remove the heat generated by the circuit, and (4) support and protect the chip from hostile environments.

Packaging is an art based on the science of establishing interconnections ranging from zero-level packages (chip-level connections), to first-level packages (either single- or multichip modules), second-level packages [e.g., printed-circuit boards (PCBs)], and third-level packages (e.g., motherboards; see Fig. 1.1). This book focuses on the chip, first, and second-level interconnections, with particular emphasis on flip chip technologies. In the context of this book, *flip chip* is defined as mounting the chip on a substrate using various interconnect materials and methods (e.g., tape-automated bonding, fluxless solder bumps, wire interconnects, isotropic and anisotropic conductive adhesives, metal bumps, compliant bumps, and pressure contacts), as long as the chip surface (circuit) is facing (oriented in the direction of) the substrate.

The requirements for smaller, more compact products and high density for high-speed circuitry drive the design of packages to higher I/O and smaller package size. In general these two design objectives are accomplished by reducing the pad size and spacing (pitch) on the chip. The presence of area array pads on the chip will, of course, compound the density, and this increased density means increased performance and power in smaller packaging systems, allowing the manufacture of lighterweight, smaller, and higher-performance products at lower cost. Some examples are videocameras and camcorders, electronic organizers, personal computers (PCs), notebook PCs, subnotebook PCs, laptop PCs, palmtop PCs, PCs with built-in portable phones, PCMCIA (Personal Computer Memory Card International Association) cards, cellular phones, pagers, portable electronic products, audiovisual products, and other multimedia products.

This chapter briefly introduces some of the necessary information for flip chip technology, including IC chip yield, known good die, and multichip module yield and presents examples for estimating thermal fatigue life of solder-bumped flip chips using both the Coffin-Manson and fracture mechanics methods. Some of the next-generation flip chip technologies are also briefly mentioned. (For more information, read Chaps. 2 through 17.) This chapter concludes with a list of 573 flip chip patents.

1.2   IC Chip Yield

1.2.1   The wafer

Figure 1.2 schematically depicts a typical silicon wafer.[51] The overall orientation of wafer cells (referred to as the *wafer step plan*) and the wafer cells themselves are the focus of wafer design and significantly impact bump mask design for wafer bumping. Typical wafer diameters are 3 in (75 mm), 4 in (100 mm), 5 in (125 mm), 6 in (150 mm), 8 in (200 mm), and the emerging 12 in (300mm); coordinates x and y are the dimensions of the chips on the wafer. A wafer usually consists of many product unit cells (PUCs), a few test unit cell (TCs), and a few combined mask alignment unit cells (CMAs) (see Fig. 1.2). (For more information, please consult Ref. 51.)

1.2.2   The number of possible chips per wafer

The physical possible number of undamaged chips (N_c) stepped from a wafer may be given by[67]

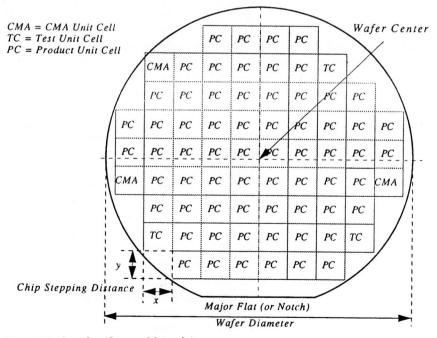

Figure 1.2   A wafer. (*Source: Motorola*)

$$N_c = \pi \frac{\left[\Phi-(1 + \Theta) \sqrt{A/\Theta}\right]^2}{4A} \tag{1.1}$$

where

$$A = xy \tag{1.2}$$

and

$$\Theta = \frac{x}{y} \geq 1 \tag{1.3}$$

In Eq. (1.1), x and y are the dimensions of a rectangular chip [in millimeters (mm)], with x no less than y; Θ is the ratio between x and y; Φ is the wafer diameter (mm); and A is the area of the chip [in square millimeters (mm^2)].

1.2.3   Some well-known IC yield formulas

For most IC manufacturing processes, it is well known[52–63] that the defect density is a random variable, varying, for example, from lot to lot, wafer to wafer, and center to edge on a wafer. To account for these variabilities in manufacturing defect densities, Murphy[52] proposed the following yield integral

$$Y = \int_0^\infty e^{-AD} f(D)\, dD \tag{1.4}$$

where Y = yield probability (or the probability of zero defects)
A = defect-sensitive area (or susceptible area, designations-sensitive area, and critical area)
D = defect density, having the units of defects per unit area
$f(D)$ = defect density distribution function

Experience with both device yield variations and spot defect counts led Murphy to suggest that the defect density distribution function should be bell-shaped, i.e., Gaussian distribution. However, with $f(D)$ = Gaussian, Eq. (1.1) cannot be integrated in closed form.

First, Murphy approximated the Gaussian distribution as a δ function (representing zero-width distribution); Eq. (1.4) then leads to the Poisson yield model

$$Y = e^{-AD_0} \tag{1.5}$$

Next, Murphy approximated the Gaussian distribution as a triangular function (representing intermediate-width distribution) and obtained

$$Y = \left(\frac{1-e^{-AD_0}}{AD_0}\right)^2 \qquad (1.6)$$

Finally, Murphy approximated the Gaussian distribution as a uniform function (representing constant and maximum-width distribution), and the yield became

$$Y = \frac{1-e^{-2AD_0}}{2AD_0} \qquad (1.7)$$

Seeds[53] assumed the $f(D)$ as an exponential function and obtained

$$Y = \frac{1}{1 + AD_0} \qquad (1.8)$$

Finally, Stapper[54-57] assumed $f(D)$ to be a gamma distribution and Okabe et al.[58] assumed $f(D)$ to be an Erlang function, and they arrived at the same form of the negative binomial yield equation:

$$Y = \frac{1}{(1 + AD_0/c)^c} \qquad (1.9)$$

For Stapper,[54-57] c is a parameter of the gamma distribution and is equal to the inverse square of the coefficient of variation of that distribution. For Okabe et al.,[58] c is an integer representing the number of process steps. In general, the value of c of the gamma distribution is smaller than the number of process steps. In Eqs. (1.5) to (1.9), D_0 is the mean of the defect density distribution, and AD_0 is the average number of faults in an IC with defect-sensitive area A.

Figures 1.3 and 1.4 are graphical representations of Eqs. (1.5) to (1.8); Eq. (1.9) was discussed and used to calculate the number of yielded chips in wafers by this author in a previous study.[1] It can be seen from Figs. 1.3 and 1.4 that Seeds's yield expression predicted much higher yields than did the others, and Seeds's own experimental data confirmed those predictions.[53] Also, Gulett's data[59] led him to a better yield prediction by averaging Seeds's and Murphy's (uniform) results.

1.2.4   A new IC yield formula

Let us consider a half-Gaussian distribution (Fig. 1.5)

$$f(D) = \frac{2}{\pi D_0} \exp\left[-\frac{1}{\pi}\left(\frac{D}{D_0}\right)^2\right], \qquad D \geq 0 \qquad (1.10)$$

Substituting Eq. (1.10) into Eq. (1.4), we have

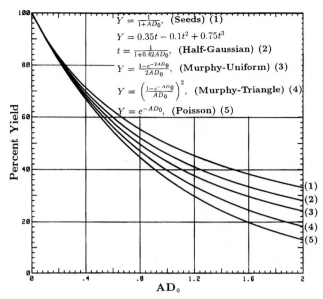

Figure 1.3 Comparison of various yield models with small AD_0.

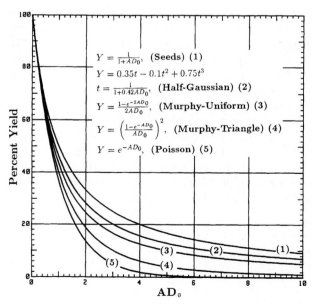

Figure 1.4 Comparison of various yield models with large AD_0.

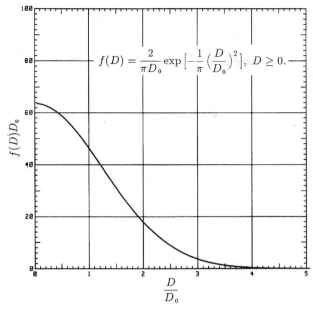

$$f(D) = \frac{2}{\pi D_0} \exp\left[-\frac{1}{\pi}\left(\frac{D}{D_0}\right)^2\right], \quad D \geq 0.$$

Figure 1.5   A half-Gaussian defect density distribution.

$$Y = \frac{2}{\pi D_0} \int_0^\infty \exp\left[-AD - \frac{1}{\pi}\left(\frac{D}{D_0}\right)^2\right] dD \qquad (1.11)$$

Let

$$z = \frac{\sqrt{\pi}\, A D_0}{2} + \frac{D}{\sqrt{\pi}\, D_0} \qquad (1.12)$$

Then, Eq. (1.11) becomes

$$Y = \frac{2}{\sqrt{\pi}} \exp\left[\frac{\pi\,(AD_0)^2}{4}\right] \int_{\sqrt{\pi}\, AD_0/2}^\infty e^{-z^2}\, dz \qquad (1.13)$$

which can be integrated in terms of an error function

$$Y = \exp\left[\frac{\pi\,(AD_0)^2}{4}\right]\left[1 - \mathrm{erf}\left(\frac{\sqrt{\pi}\, AD_0}{2}\right)\right] \qquad (1.14)$$

It should be noted that Eq. (1.14) is different from Eq. (1.12) (of Ref. 60), and provides an opportunity to obtain a couple of approximated closed-form solutions. Noting that[173]

$$\text{erf}\left(\frac{\sqrt{\pi}\,AD_0}{2}\right) = 1 - (a_1 t + a_2 t^2 + a_3 t^3)\exp\left[-\frac{\pi\,(AD_0)^2}{4}\right] + \epsilon \quad (1.15)$$

where

$$|\epsilon| \le 0.000025 \quad (1.16)$$

$$a_1 = 0.34802 \quad (1.17)$$

$$a_2 = -0.09588 \quad (1.18)$$

$$a_3 = 0.74786 \quad (1.19)$$

and

$$t = \frac{1}{1 + 0.42\,AD_0} \quad (1.20)$$

and substituting Eq. (1.15) into Eq. (1.14) and dropping the $|\epsilon| \le 0.000025$ term (which is very insignificant for most IC applications), we have a new yield formula:

$$Y = a_1 t + a_2 t^2 + a_3 t^3 \quad (1.21)$$

Equation (1.21) is plotted in Figs. 1.3 and 1.4 on curve 2. It can be seen that the present results are very close to the average values of Seeds' and Murphy's (uniform) and Gulett's data[59] (Fig. 1.6). Furthermore, Eq. (1.21) can be simplified to

$$Y = 0.35t - 0.1t^2 + 0.75t^3 \quad (1.22)$$

Equation (1.22) is also plotted in Figs. 1.3 and 1.4 on curve 2. It can be seen that Eqs. (1.21) and (1.22) are virtually identical. Thus, for a given AD_0, the yield can be readily determined by Eqs. (1.22) and (1.20).

Since Eq. (1.22) is a cubic equation, the closed-form solution of AD_0 for a given Y can be determined by

$$AD_0 = \frac{0.96 - S_1 - S_2}{0.42(0.04 + S_1 + S_2)} \quad (1.23)$$

where

$$S_1 = \left(r + \sqrt{r^2 + 0.0035}\right)^{1/3} \quad (1.24)$$

$$S_2 = -\left(-r + \sqrt{r^2 + 0.0035}\right)^{1/3} \quad (1.25)$$

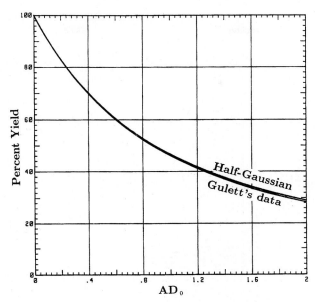

Figure 1.6 Comparison of the present new yield model with Gulett's data.

and

$$r = -0.01 + 0.67Y \qquad (1.26)$$

It should be emphasized that Eqs. (1.22) and (1.23) are simple algebraic equations. They can be easily programmed on a calculator or computer for IC process and package assembly simulation and cost modeling.

1.2.5   Number of yielded chips per wafer

Once the IC chip yield is determined and Lau's chip yield formula is used, the total number of yielded unassembled chips per wafer (N_y) can be obtained by

$$N_y = \pi \frac{\left[\Phi - (1 + \Theta)\sqrt{A/\Theta}\right]^2}{4A} (0.35t - 0.1t^2 + 0.75t^3) \qquad (1.27)$$

Figures 1.7 through 1.12 show the number of yielded chips for various wafer sizes (5 in or 125 mm, 6 in or 150 mm, and 8 in or 200 mm) and mean defect densities ($D_0 = 0.01, 0.02, 0.03, 0.05, 0.1, 0.2$).

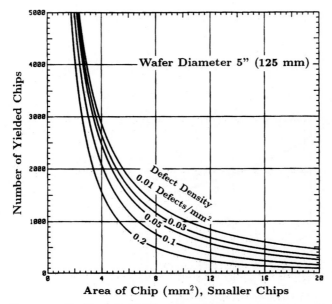

Figure 1.7 Number of yielded chips in a 125-mm wafer (small chip area).

1.3   IC MCM Yield

1.3.1   MCM yield equations

Prediction of the overall probable yield for an assembled multichip module (MCM) is very complicated and difficult.[69–81] It depends on the known good probabilities for the chip, substrate, chip interconnects, system design, manufacturing, and other parameters:

$$Y_{MCM} = Y_C \, Y_S \, Y_I \, Y_D \, Y_M \cdots \qquad (1.28)$$

where Y_{MCM} = overall MCM yield
Y_C = a function of the chip yield
Y_S = substrate yield
Y_I = interconnect yield
Y_D = design yield
Y_M = manufacturing yield

In Eq. (1.28), all the terms on the right-hand side (RHS) are important for the overall MCM yield. For example, even if you have a 100 percent chip yield (all the chips are known good), 100 percent substrate yield (the substrate is perfect), 100 percent interconnect yield

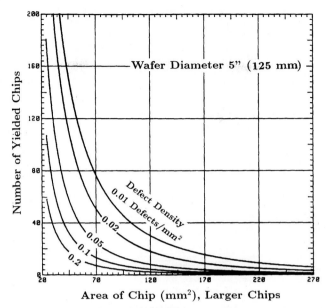

Figure 1.8 Number of yielded chips in a 125-mm wafer (large chip area).

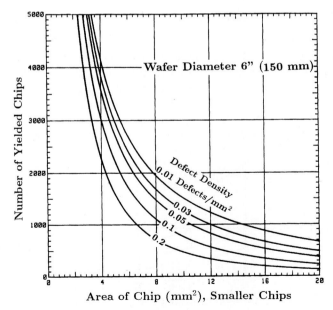

Figure 1.9 Number of yielded chips in a 150-mm wafer (small chip area).

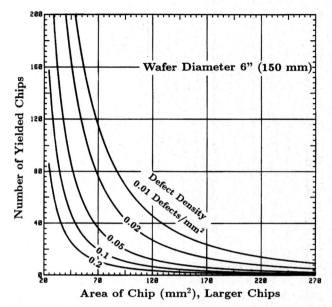

Figure 1.10 Number of yielded chips in a 150-mm wafer (large chip area).

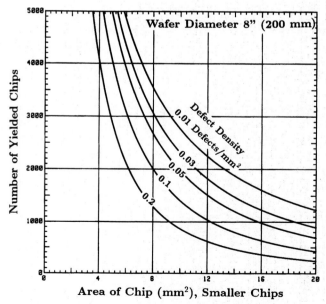

Figure 1.11 Number of yielded chips in a 200-mm wafer (small chip area).

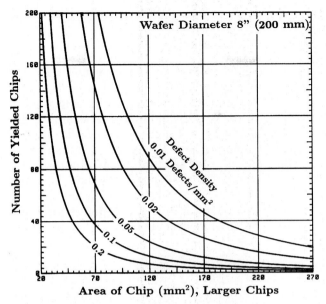

Figure 1.12 Number of yielded chips in a 200-mm wafer (large chip area).

(all the interconnects between the chips and substrate are mechanically solid and electrically sound), and 100 percent manufacturing yield ($\infty\sigma$), the overall MCM yield will be zero if the design yield is zero (e.g., short-circuits).

Although all RHS terms in Eq. (1.28) are critical for overall MCM yield, most of them can be determined at an early preproduction stage and achieved at very high yield. For example, the MCM design, substrate construction, and manufacturing process can be developed, modified, and qualified in house prior to production. (Usually, for a mature MCM assembly process, $Y_D > 99$ percent, $Y_I > 90$ percent, and $Y_M > 95$ percent.) However, Y_C, which is usually out of the control of the system house and thus becomes more critical to the overall MCM yield. This chapter emphasizes the relationship between chip yield and MCM yield.

In direct-chip-attach (DCA) and MCM assemblies, there are usually more than one chip on the substrate. The MCM yield (not the overall MCM yield) depends on the chip yield and the number of chips on the substrate (e.g., see Refs. 1, 2, and 69):

$$Y_C\ (\%) = 100Y^{N_c} \tag{1.29}$$

where Y_C = MCM yield
    Y = chip yield
    N_c = number of chips on the substrate

Equation (1.29) is plotted in Figs. 1.13 and 1.14 for different values of chip yield and chip number. For a five-chip MCM with chip yield $Y = 60$ percent, the MCM yield is $Y_C = 8$ percent. However, if the chip yield is increased from 60 to 95 percent, the MCM yield increases dramatically to 77 percent. It can be seen that "good" or "high" chip yield is very important for MCM and DCA [1] applications. [*Note:* Eq. (1.29) assumes that all the chips have the same IC yield and were not fully tested at high speed and burned in at elevated temperature for known good die prior to MCM assembly.]

If the average number of faults (AD_0) in an IC with defect-sensitive area A is known, and Lau's chip yield formula [Eq. (1.22)] is used, Eq. (1.29) can be written as

$$Y_C (\%) = 100(0.35t - 0.1t^2 + 0.75t^3)^{N_c} \tag{1.30}$$

and t is as given in Eq. (1.20). Equation (1.30) is plotted in Figs. 1.15 and 1.16 for various values of AD_0 and N_c. For example, for a four-chip MCM in which the average number of faults in the chip is $AD_0 = 0.13$, the MCM yield will be 60 percent.

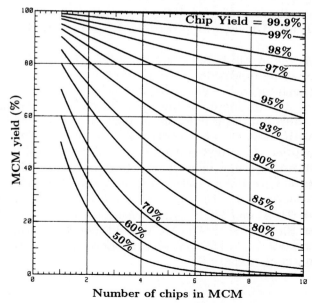

Figure 1.13 MCM yield as a function of chip yield (small number of chips).

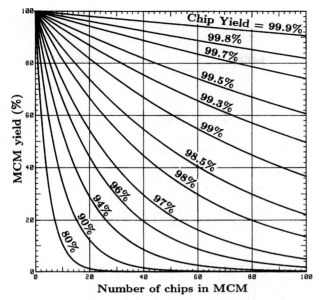

Figure 1.14 MCM yield as a function of chip yield (large number of chips).

If a MCM consists of more than one chip type, the MCM yield is given by (see, e.g., Refs. 2, 67, and 69)

$$Y_C\,(\%) = 100\,(Y_A^{N_A}\ Y_B^{N_B}\ \cdots\ Y_I^{N_I}) \qquad (1.31)$$

where Y_C = MCM yield
$\quad\ \ Y_I$ = chip yield for chip type I
$\quad\ \ N_I$ = number of type I chips on the substrate

For example, assume that there is a six-chip MCM with one microprocessor chip, two ASIC (application-specific IC) chips, and three memory chips. In early production, the chip yield for the microprocessor, ASIC, and memory are, respectively, 55, 50, and 60 percent. Then, the MCM yield is $Y_C = 100\,(0.55)^1\,(0.5)^2\,(0.6)^3 = 3$ percent. Later in mature products, the chip yield for the microprocessor, ASIC, and memory are increased to, respectively, 90, 85, and 99 percent and the MCM yield becomes $Y_C = 100\,(0.9)^1\,(0.85)^2\,(0.99)^3 = 63$ percent.

Now, let us assume that the overall MCM yield (Y_{MCM}) is based solely on the chip yield; then we have

$$Y_{MCM} = Y_C \qquad (1.32)$$

In this case, the overall MCM yield or simply MCM yield (Y_C) in ei-

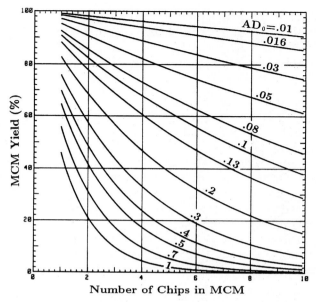

Figure 1.15 MCM yield as a function of AD_0 (small number of chips).

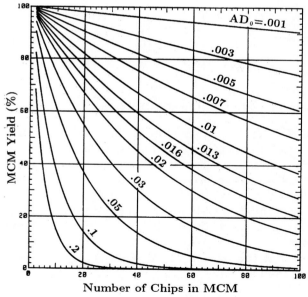

Figure 1.16 MCM yield as a function of AD_0 (large number of chips).

ther Eq. (1.29), (1.30), or (1.31) will become a shipped defect if the test method used for the MCM assembles cannot identify the defect for successful rework.[1,2,67,69] The resultant shipped MCM yield (Y_{ms}), i.e., the percentage of MCMs shipped which pass final MCM test and are fault-free, is given by

$$Y_{ms} (\%) = 100 Y_C^{(1 - \eta)} \tag{1.33}$$

Here, $0 \leq \eta \leq 100$ percent is the test fault coverage (in percent) and is defined as the ability of the tester or inspection method to identify defects. When $\eta = 0$, the test cannot detect any faults even with all the possible faults present. On the other extreme, when $\eta = 100$ percent, the test detects all the possible faults whenever they are present. Equation (1.33) is plotted in Fig. 1.17 for various values of MCM yield (Y_C) and test fault coverage (η).

On the other hand, the resultant shipped MCM defect level D_m (%) (the percentage of MCMs shipped which pass final MCM test but actually may be faulty) is given by ($1 - Y_{ms}$), or

$$D_m (\%) = 100 [1 - Y_C^{(1 - \eta)}] \tag{1.34}$$

Equation (1.34) is plotted in Fig. 1.18 for various values of Y_C and η. With Figs. 1.13 through 1.18, tradeoffs between the number of chips,

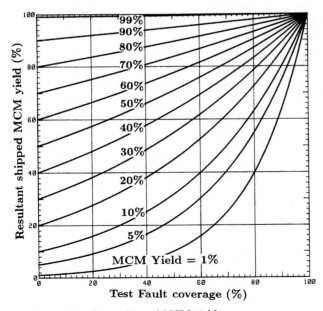

Figure 1.17 Resultant shipped MCM yield.

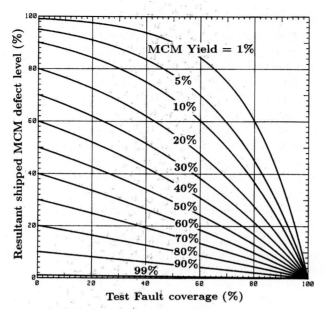

Figure 1.18   Resultant shipped MCM defect level.

defect density, chip size, chip yield, MCM yield, MCM rework, resultant shipped MCM yield, and resultant shipped MCM defect level can be made.

For example, for a 10-chip MCM having an average number of chip faults of $AD_0 = 0.115$ and test fault coverage $\eta = 95$ percent, the chip yield will be 90 percent [Eq. (1.22) or Figs. 1.3 and 1.4], MCM yield will be 35 percent [Eq. (1.30) or Figs. 1.15 and 1.16], the resultant shipped MCM yield is 95 percent [Eq. (1.33) or Fig. 1.17] of the 35 percent-yield MCMs shipped, and the resultant shipped MCM defect level will be 5 percent [Eq. (1.34) or Fig. 1.18] of the 35 percent-yield MCMs shipped. In this case, 65 percent of the MCM will require at least one rework.

On the other hand, if the average number of faults in a chip (AD_0) is decreased from 0.115 to 0.0115 and the test fault coverage remains the same (95 percent), the chip yield will increase from 90 to 99 percent, the MCM yield will be 90 percent, the resultant shipped MCM yield will be 99.5 percent of the 90 percent-yield MCMs shipped, the resultant shipped MCM defect level will be 0.5 percent of the 90 percent-yield MCMs shipped, and only 10 percent of the MCM will require rework. This indicates the importance of known good dies (KGDs) and the logic that the more bad chips (higher average number of faults in a chip or lower chip yield) there are, the greater the likelihood that they will escape into the MCM assembly.

1.3.2 Known good die (KGD)

The importance of KGD for MCM and DCA applications was shown in Sec. 1.3.1 with respect to the relationship between chip yield and MCM yield. There are basically two ways to test and burn in for the KGDs: (1) at the wafer level and (2) at the individual chip level. Because of the power distribution, cooling, and wafer contact problems, test at high speed and burn-in of chip at wafer level poses a technical challenge. Also, test at high speed and burn-in for individual bare chip using sockets, probe cards, anisotropic conductive compliant films, etc. may damage the pads on the chip, limit high-frequency capability, and increase cost. Built-in self-testing (BIST) chips and boundary scan test of interchip connections are under active investigation. For more information about test and burn-in for KGDs, please read Chaps. 16 and 17 of this book.

If tests (e.g., parametric, functional, full-speed, elevated-temperature, and accelerated burn-in) are performed at the individual bare chip or, preferably, the whole-wafer level for the KGDs prior to MCM assembly, then the chip yield Y, e.g., Eq. (1.22), will become a shipped defect if the test methods used for testing KGD cannot identify the defect. In general, the resultant shipped chip yield Y_{cs} (the percentage of chips shipped which pass final tests and are fault-free) is given by

$$Y_{cs}(\%) = 100Y^{(1-\xi)} \qquad (1.35a)$$

If Lau's chip yield formula is used, then we have

$$Y_{cs}(\%) = 100(0.35t - 0.1t^2 + 0.75t^3)^{(1-\xi)} \qquad (1.35b)$$

In Eqs. (1.35a) and (1.35b), $0 \le \xi \le 100$ percent is the test fault coverage (%) and is defined as the ability of the tester or inspection method to identify the IC defects, Y is the chip yield, and t is as given in Eq. (1.20). When $\xi = 0$, the test cannot detect any faults even with all the possible faults on the IC present. On the other extreme, when $\xi = 100$ percent, the test detects all the possible faults on the IC whenever they are present. Equation (1.35a) is plotted in Fig. 1.19 and Eq. (1.35b) is plotted in Figs. 1.20 and 1.21 for various values of chip yield (Y) and test fault coverage (ξ) and the average number of faults in a chip (AD_0) and ξ.

On the other hand, the resultant shipped chip defect level D_c (%) (the percentage of chips shipped which pass final IC test but may actually be faulty) is given by $(1 - Y_{cs})$, or

$$D_c(\%) = 100\,[1 - Y^{(1-\xi)}] \qquad (1.36a)$$

or

$$D_c(\%) = 100\,[1 - (0.35t - 0.1t^2 + 0.75t^3)^{(1-\xi)}] \qquad (1.36b)$$

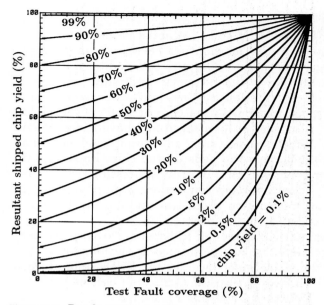

Figure 1.19   Resultant shipped chip yield as a function of chip yield.

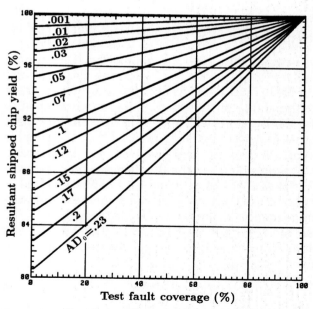

Figure 1.20   Resultant shipped chip yield as a function of small AD_0.

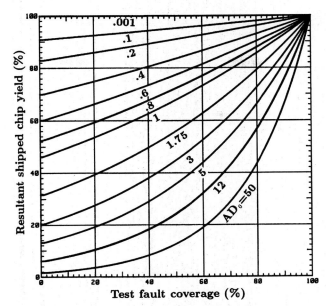

Figure 1.21   Resultant shipped chip yield as a function of large AD_0.

if Lau's chip yield formula is used. Equation (1.36a) is plotted in Fig. 1.22 and Eq. (1.36b) in Figs. 1.23 and 1.24 for various values of Y and ξ and for AD_0 and ξ. For example, if the chip yield of an IC is $Y = 85$ percent and the IC test fault coverage is $\xi = 90$ percent, the resultant shipped chip yield will be $Y_{cs} = 98$ percent [Eq. (1.35a) or Fig. 1.19] of the 85 percent-yield chips shipped, and the resultant shipped chip defect level will be $D_c = 2$ percent [Eq. (1.36a) or Fig. 1.22] of the 85 percent-yield chips shipped.

If the average number of faults in a chip is $AD_0 = 3$ and the test fault coverage is $\xi = 80$ percent, the chip yield will be 20 percent [Eq. (1.22) or Fig. 1.4] and the resultant shipped chip yield will be 72 percent [Eq. (1.35b) or Fig. 1.21] of the 20 percent-yield chips shipped, and the resultant shipped chip defect level will be 28 percent [Eq. (1.36b) or Fig. 1.24] of the 20 percent-yield chips shipped.

The KGD MCM yield (Y_{KGD}) is given by

$$Y_{KGD} (\%) = 100 Y_{cs}^{N_c} = 100 Y^{(1 - \xi)N_c} \qquad (1.37a)$$

or

$$Y_{KGD} (\%) = 100(0.35t - 0.1t^2 + 0.75t^3)^{(1 - \xi)N_c} \qquad (1.37b)$$

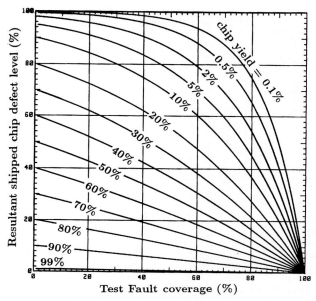

Figure 1.22 Resultant shipped chip defect level as a function of chip yield.

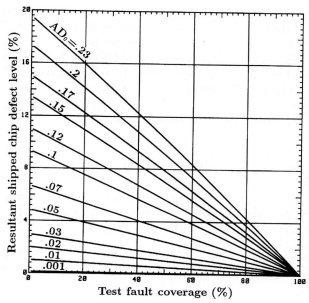

Figure 1.23 Resultant shipped chip defect level as a function of small AD_0.

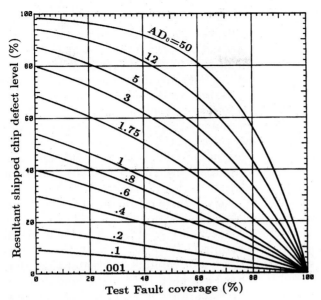

Figure 1.24   Resultant shipped chip defect level as a function of large AD_0.

if Lau's chip yield formula is used. N_c is the number of chips in the KGD MCM assembly. Again, Eq. (1.37) will become a shipped defect if the test method used for the KGD MCM assembly cannot identify the defect for successful rework. The resultant shipped KGD MCM yield (S_{KGD}), i.e., the percentage of KGD MCM shipped which pass final MCM test and are fault free, is given by

$$S_{KGD} (\%) = 100Y_{KGD}^{(1-\eta)} = 100Y^{(1-\xi)(1-\eta)N_c} \qquad (1.38a)$$

or

$$S_{KGD} (\%) = 100(0.35t - 0.1t^2 + 0.75t^3)^{(1-\xi)(1-\eta)N_c} \qquad (1.38b)$$

Also, the resultant shipped KGD MCM defect level (D_{KGD}), i.e., the percentage of KGD MCM shipped which pass final MCM test but may actually be faulty is given by

$$D_{KGD} = 100\,[1 - Y_{KGD}^{(1-\eta)}] = 100\,[1 - Y^{(1-\xi)(1-\eta)N_c}] \qquad (1.38c)$$

or

$$D_{KGD} = 100\,[1 - (0.35t - 0.1t^2 + 0.75t^3)^{(1-\xi)(1-\eta)N_c}] \qquad (1.38d)$$

where ξ and η are, respectively, the IC and MCM test fault coverages.

1.3.3 Unknown bad die (UBD)

It was shown in Sec. 1.3.2 that unless the test system provides a 100 percent test fault-coverage, some bad bare chips will "escape" the test system and become unknown bad die (UBD). However, with today's complex MCM and DCA systems, a 100 percent test effectiveness system for bare chips is almost impossible or extremely expensive. Thus, UBD is a critical issue for MCM and DCA applications.

For semiconductor producers, shipping KGD at a reasonable price will enhance customer relations; shipping UBD will create many problems for the customers and eventually delete the clientele. Thus, in the past, most semiconductor producers did not provide bare wafers or chips to their customers because they did not want to be liable for UBD and also did not want to disclose their confidential wafer yields and reduce their profit margins because of nonpackaging of chips.

Today, more semiconductor producers are willing to learn how to sell KGD and minimize the risk in selling UBD. A few semiconductor producers are actually selling KGD at a cost equal to or greater than that of conventional packaged ICs that have been fully tested and guaranteed. However, the total bare-chip sales still represent a negligible percentage of total semiconductor revenues, and KGDs represent only a small percentage of those sales. Thus, the incentive for semiconductor producers to invest significant resources in developing a 100 percent test fault coverage system ($\xi = 1$) for bare-die testing in order to avoid UBD is negligible. The UBD problem will pass to the next level of interconnect, i.e., MCM and DCA assembly and tests. Chapters 16 and 17 of this book will provide more information on KGD and UBD.

1.3.4 Known good substrate (KGS)

For MCM and DCA applications, assembling some KGDs on a bad substrate could lead to either discarding the whole assembly or very expensive removal, rework, replacement, and repair or would increase the material and manufacturing costs of the product. On the other hand, extensive electrical tests of substrates prior to MCM or DCA assembly will also increase the product costs. Thus, there is a tradeoff between the objective of KGS, KGD, and the business (profit-margin) needs.

Testing of MCM and DCA substrates is very complex and costly. The test equipment alone will cost three to five times more than the standard surface-mount technology (SMT) equipment. A test fixture that is unique to the board must be constructed to interface the selected product test locations with specific test points of the test system. There is a further requirement for sophisticated software specifying the manner in which these test points should be observed for interconnection to a properly fabricated product. A major commitment for better-qualified

TABLE 1.1   Electrical Test of Printed-Circuit Boards

Electrical test	IPC-ET-652	MIL-55110D (production)
Maximum continuity resistance pass/fail threshold test	Class 1: general electronic = 50 V Class 2: dedicated service = 20 V Class 3: high reliability = 20 V	= 10 V
Maximum continuity current test	Undefined	Per MIL-STD-275
Isolation resistance pass/fail threshold test	Class 1: general electronic = 500 kV Class 2: dedicated service > 2 MV Class 3: high reliability > 2 MV	>2 MV
Apply voltage to passing networks during isolation test	High enough to provide sufficient current for the measurement in question, but low enough to prevent arcover	40 V, or twice the maximum rated voltage on the board, whichever is greater

test and fixture personnel to implement the more complex technologies and operate the more sophisticated equipment must be made before this process is implemented for bare-board testing.

Table 1.1 summarizes two published electrical tests and requirements of PCBs for laminate MCM (MCM-L) and DCA on PCB applications. It can be seen that the bare board is first subjected to the continuity test (for open circuits), and during the continuity test, a current is allowed to pass through the conductor under test. The isolation test (for short circuits or leaks) will verify that each network is well isolated from the rest of the board, and during the isolation test of each network, a small amount of current is injected into the network. For some products, if the characteristic impedance Z_0 is required, it can be tested off line using specialized radio-frequency (RF) equipment.

1.4   Classical Flip Chip Technologies

As mentioned earlier in this chapter, a *flip chip* is defined as a chip mounted on the substrate with various interconnect materials and methods, such as tape-automated bonding, fluxless solder bumps, wire interconnects, isotropic and anisotropic conductive adhesives, metal bumps, compliant bumps, and pressure contacts, as long as

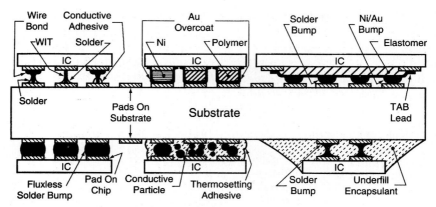

Figure 1.25 Various flip chip technologies.

the chip surface (active area or I/O side) is facing the substrate (see Fig. 1.25).[81–163]

Theoretically, *wafer scale integration* (WSI), which builds the entire system or subsystems on a single wafer, is the most ideal electronic assembly. However, because of the poor yield of WSI, wafers are usually broken up into individual chips. These chips are then packaged on a substrate (chip on board) or in a carrier (either single-chip or multi-chip modules). There are at least three popular methods for interconnecting the chip(s) on the substrate: face-up wire bonding, face-up tape-automated bonding, and flip chip. Among these three methods, flip chip provides the shortest possible leads, lowest inductance, highest frequency, best noise control, highest density, greatest number of I/Os, smallest device footprints, and lowest profile.[81–163]

In this section, the classical solder-bumped flip chip technologies will be briefly discussed and two different methods of predicting the thermal fatigue life of the flip chip solder joints presented.

1.4.1 Solder-bumped flip chip technologies

The solder-bumped flip chip was introduced by IBM in the early 1960s for their *solid logic technology* (SLT), which became the logical foundation of the IBM System/360 computer line. The so-called C4 (controlled-collapse chip connection) technology utilizes solder bumps deposited on wettable metal terminals on the chip and a matching footprint of solder wettable terminals on the substrate. The solder-bumped flip chip is aligned to the substrate, and all solder joints are made simultaneously by reflowing the solder.

The solder materials IBM used are usually the 5wt%Sn/95wt%Pb solder, which has solid and liquid temperatures of 308 and 312°C, respectively. Sometimes, IBM also used 3wt%Sn/97wt%Pb solder, which has solid and liquid temperatures of 314 and 320°C, respectively. The substrate materials IBM used are usually the ceramics which have a thermal expansion coefficient of about 4 to 6×10^{-6} °C^{-1}. During reflow, because of the surface tension of the molten solder, the chip will self-align to the substrate; this leads to a very high-yield manufacturing process. Also, the C4 technology allows for availability of the whole chip surface for I/O interconnections.

The C4 technology consists of two basic assembly tasks: wafer bumping and substrate preparation. A detailed description of these two tasks and flip chip assembly is beyond the scope of this chapter; interested readers are encouraged to read Chaps. 2 through 4 of this book and the work published by IBM[15,16,82–113] and others.[114–163] However, the most important features of these two tasks are briefly described in the following paragraphs.

The most important task in wafer bumping is to put down the ball limiting metallurgy (BLM) or underbump metal (UBM) on the aluminum pads. The BLM defines the region of terminal metallurgy on the top surface of the chip that is wetted by the solder which is deposited by either electroplating[118,119,149,150] or evaporation[84–90,99–101] methods. The BLM usually consists of three layers (Fig. 1.26a and 1.26b): adhesion and/or barrier layer, wetting layer, and oxidation barrier layer. For most of the C4 bump structures, the adhesion and/or barrier layer consists of a chromium layer (~0.15 μm) and a phased 50/50 CrCu layer (~0.15 μm). The function of this layer is to form a strong bond with the Al pad and with the passivation layer (Fig. 1.26a and 1.26b). The wetting layer is made of copper (~1 μm) which must remain at least partially intact through all the remaining reflow processes, e.g., chip level interconnect, chip carrier interconnect, PCB assembly and rework. The oxidation barrier layer (~0.15 μm) is made of Au, which protects the Cu from oxidation. It is recommended in Ref. 51 that in order to have a reliable C4 solder joint, the dimensions (diameters) of the chip pad on the wafer are: 50 μm ± 1 μm for the terminal via contact, 64 μm ± 4 μm for the passivation opening (i.e., the step is 7 μm ± 1.5 μm) and 75 μm ± minimum for the final metal (Al) pad (Fig. 1.26a and 1.26b).

The most important step in preparing the ceramic substrate is to put down the *topside* (or *top surface*) *metallurgy* (TSM), which defines the terminal metallurgy on the substrate to which the solder-bumped flip chips are joined (Fig. 1.26a and 1.26b). The formation of TSM can be achieved by either thick- or thin-film technology. Thin-film contact technology is similar to the BLM; however, thick-film technology involves development of wettable surfaces by plating Ni and Au over generally nonwettable surfaces. Solder flow is restricted by the use of glass or chromium dams where necessary. A few substrate TSM structures that are typically used are shown in Fig. 1.27.[16]

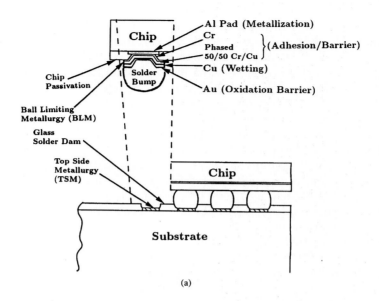

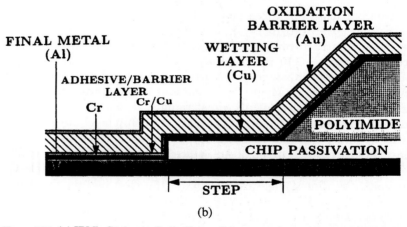

Figure 1.26 (*a*) IBM's C4 (controlled collapse chip connection) technology; (*b*) IBM's ball-limiting metallurgy structure.

1.4.2 Low-cost solder-bumped flip chip technologies

IBM at Yasu, Japan, has been assembling solder-bumped flip chips on organic PCB since 1990.[98] Their applications include personal computers, PCMCIA cards, and token-ring LAN (local area network) adapter cards. Their results showed that solder-bumped flip chip assembly technology is applicable for low-cost PCB as well as high-cost substrates, e.g., ceramics.

IBM's (at Yasu) DCA technology[95–98] is based on IBM's C4 technology. In order to be SMT-compatible, the Cu pads on the PCB are coat-

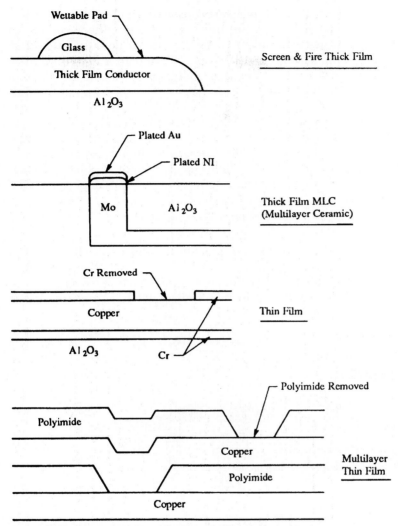

Figure 1.27   IBM's various ceramic substrate pad structures.[16]

ed with 63Sn/37Pb solder using IBM's molten solder injection machine. The problem of large thermal expansion mismatch between the silicon chip and the PCB is reduced by filling the gap between the chip and PCB with an epoxy resin. Because of capillary action, this resin "completely" fills the space underneath the chip. The chip and PCB are then firmly bonded by curing the resin (see Fig. 1.28).

Unlike IBM's composite solder interconnects, Sharp's wafers are bumped with eutectic 60Sn/40Pb solder by electroplating[118,119] (Fig.

1.29a). Their substrates for the solder-bumped flip chips are standard
FR-4 PCBs with a CuNiAu metal finish (Fig. 1.29b). Sharp conducted
thermal cycling tests (-45 to 100°C and 1 cycle per hour) of five different square flip chip assemblies (3-, 6-, 9-, 12-, and 15-mm) with underfill resin; no failures were observed at 1000 cycles. For the smaller flip
chip (3-, 6-, and 9-mm) assemblies with underfill encapsulant, no failures were observed even after 9000 cycles[118,119] (Fig. 1.30).

   In the next section, a nonlinear finite element method will be used to
determine the thermal stresses and plastic strains in an encapsulated
low-cost flip chip solder joint. The average thermal fatigue life of the
solder joint is then estimated on the basis of calculated plastic strains,
Coffin-Manson law, and isothermal fatigue data of the solder material.

1.4.3   Thermal fatigue life prediction of encapsulated flip chip solder joints (Coffin-Manson method)

Figure 1.31 schematically depicts a solder-bumped flip chip on PCB
with underfill epoxy encapsulant. It can be seen that the silicon chip
is 5.5 mm in length. (The other dimensions of the chip are assumed to
be 0.635 mm thickness and 2.75 mm width.) The thermal coefficient

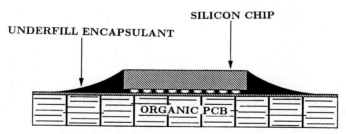

Figure 1.28   Solder-bumped flip chip on low-cost PCB with underfill
encapsulant.

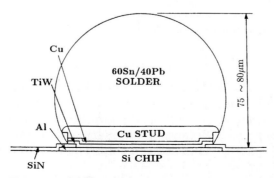

Figure 1.29   (a) Sharp's electroplated solder bump.

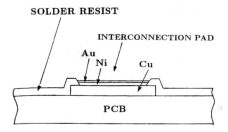

SOLDER RESIST

INTERCONNECTION PAD

Au
Ni   Cu

PCB

Figure 1.29 (*b*) Sharp's PCB for low-temperature solder-bumped flip chip applications.

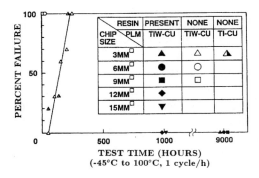

CHIP SIZE \ RESIN PLM	PRESENT TIW-CU	NONE TIW-CU	NONE TI-CU
3MM□	▲	△	▲
6MM□	●	○	
9MM□	■	□	
12MM□	◆		
15MM□	▼		

PERCENT FAILURE

TEST TIME (HOURS)
(-45°C to 100°C, 1 cycle/h)

Figure 1.30 Sharp's flip chip on-board thermal cycling results.

of expansion of the silicon is 2.8×10^{-6} °C^{-1}. The Young modulus of the silicon is 131 GPa, and the Poisson ratio is 0.3 (Table 1.2).

The thermal coefficient of expansion of the epoxy encapsulant is 30×10^{-6} °C^{-1}. The Young modulus of the epoxy is 6 GPa, and the Poisson ratio is 0.35 (Table 1.2). (The epoxy encapsulant is assumed to be 0.102 mm thick.)

The thickness of the PCB is 1.524 mm and is made of FR-4 epoxy/glass. The thermal coefficient of linear expansion of the FR-4 material is 18×10^{-6} °C^{-1} in the x and y (i.e., horizontal) directions and is 70×10^{-6} °C^{-1} in the z (i.e., vertical) direction. The Young modulus of the PCB is 22 GPa, and the Poisson ratio is 0.28 (Table 1.2).

There are at least three major systems of thermal stresses and strains acting at the solder joints and encapsulant, which are attribut-

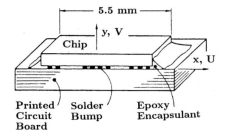

5.5 mm

y, V

Chip

x, U

Printed Circuit Board   Solder Bump   Epoxy Encapsulant

Figure 1.31 A low-cost solder-bumped flip chip assembly.

TABLE 1.2   Material Properties of Low-Cost Flip Chip Assemblies

		Young's Modulus, GPa	Poisson's ratio	Thermal coefficient of linear expansion 10^{-6} (m/m-°C)
Solder		10	0.40	21
Encapsulant		6	0.35	30
Silicon		131	0.30	2.8
FR-4 PCBs	x and y	22	0.28	18
	z	22	0.28	70
Copper		121	0.35	17

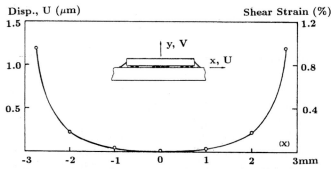

Figure 1.32 Relative displacement in the x direction between the chip and the PCB, and the distribution of average shear strains in the interconnection.[155]

able to the (1) local thermal expansion mismatch between the silicon chip, solder joint, epoxy encapsulant, and the FR-4 PCB; (2) global thermal expansion mismatch between the silicon chip and the FR-4 PCB; and (3) overall bending of the solder-bumped flip chip assembly.

The global thermal expansion mismatch and bending effects of this assembly have been studied by Guo et al.[155] Using the Moiré interferometry method, they found that, even with the help of the epoxy resin, the thermal expansion mismatch (relative displacement) distribution in the flip chip assembly is not uniform and the maximum relative displacement occurs at the corner of the chip (Fig. 1.32). In this chapter the total (local, global, and bending) thermal response of this solder-bumped flip chip assembly will be addressed by a local finite element analysis model[156] with the global thermal expansion mismatch and bending results given by Ref. 155 as the imposed boundary conditions.

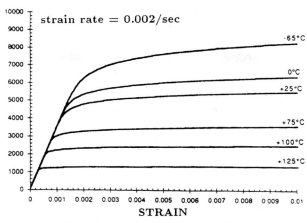

Figure 1.33   Stress–strain curves for 63/37 SnPb solder.

Because of the low yield strength and high ductility of the 63Sn/37Pb solder,[7] this solder is assumed to be an elastoplastic material. The Young modulus of the solder is 10 GPa, and the Poisson ratio is 0.4. The thermal coefficient of linear expansion of the solder is 21×10^{-6} °C^{-1}. To obtain a conservative estimate of the fatigue life (based on the Coffin-Manson law and plastic strain calculation), the stress–strain curve at 120°C (yield stress = 8.3 MN/m^2, yield strain = 0.0008, strain-hardening parameter = 0.1[164–166] has been used (Fig. 1.33). The nominal dimension of the solder joint is about 0.125 mm.[156]

In the present study, the yield surface (i.e., yield criterion) of the solder is assumed to obey the distortion energy theory (i.e., Von Mises yield criterion), and the plastic flow rule (i.e., plastic stress–strain relationship) is assumed to obey the incremental deformation theory (i.e., Prandtl-Reuss theory). The fundamental equations governing the present boundary-value problem are the *equilibrium equations* [Eq. (1.39)], *strain-displacement equations* [Eq. (1.40)], and *constitutive equations* [Eq. (1.41)].

$$\frac{\partial \sigma_{ij}}{\partial x_j} = 0 \tag{1.39}$$

$$\epsilon_{ij} = \frac{1}{2} \left(\frac{\partial u_i}{\partial x_j} + \frac{\partial u_j}{\partial x_i} \right) \tag{1.40}$$

$$d\epsilon_{ij} = d\epsilon_{ij}^e + d\epsilon_{ij}^p \tag{1.41}$$

where

$$de_{ij}^{\ e} = \frac{dS_{ij}}{2G} + (1-2v)\delta_{ij}\frac{d\sigma_{ij}}{3E} \qquad (1.42)$$

$$d\epsilon_{ij}^{\ p} = d\lambda\,\frac{\partial f}{\partial \sigma_{ij}} \qquad (1.43)$$

and

$$S_{ij} = \sigma_{ij} - \frac{1}{3}\,\sigma_{\beta\beta}\delta_{ij} \qquad (1.44)$$

If

$$f = \frac{1}{2}\,S_{ij}S_{ij} - \frac{\sigma_0^{\ 2}}{3} \qquad (1.45)$$

$$\bar{\sigma} = \sqrt{\frac{3}{2}\,S_{ij}S_{ij}} \qquad (1.46)$$

and

$$\overline{d\epsilon^p} = \sqrt{\frac{2}{3}\,d\epsilon_{ij}^{\ p}d\epsilon_{ij}^{\ p}} \qquad (1.47)$$

then

$$d\lambda = \frac{3}{2}\frac{\overline{d\epsilon^p}}{\bar{\sigma}} \qquad (1.48)$$

and $\bar{\sigma}$ and $\overline{\epsilon^p} = \int\overline{d\epsilon^p}$ coincide with the uniaxial stress–strain curve (Fig. 1.33). In these equations, σ_{ij} is the stress tensor; x_i are the coordinates of each particle in the original configuration; ϵ_{ij} is the strain tensor: u_i are the displacement components; $d\epsilon_{ij}$ is the incremental strain tensor, $d\epsilon_{ij}^{\ e}$ is the incremental elastic strain tensor; $d\epsilon_{ij}^{\ p}$ is the incremental plastic strain tensor; S_{ij} is the stress deviation tensor; G is the shear modulus; v is the Poisson ratio for elastic deformation; δ_{ij} is the Kronecker delta; $d\sigma_{ij}$ is the incremental stress tensor; E is the modulus of elasticity; f is the yield function (or loading function); $d\lambda$ is a nonnegative function which depends on stress, strain, and strain history; $\bar{\sigma}$ is the effective stress; $\overline{d\epsilon^p}$ is the incremental effective plastic strain; $\overline{\epsilon^p} = \int\overline{d\epsilon^p}$ is the accumulated effective plastic strain; and σ_0 is the uniaxial yield stress. All the indices range over 1, 2, 3, and the summation convention for repeated indices is used.

The present boundary-value problem is to solve the preceding equations for the elastoplastic stresses and strains of the corner solder joint (Fig. 1.34) when it is subjected to an incrementally decreased temperature from 82 to 22°C ($\Delta T = -60$°C). The displacement boundary conditions are as follows: (1) at the left-hand side (LHS) of the silicon chip all points cannot move in the x direction, and (2) at the bottom of the FR-4 PCB all points move 0.00134 mm in the negative x direction and 0 mm

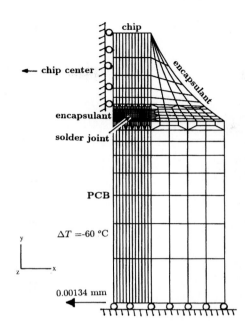

Figure 1.34 Boundary-value problem of the solder-bumped flip chip assembly.

(fixed) in the y direction. It should be noted that the measured relative displacement in the x direction (0.0012 mm) (Fig. 1.32 and Ref. 155), as a result of the effects of bending and global thermal expansion mismatch between the chip and PCB, has been superposed to that in the z direction (0.0006 mm), i.e., $\sqrt{(0.0012)^2 + (0.0006)^2} = 0.00134$ mm.

Figure 1.34 shows the finite element model for the analysis of the corner portion (which includes the chip, solder joint, encapsulant, and PCB) of the 63Sn/37Pb solder bumped flip chip assembly. A high-order two-dimensional (2D) plane stress element has been used for the model. Each element has eight nodes. Each node has two degrees of freedom. The temperature and displacements are applied to the flip chip assembly decrementally. A total of seven decrements are applied with about 11 iterations per decrement.

The whole-field displacements (deflections) of the flip chip assembly are shown in Fig. 1.35. The dotted lines represent the original mesh and the solid lines, the displaced mesh at $\Delta T = -60°C$. It can be seen that the displacement of the FR-4 PCB in the vertical direction is larger than that in the horizontal directions. This is because the coefficient of thermal expansion in the vertical direction (70×10^{-6} °C^{-1}) of the PCB is several times greater than that in the horizontal directions (18×10^{-6} °C^{-1}). It can also be seen that the deformations of the epoxy encapsulant (Fig. 1.36) and solder joint (Fig. 1.37) are dominated by shear.

The von Mises (effective) stress [$\bar{\sigma}$, in megapascals (MPa)] distribu-

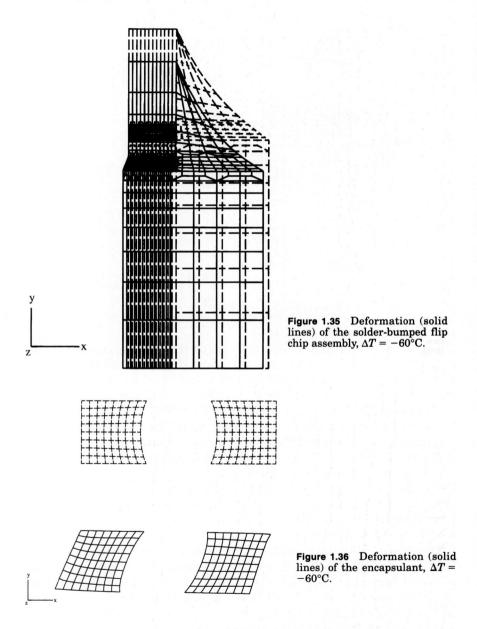

Figure 1.35 Deformation (solid lines) of the solder-bumped flip chip assembly, $\Delta T = -60°C$.

Figure 1.36 Deformation (solid lines) of the encapsulant, $\Delta T = -60°C$.

tion, defined by Eq. (1.46), in the solder joint is shown in Fig. 1.38. It can be seen that the von Mises stress distribution in the solder joint is quite uniform. The difference between the largest and smallest stresses is less than 10 percent. The higher stresses occur near the corners of the solder joint. Figure 1.39 shows the von Mises stress

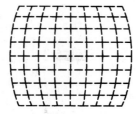

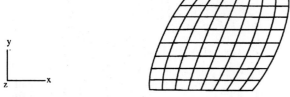

Figure 1.37 Deformation (solid lines) of the solder joint, $\Delta T = -60°C$.

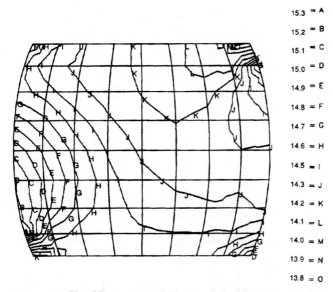

15.3	= A
15.2	= B
15.1	= C
15.0	= D
14.9	= E
14.8	= F
14.7	= G
14.6	= H
14.5	= I
14.3	= J
14.2	= K
14.1	= L
14.0	= M
13.9	= N
13.8	= O

Figure 1.38 Von Mises stress (MPa) in the solder joint (nonlinear solution).

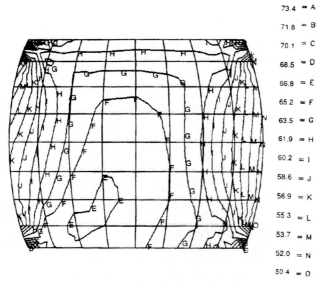

73.4	= A
71.8	= B
70.1	= C
68.5	= D
66.8	= E
65.2	= F
63.5	= G
61.9	= H
60.2	= I
58.6	= J
56.9	= K
55.3	= L
53.7	= M
52.0	= N
50.4	= O

Figure 1.39 Von Mises stress (MPa) in the solder joint (linear solution).

(MPa) distribution in the solder joint from a linear analysis (i.e., assuming a linear stress–strain relation for the solder joint). It can be seen that the magnitudes of the stress predicted by the linear analysis are more than four times (unreasonably) larger than those obtained by nonlinear analysis.

The accumulated effective plastic strain ($\overline{\epsilon^p} = \int \overline{d\epsilon^p}$) with $\overline{d\epsilon^p}$ defined in Eq. (1.47) can be written as follows and is shown in Fig. 1.40:

$$\overline{\epsilon^p} = \int \overline{d\epsilon^p} = \frac{\sqrt{2}}{3} \int \sqrt{(d\epsilon_x^p - d\epsilon_y^p)^2 + (d\epsilon_x^p)^2 + (d\epsilon_y^p)^2 + \frac{3}{2}(d\gamma_{xy}^p)^2} \quad (1.49)$$

where $\int d\epsilon_x^p$ = accumulated plastic normal strain acting in the x direction (Fig. 1.41)

$\int d\epsilon_y^p$ = accumulated plastic normal strain acting in the y direction (Fig. 1.42)

$\int d\gamma_{xy}^p$ = accumulated plastic shear strain acting in the y direction of the plane normal to the x axis (Fig. 1.43)

It can be seen that the values of $\int d\epsilon_x^p$ and $\int d\epsilon_y^p$ are much smaller than those of $\int d\gamma_{xy}^p$. Thus, the solder joint is dominated by shear action.

In general, the accumulated effective plastic strain $\int \overline{d\epsilon^p}$ is larger than all the accumulated plastic strain components ($\int d\epsilon_x^p$, $\int d\epsilon_y^p$, and $\int d\gamma_{xy}^p$). For shear-deformation-dominant cases, however, the values of

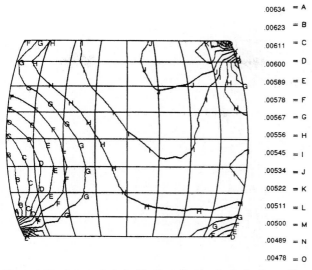

.00634	= A
.00623	= B
.00611	= C
.00600	= D
.00589	= E
.00578	= F
.00567	= G
.00556	= H
.00545	= I
.00534	= J
.00522	= K
.00511	= L
.00500	= M
.00489	= N
.00478	= O

Figure 1.40 Accumulated effective plastic strain in the solder joint.

the accumulated plastic shear strain component ($\int d\gamma_{xy}{}^{\mathrm{p}}$, Fig. 1.43) is larger than that of the accumulated effective plastic strain ($\int d\epsilon^{\mathrm{p}}$, Fig. 1.40). Thus, for a conservative estimate of the thermal fatigue life of the flip chip solder joint, the values of $\int d\gamma_{xy}{}^{\mathrm{p}}$ shown in Fig. 1.43 should be used.

Figure 1.44 shows the total shear strain (elastic and plastic) distribution in the solder joint by the present nonlinear analysis. Comparing Figs. 1.44 and 1.43 (the accumulated plastic shear strain), it can be seen that most of the total shear strain is accumulated plastic shear strain and the total elastic shear strain is very small.

In Coffin-Manson's relation[174]

$$N_f = \theta(\Delta\gamma_p)^\phi \qquad (1.50)$$

where N_f = number of cycles to failure
$\Delta\gamma_p$ = plastic strain range
θ, ϕ = material constants

For 60Sn/40Pb or 63Sn/37Pb solders, θ and ϕ were determined by Solomon[168] at −50, 35, and 125°C. The average values were θ = 1.2928 and ϕ = −1.96.

The maximum accumulated plastic shear strain in the solder joint of the flip chip assembly has been determined to be 0.0107 (Fig. 1.43). Thus, from Eq. (1.50) and Solomon's solder fatigue data, the solder

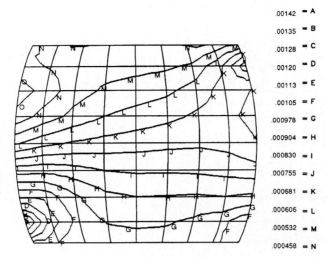

.00142	▬ A
.00135	▬ B
.00128	▬ C
.00120	▬ D
.00113	▬ E
.00105	= F
.000978	▬ G
.000904	▬ H
.000830	= I
.000755	= J
.000681	= K
.000606	= L
.000532	▬ M
.000458	▬ N

Figure 1.41 Normal plastic strain in the x direction.

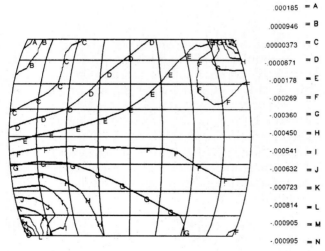

.000185	= A
.0000946	= B
.00000373	= C
-.0000871	= D
-.000178	▬ E
-.000269	= F
-.000360	= G
-.000450	▬ H
-.000541	= I
-.000632	= J
-.000723	= K
-.000814	▬ L
-.000905	= M
-.000995	▬ N

Figure 1.42 Normal plastic strain in the y direction.

joint should have an average life of 9420 cycles ($\Delta\gamma_p = 0.0107$). In practical applications, a safety factor should be applied.

The von Mises stress (MPa) distribution in the flip chip encapsulant is shown in Fig. 1.45. It can be seen that the maximum stress (107 MPa) occurs at the left-hand free-edge corners of the local finite element model. It should be noted, however, that this large stress is due to the limitation of the present boundary-value problem (very

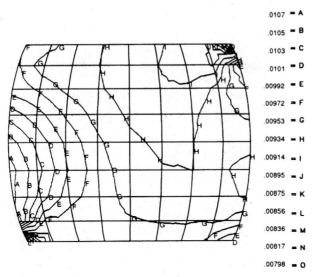

.0107	= A
.0105	= B
.0103	= C
.0101	= D
.00992	= E
.00972	= F
.00953	= G
.00934	= H
.00914	= I
.00895	= J
.00875	= K
.00856	= L
.00836	= M
.00817	= N
.00798	= O

Figure 1.43 Plastic shear strain in the xy plane.

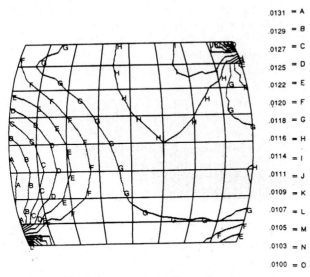

.0131	= A
.0129	= B
.0127	= C
.0125	= D
.0122	= E
.0120	= F
.0118	= G
.0116	= H
.0114	= I
.0111	= J
.0109	= K
.0107	= L
.0105	= M
.0103	= N
.0100	= O

Figure 1.44 Total shear strain in the xy plane.

large thermal stresses occur at the free edge of the interface of any composite structures[169,170]) and thus will not happen in real conditions. Thus, the maximum stress (63.2 MPa) occurs at point G, the interface between the solder joint and the encapsulant. This stress acts

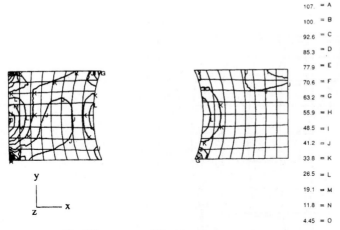

107. = A
100 = B
92.6 = C
85.3 = D
77.9 = E
70.6 = F
63.2 = G
55.9 = H
48.5 = I
41.2 = J
33.8 = K
26.5 = L
19.1 = M
11.8 = N
4.45 = O

y

z ——— x

Figure 1.45   Von Mises stress (MPa) in the epoxy encapsulant.

at a very small area (Fig. 1.45) as a result of stress concentration and decreases rapidly to a much smaller value (41.2 MPa at point J). This stress is lower than the yield stress (45 MPa) of the encapsulant suggested by the manufacturer's data. It should be noted that the epoxy encapsulant is highly temperature- and rate-dependent.[117,160,161,163] These variables have to be included in the analysis for an accurate estimation of the stresses in the encapsulant.

The von Mises stress (MPa) distribution in the lower portion of the silicon chip is shown in Fig. 1.46. It can be seen that the maximum stress (129 MPa) occurs at the bottom corners of the chip. This large stress is due to the large thermal expansion mismatch between the silicon chip (2.8×10^{-6} °C^{-1}) and the epoxy encapsulant (30×10^{-6} °C^{-1}) and solder joint (21×10^{-6} °C^{-1}). This large stress decreases rapidly to a much smaller value (54.2 MPa at point J) only a short distance away from the bottom of the chip.

1.4.4   Thermal fatigue life prediction of flip chip solder joints (fracture mechanics method)

In this section, the fracture mechanics method[167] will be used to predict the thermal fatigue life of the self-stretching flip chip (SSFC) solder joints connecting a silicon chip and an Al$_2$O$_3$ substrate[162] (Fig. 1.47). This flip chip assembly was tested by Satoh[162] and subjected to the thermal cycling condition (-50 to $+150$°C and one cycle per hour). The results of a corner SSFC solder joint are shown in Fig. 1.48. It can be seen that the corner SSFC solder joint cracked through its

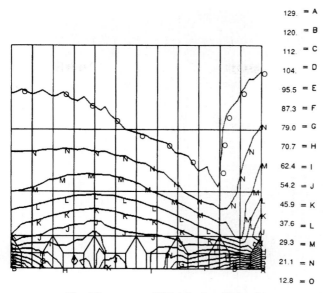

129.	= A
120.	= B
112.	= C
104.	= D
95.5	= E
87.3	= F
79.0	= G
70.7	= H
62.4	= I
54.2	= J
45.9	= K
37.6	= L
29.3	= M
21.1	= N
12.8	= O

Figure 1.46   Von Mises stress (MPa) in the silicon chip.

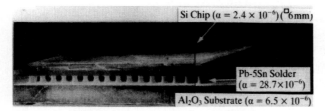

Figure 1.47   Self-stretching flip chip solder joint.[162]

cross section at about 3000 cycles. The purpose of this section is to model (by means of fracture mechanics) the thermal fatigue crack propagation of the corner SSFC solder joint.

The horizontal dimensions of the Si chip are about 6×6 mm (Fig. 1.47). The coefficient of thermal expansion is 2.4×10^{-6} °C^{-1} for the Si chip and is 6.5×10^{-6} °C^{-1} for the Al$_2$O$_3$ substrate. The geometry of the SSFC solder joint is shown in Figs. 1.48, 1.49, and 1.50. It can be seen that the joint is 0.32 mm high. The diameters of the bottom and top surfaces are 0.38 and 0.29 mm, respectively. The solder joint is made of 5/95 SnPb and has a Young modulus of $E = 2600$ MPa, yield stress of $\sigma_y = 18$ MPa, Poisson ratio of $\nu = 0.3$, and coefficient of ther-

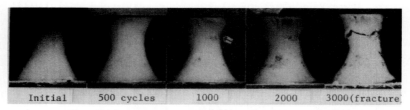

Figure 1.48 Thermal fatigue crack propagation in the self-stretching flip chip solder joint. Test condition: −50 to + 150°C, one cycle per hour.[162]

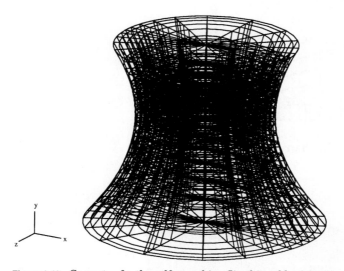

Figure 1.49 Geometry for the self-stretching flip chip solder joint.

mal expansion of $\alpha = 28.7 \times 10^{-6}$ °C^{-1}.

During the temperature cycling (−50 to + 150°C or $\Delta T = 200°C$), because of the assembly global thermal expansion mismatch between the Si chip (4.24 × 200 × 0.0000024 = 0.002 mm) and the Al$_2$O$_3$ substrate (4.24 × 200 × 0.0000065 = 0.0055 mm), and the thermal expansion of the solder joint (0.32 × 200 × 0.0000287 = 0.00184 mm), the corner solder joint is subjected to a complex state of stress and strain. These stresses and strains produce the driving force for solder joint failures. Since most of the thermal fatigue life of ductile materials is spent in propagating the crack (i.e., fatigue crack growth),[7] the strains and stresses around the crack tip (J integral and stress intensity factor) for different crack lengths in the solder joint are of utmost interest. In this section, seven different crack lengths have been studied. The boundary conditions imposed on the corner SSFC solder joint are shown in Fig. 1.50. It can be seen that the bottom surface of the

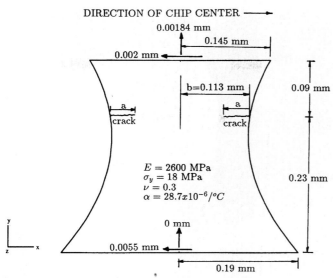

DIRECTION OF CHIP CENTER ⟶

Figure 1.50 Material properties, boundary conditions, and geometry of the self-stretching flip chip solder joint.

joint is subjected to a 0.0055-mm displacement moving to the left and the top surface of the joint is subjected to a 0.002-mm displacement moving to the left and a 0.00184-mm displacement moving in the upward direction.

The finite element model (dotted lines) and deformation (solid lines) of the corner SSFC solder joint without crack subjected to a $\Delta T = 200°C$ are shown in Fig. 1.51. Because of the symmetry of the problem, only half of the solder joint is modeled. A high-order, three-dimensional (3D) solid element has been used for the model. Each element has 20 nodes. Each node has three degrees of freedom. It can be seen from Fig. 1.51 that the corner SSFC solder joint moves away from the center of the chip because of the thermal expansion of the Si chip and Al_2O_3 substrate. It can also be seen that the lower part of the solder joint moves more than does the upper part.

The von Mises stress (MPa) contours in the SSFC solder joint is shown in Fig. 1.52. It can be seen that the maximum stress (42.1 MPa) occurs near the throat of the solder joint (see also Fig. 1.48). Thus, any solder joint cracks should begin at this location. In this study, we will model the crack propagation near the throat of the solder joint.

The 3D calculation of the stress intensity factor and J integral for various crack lengths of the SSFC solder joint (Figs. 1.48 to 1.50) is very time-consuming. Thus, in the present study the 2D plane stress theory is used. Figure 1.53 shows the von Mises stress (MPa) con-

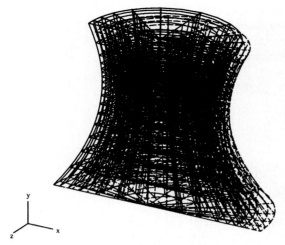

Figure 1.51 Deformation (solid lines) of the corner self-stretching flip chip solder joint (no crack).

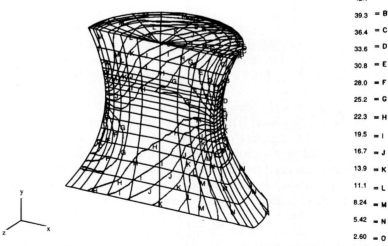

42.1	= A
39.3	= B
36.4	= C
33.6	= D
30.8	= E
28.0	= F
25.2	= G
22.3	= H
19.5	= I
16.7	= J
13.9	= K
11.1	= L
8.24	= M
5.42	= N
2.60	= O

Figure 1.52 3D von Mises stress distribution (MPa) in the self-stretching flip chip solder joint (no crack).

tours in the solder joint. A high-order 2D plane stress element has been used for the model. Each element has eight nodes. Each node has two degrees of freedom. It can be seen from Figs. 1.52 and 1.53 that the stress distribution for the 3D and 2D cases is very similar except for the stress magnitude (the 2D results are 11 percent less than the 3D results).

Figure 1.54 shows the deformed shape of the corner SSFC solder

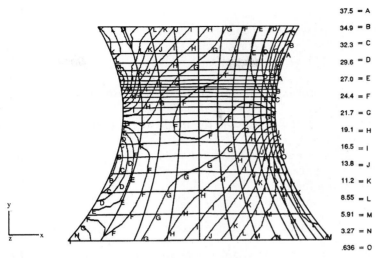

37.5	= A
34.9	= B
32.3	= C
29.6	= D
27.0	= E
24.4	= F
21.7	= G
19.1	= H
16.5	= I
13.8	= J
11.2	= K
8.55	= L
5.91	= M
3.27	= N
.636	= O

Figure 1.53 2D von Mises stress distribution (MPa) in the self-stretching flip chip solder joint (no crack).

$$\eta = a/b = 0.05$$
$$\Delta J = 0.0088 \text{ N/mm}$$
$$\Delta K = 4.78 \text{ MPa}\sqrt{\text{mm}}$$

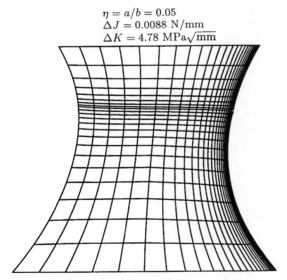

Figure 1.54 Deformation of the corner self-stretching flip chip solder joint with crack length $a = 0.0057$ mm.

joint with a crack near the throat. The crack length is $a = 0.00565$ mm. The element type and boundary conditions are exactly the same as those of the crack-free corner SSFC solder joint. In this analysis, ΔJ has been calculated at three contours around the crack tip. The average value is shown in Fig. 1.54 ($\Delta J = 0.0088$ N/mm). Once ΔJ is determined, the stress intensity factor ΔK can be determined by[175,176]

$$\Delta J = \frac{\Delta K_{\text{I}}^2 + \Delta K_{\text{II}}^2}{E} + \frac{\Delta K_{\text{III}}^2}{2G} \tag{1.51}$$

where ΔK_{I} = stress intensity factor for mode I (opening) type of fracture

ΔK_{II} = stress intensity factor for mode II (in-plane shear) type of fracture

ΔK_{III} = stress intensity factor for mode III (transverse shear) type of fracture

E = Young modulus

G = shear modulus

In this study, $\Delta K_{\text{II}} = \Delta K_{\text{III}} = 0$; then

$$\Delta K_{\text{I}} = \Delta K = \sqrt{E\,\Delta J} \tag{1.52}$$

Thus, for a crack length $a = 0.00565$ mm, $\Delta K = 4.78$ MPa$\sqrt{\text{mm}}$ (Fig. 1.54).

Figures 1.55 through 1.60 show the deformed shapes of the corner SSFC solder joint with crack length $a = 0.0113, 0.0226, 0.0339, 0.0565, 0.0791$, and 0.1017 mm, respectively. It can be seen that for

$$\eta = a/b = 0.1$$
$$\Delta J = 0.0153 \text{ N/mm}$$
$$\Delta K = 6.31 \text{ MPa}\sqrt{\text{mm}}$$

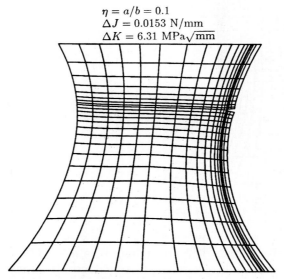

Figure 1.55 Deformation of the corner self-stretching flip chip solder joint with crack length $a = 0.0113$ mm.

$$\eta = a/b = 0.2$$
$$\Delta J = 0.0228 \text{ N/mm}$$
$$\Delta K = 7.70 \text{ MPa}\sqrt{\text{mm}}$$

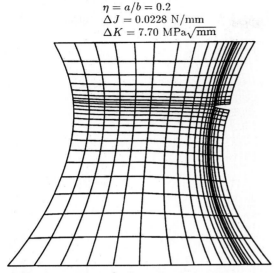

Figure 1.56 Deformation of the corner self-stretching flip chip solder joint with crack length $a = 0.0226$ mm.

$$\eta = a/b = 0.3$$
$$\Delta J = 0.0262 \text{ N/mm}$$
$$\Delta K = 8.25 \text{ MPa}\sqrt{\text{mm}}$$

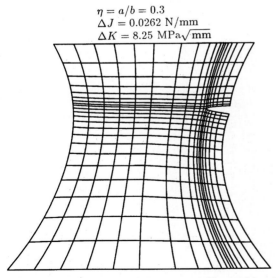

Figure 1.57 Deformation of the corner self-stretching flip chip solder joint with crack length $a = 0.0339$ mm.

$$\eta = a/b = 0.5$$
$$\Delta J = 0.0263 \text{ N/mm}$$
$$\Delta K = 8.25 \text{ MPa}\sqrt{\text{mm}}$$

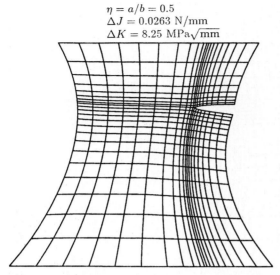

Figure 1.58 Deformation of the corner self-stretching flip chip solder joint with crack length $a = 0.0565$ mm.

$$\eta = a/b = 0.7$$
$$\Delta J = 0.0264 \text{ N/mm}$$
$$\Delta K = 8.25 \text{ MPa}\sqrt{\text{mm}}$$

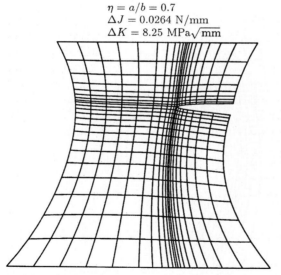

Figure 1.59 Deformation of the corner self-stretching flip chip solder joint with crack length $a = 0.0791$ mm.

$\eta = a/b = 0.9$
$\Delta J = 0.0265$ N/mm
$\Delta K = 8.25$ MPa$\sqrt{\text{mm}}$

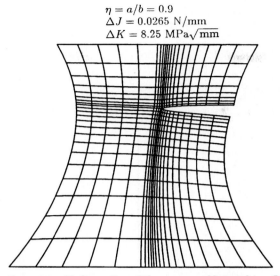

Figure 1.60 Deformation of the corner self-stretching flip chip solder joint with crack length $a = 0.1017$ mm.

$a = 0.0113$ mm, $\Delta J = 0.0153$ N/mm and $\Delta K = 6.31$ MPa$\sqrt{\text{mm}}$; for
$a = 0.0226$ mm, $\Delta J = 0.0228$ N/mm and $\Delta K = 7.70$ MPa$\sqrt{\text{mm}}$; for
$a = 0.0339$ mm, $\Delta J = 0.0262$ N/mm and $\Delta K = 8.25$ MPa$\sqrt{\text{mm}}$; for
$a = 0.0565$ mm, $\Delta J = 0.0263$ N/mm and $\Delta K = 8.25$ MPa$\sqrt{\text{mm}}$; for
$a = 0.0791$ mm, $\Delta J = 0.0264$ N/mm and $\Delta K = 8.25$ MPa$\sqrt{\text{mm}}$; and
for $a = 0.1017$ mm, $\Delta J = 0.0265$ N/mm and $\Delta K = 8.25$ MPa$\sqrt{\text{mm}}$.
These values are plotted in Figs. 1.61 and 1.62 for ΔJ and ΔK, respectively. The best-fit equations for ΔJ and ΔK are, respectively

$$\Delta J = 0.01(-.04 + 21.21\eta - 63.20\eta^2 + 88.07\eta^3 - 57.21\eta^4 + 13.70\eta^5) \tag{1.53}$$

$$\Delta K = 2.62 + 52.18\eta - 187.91\eta^2 + 330.18\eta^3 - 283.39\eta^4 + 95.03\eta^5 \tag{1.54}$$

where

$$\eta = \frac{a}{b} \tag{1.55}$$

In Eq. (1.55), a is the crack length and $b = 0.113$ mm (Fig. 1.50). Figure 1.63 shows the von Mises stress distribution (MPa) in the SSFC solder joint with crack length $a = 0.0339$ mm (Fig. 1.58). It can be seen that very large stress concentrates around the crack tip (Fig.

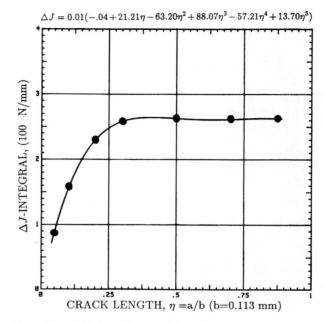

Figure 1.61 ΔJ integral vs. crack length.

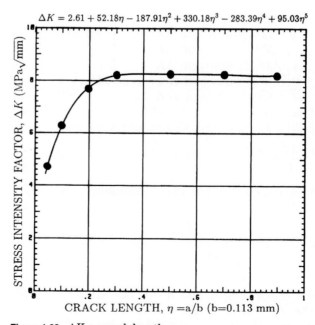

Figure 1.62 ΔK vs. crack length.

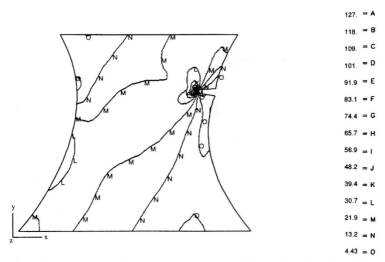

127.	= A
118.	= B
109.	= C
101.	= D
91.9	= E
83.1	= F
74.4	= G
65.7	= H
56.9	= I
48.2	= J
39.4	= K
30.7	= L
21.9	= M
13.2	= N
4.43	= O

Figure 1.63 2D von Mises stress distribution (MPa) in the flip chip solder joint with crack length $a = 0.0339$ mm.

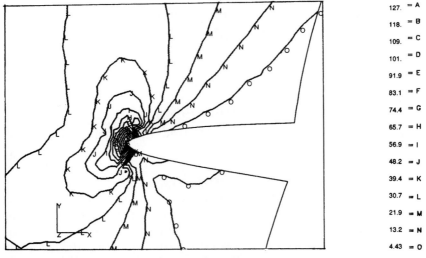

127.	= A
118.	= B
109.	= C
101.	= D
91.9	= E
83.1	= F
74.4	= G
65.7	= H
56.9	= I
48.2	= J
39.4	= K
30.7	= L
21.9	= M
13.2	= N
4.43	= O

Figure 1.64 2D von Mises stress distribution (MPa) around the flip chip solder joint crack tip with crack length $a = 0.0339$ mm.

1.64). This stress, however, decreases away from the crack tip. As a matter of fact, except near the crack tip, the stresses in the cracked solder joint (Figs. 1.63 and 1.64) are smaller than those in the un-cracked solder joint (Fig. 1.53).

Paris' law[175,176] is expressed as

$$\frac{da}{dN} = \gamma \Delta K^{\text{ß}} \tag{1.56}$$

where a = crack length
  N = number of cycles to failure
  $\gamma, \text{ß}$ = material constants

For 63Sn/37Pb solder, γ and ß have been determined by Logsdon et al.[171] at 24°C and 10 Hz (because of lack of fatigue crack growth data of 5Sn/95Pb solder, which is assumed to be equal to that of 63Sn/37Pb solder). Substituting their values for $\gamma = 2.77 \times 10^{-7}$ and ß = 3.26 into Paris' equation, we have

$$N = \int \frac{10^7 da}{2.77(\Delta K)^{3.26}} \tag{1.57}$$

In this equation, a is in centimeters, N is in cycles, and ΔK is in MPa$\sqrt{\text{m}}$. Thus, fatigue life of a solder joint can be estimated as soon as ΔK is known to be a function of crack length a.

Substituting Eqs. (1.54) in Eq. (1.57) and noting the units of the crack length a and ΔK, we have

$$N = \int_{0.05}^{0.9} \frac{0.113(10^7)d\eta}{(10)2.77(0.03162)^{3.26}(\Delta K)^{3.26}} \tag{1.58}$$

or

$$N =$$

$$\int_{0.05}^{0.9} \frac{3.1675(10^9)d\eta}{[2.62 + 52.18\eta - 187.91\eta^2 + 330.18\eta^3 - 283.39\eta^4 + 95.03\eta^5]^{3.26}} \tag{1.59}$$

After integration, we have $N = 3,460,000$ cycles.

As mentioned earlier, the solder material constants determined by Logsdon et al.[171] are at 10 Hz and 24°C. However, the thermal cycling results obtained by Satoh et al.[162] are at one cycle per hour and -50 to $+150$°C. Thus, a factor which captures the frequency and temperature effects is needed to transfer the lower stress condition to the higher stress condition. In this study, we assume the thermal fatigue life (N_f) of the corner SSFC solder joint to be

$$N_f = \omega N \tag{1.60}$$

where

$$\omega = \left(\frac{f_0}{f_t}\right)^{1/3} \left(\frac{\max T_t}{\max T_0}\right)^2 \qquad (1.61)$$

For the present study, $f_0 = 1$ cycle/h, $\max T_0 = 150°C$, $f_t = 10$ Hz, and $\max T_t = 24°C$. Then

$$\omega = (1/36,000)^{1/3} \, (24/150)^2 = 0.00078 \qquad (1.62)$$

$$N_f = (0.00078) \, (3,460,000) = 2700 \qquad (1.63)$$

This value compares very well with that (3000) of Satoh et al.[162]

On the basis of this study, the following conclusions and recommendations are made:

1. The finite element analysis showed that the maximum stress occurred near the throat of the corner SSFC solder joint. This result agreed very well with the observed failure location.

2. The ΔJ and ΔK values have been calculated for seven different crack lengths of the corner SSFC solder joint and have been best-fitted into functions of the crack length.

3. The thermal fatigue life ($\Delta T = 200°C$, 1 cycle/h) of the corner SSFC solder joint has been estimated by the calculated ΔK, Paris' law, and solder fatigue crack growth rate data to be 2700 cycles. This compared very well with experimental results (3000 cycles).

4. On the basis of all these assumptions, the present fracture mechanics approach to predict the thermal fatigue life of solder joints represented only a very small step beyond the Coffin-Manson approach. However, it is the author's hope that the present study has opened the door for advanced solder joint fatigue analysis by the fracture mechanics method.

5. The fatigue of materials is incompletely understood. This has been especially true for solder alloys because of the small amount of experimental data. In order to better understand and predict the solder joint thermal fatigue life, additional mechanical properties of bulk solders and joints are needed.[7,172] These are the true relationships between stress, strain, temperature, frequency, and time relation; fracture toughness and fatigue crack growth rate at various temperatures and frequencies; and the peeling and shearing fracture initiation strength at various temperatures.

1.5   Next-Generation Flip Chip Technologies

The strength of this book is in providing readers some of the advanced flip chip technologies which could lead to lower cost and higher performance applications. These include the solder-bumped flip chip technology, conductive polymer flip chip technology, anisotropic conductive flip chip technology, wired flip chip technology, TAB flip chip technology, metallurgy bumped flip chip technology and KGD methods.

1.5.1 Solder-bumped flip chip technology

In most of the standard or traditional fine-pitch solder bumped flip chip technologies, flux residues are entrapped underneath the chip and are very difficult to clean. These residues could degrade the flip chip assembly reliability with corrosion effects, gross solder voids, passivation delaminations, etc. Thus, a technique that will eliminate both flux and postsolder cleaning is desperately needed for low-cost and reliable flip chip applications. The objective of these techniques is to address the problem of the surface oxides covering the soldering surfaces that prevent joining from taking place.

A new fluxless process called *plasma-assisted dry soldering* (PADS) developed by MCNC[124,125] is presented by Koopman and Nangalia in Chap. 2 of this book. PADS is a dry pretreatment that converts the surface oxides to oxyfluorides. This conversion film passivates the solder surface and breaks when the solder melts in an inert (nitrogen) oven or even in oxidizing (air) ambients.

Some important aspects of the electrical and thermal performance of solder-bumped flip chip assemblies[133,134] are discussed by Sharma and Subrahmanyan in Chap. 3. Also, on the basis of a damage integral approach, they outlined a methodology to develop accelerated test strategies for evaluating new solder bumped flip chip technologies.

Flip chip technology, especially for very large and high-I/O chips, has been recognized as one of the most important technologies necessary to meet the stringent requirements of workstation computer applications.[130] In Chap. 4, Chung, Dolbear, and Nelson examine the objectives, approaches, key technology elements, and some of the test results of the thermal management, manufacturability, and reliability of very large and high-I/O flip chip technology.

The needs for increased interconnection I/O and performance have led AT&T to develop a platform which consists of a composite of substrate attachment, and packaging that they call *microinterconnect technology*.[131,132] It exploits the unique capabilities of the silicon MCM-D interconnection architectures. In Chap. 5, Degani, Dudderar, Frye, Tai, Lau, and Han describe the design, development, and optimization of these manufacturing processes for the cost-effective large-volume assembly of flip chip silicon-on-silicon MCM-D through the use of a custom solder-bumped flip chip technology and high-speed automated assembly.

1.5.2 Conductive polymer flip chip technology

Conductive adhesive polymer materials have been used by the electronics industry for a long time. However, these materials are only now beginning to be recognized for high volume and commercial flip chip applications.[135] In Chap. 6, Estes and Kulesza present an overview of various conductive adhesive polymer materials and their applications to flip chip technology. Some of the design, process, electrical, and mechanical reliability data are also provided.

Some drawbacks of conductive adhesive flip chip technology are nonreworkability and movement in the z axis caused by the thermal expansion coefficient and water absorption of the adhesive. With these in mind, MCC (Microelectronics and Computer Technology Corporation) developed new compliant polymer-core bumps with a metal coating.[137] Breen, Duane, German, Keswick, and Nolan (Chap. 7) describe the polymer bump formation and metallization, assembly and rework processes, and fabrication cost pictures. Also, some reliability data are presented.

1.5.3 Anisotropic conductive flip chip technology

Anisotropic conductive adhesive materials are available in two different forms: anisotropic conductive film (ACF) and anisotropic conductive adhesive (ACA) paste. They are also called *z-axis conductive adhesives* because they are designed to be conductive between the electrodes of the chip and substrate in the vertical direction and nonconductive (no short circuits) in the horizontal direction.

Fujitsu developed a new ACA material for fine-pitch interconnection between flip chip and PCB.[148] Their ACA material consists of conductive particles (usually Ag) coated with a thin dielectric resin. During bonding, the high pressure between the pad on the chip and the corresponding pad on the PCB destroys the resin layer between the pads. The electrical connection is established through the Ag conductive particles. In Chap. 8, Date, Hozumi, Tokuhira, Usui, Horikoshi, and Sato present the characteristics of their ACA material and the reliability data of the ACA interconnection between the chip and PCB.

A few ACF materials have been developed by Hitachi for flip chip applications.[158,159] In Chap. 9, Watanabe, Shiozawa, Takemura, and Ohta describe the principle of ACF interconnects and the role of conducting particles dispersed in ACF through voltage, current, and resistance measurements. Also, they disclose a double-layer ACF material which consists of a nonfilled adhesive layer and a conducting particle-filled adhesive layer.

The applications of ACA and ACF materials to flip chip-on-glass (COG) technology[1,136] are presented in Chap. 10. The processes and materials for several flip COG assembly methods using ACA or ACF have been discussed along with each method's technical merits and demerits by Lee.

1.5.4   Wired flip chip technology

More than 90 percent of the chips used today are face-up wire-bonded on the substrate. Thus, making the Au stud with a wire bonder should be very straightforward and the reliability data are already available. However, as mentioned earlier, face-down chip technology has many advantages over the face-up chip technology. Thus, a few wired flip chip technologies are presented in this book.

A few years ago, a new assembly process called *bonded interconnect pin* (BIP) technology was developed by Raychem Corporation.[143] It utilizes thermosonic ball bonding and reflow soldering to create electrical interconnections between the flip chip and the substrate. In Chap. 11, Baba and Carlomagno present the design, fabrication, and assembly process of the BIP technology and also provide some reliability data on BIP assembly under temperature, mechanical, shock and vibration, and stabilization bake conditions.

Fujitsu developed a slightly different wired flip chip technology and put it into productions for their low-end computer products. They use wire bonder to make the stud bumps and conductive paste to attach the stud bumps on the PCB. Reliability of the assembly is enhanced by encapsulant.[146] Tsunoi, Kusagaya, and Kira (Chap. 12) describe the structure, stud-bump formation and leveling, conductive paste application, assembly process, and rework. Reliability testing results are also presented.

Flip chip technologies using the wire bonder can apply only to peripheral array pads. For area array pads, FCPT (Fujitsu Computer Packaging Technologies) developed a new wire interconnect technology (WIT) for high-density and high-performance flip chip applications.[144] WIT "grows" very small metal posts (usually 10 μm in diameter and 50 μm long) from the chip pad, and the other end of the posts are soldered on the substrate. In Chap. 13, Moresco, Love, Chou, and Holalkere disclose their WIT fabrication and assembly process, and reliability testing results. The flip chip WIT technical characteristics are also presented through electrical, mechanical stress, heat transfer, α-particle radiation, and routing analyses and measurements.

1.5.5   TAB flip chip technology (CSP)

The area array tape-automated bonding (TAB) technology plays a very important role in electronics packaging. By combining the most advantages of area array TAB, wire bonding, and flip chip technologies, Tessera developed a small, thin, high-performance chip size package (CSP) called micro–ball grid array (μBGA).[138–141] It allows a chip pitch as small as 50 μm (which is competitive with advanced TAB pad pitch), and the flexible gold beam lead is assembled to the chip pad with conventional thermosonic wire-bonding equipment. The leads fan inwardly from the chip pads to a NiAu (or solder) bump array that serves to make connection to the next-level package. High level of reliability can be achieved with the elastomer compliant layer, which also provides the opportunity for full test and burn-in for KGD prior to final assembly. In Chap. 14, DiStefano and Fjelstad present the design, material, bump metallurgy, and assembly process of their μBGA. Electrical and thermal performance, reliability testing, and burn-in of the μBGA are also provided.

1.5.6 Metallurgy-bumped flip chip technology

As mentioned earlier, a fluxless and no-clean process is desperately needed for low-cost and reliable flip chip applications[126–128]. In Chap. 15, Zakel and Reichl describe their bumping processes and characteristics for the Au (by electroplating), AuSn (by electroplating), and NiAu (by electroless) metallurgies, and Au-stud and solder-stud bumps and also provide the assembly processes and test data of flip chip assemblies using Au, AuSn, NiAu, Au-stud and solder-stud bumps.

1.5.7 KGD for flip chip technology

The importance of KGD for flip chip DCA and MCM applications has been shown in Sec. 1.3. In Chap. 16, Gilg addresses the KGD definition and standards, and the cost for producing KDG and also present various methods (such as process control, statistical sampling, testing every die, test & burn-in carriers, and wafer-level burn-in) of assuring KGD for flip chip applications.

In Chap. 17, Rinne explains the principles of burn-in for clip chip ICs and also provides some key flip chip burn-in considerations such as alignment, contact area, differential expansion, oxidation, solid-state diffusion, bump integrity, and plastic and creep deformation of the bump. Table 1.3 shows the flowchart of this book.

1.6 Summary

A brief introduction to IC chip yield, KGD, MCM yield, and some of the necessary information for flip chip technologies was presented. Two methods for estimating the thermal fatigue life of flip chip solder joints were provided. Next-generation flip chip technologies such as the fluxless solder bump, conductive polymer, anisotropic conductive

TABLE 1.3   Flowchart for Chapters in This Book*

Introduction to flip chip technologies for multichip modules (1)

Solder-bumped flip chip:
　　Fluxless joining (2)
　　Material, process, and reliability (3)
　　Large, high-I/O chips (4)
　　Cost-effective flip chip MCMs (5)

Conductive polymer flip chip:
　　Conductive adhesive polymers (6)
　　Compliant bumps (7)

Anisotropic conductive flip chips:
　　ACA (8)
　　ACF (9)
　　COG (10)

Wired flip chip:
　　BIP (11)
　　Stud bumps (12)
　　Wire interconnects (13)

TAB flip chip:
　　μBGA chip size package (14)

Metallurgy flip chip:
　　Au, AuSn, NiAu bumps (15)

KGD for flip chip applications:
　　Assurance technologies for KGD (16)
　　Test for KGD (17)

Authors' biographies

Index

*Numbers in parentheses represent chapter numbers.

adhesive, wire interconnect, chip size package, metallurgy bump, and KGD testing were briefly discussed.

As the trend in the electronics industry toward streamlining products—making them more compact in size, faster (shorter cycle time), and yet more user-friendly, functional, powerful, reliable, and less expensive—continues, it seems clear that the use of flip chip technologies will be increased in order to satisfy these needs. It is our hope that the information presented in this book may assist in removing roadblocks, avoiding unnecessary false starts, and accelerating design, material, and process development of flip chip technologies.

1.7 Flip Chip Patents

The following flip chip patents have been arranged in two groups: U.S. patents and non-U.S. (world) patents. All the U.S. patents are listed with their exact title, patent number, and issue date. All the non-U.S. (world) patents are listed with their World Patent Index Accession Number (abbreviated WPI Acc. in list), and a very brief description. (The first two digits of the WPI Accession Number represent the issue year.) Failure to include any particular flip chip patent indicates only that the author either had not read it or was unaware of it when this book went to press.

1.7.1 Some U.S. flip chip patents

1. "Flip Chip MICROFUS," U.S. Patent 5,363,082, Nov. 8, 1994.
2. "Flip Chip Package and Method of Making," U.S. Patent 5,352,926, Oct. 4, 1994.
3. "Method for Forming a Flip-Chip Bond from a Gold-Tin Eutectic," U.S. Patent 5,346,857, Sept. 13, 1994.
4. "Direct Chip Attach Module (DCAM)," U.S. Patent 5,313,366, May 17, 1994.
5. "Backplane Grounding for Flip-Chip Integrated Circuit," U.S. Patent 5,311,059, May 10, 1994.
6. "Method and Apparatus for Isolation of Flux Materials in Flip-Chip Manufacturing," U.S. Patent 5,299,730, April 5, 1994.
7. "Packaging Method for Flip-Chip Type Semiconductor Device," U.S. Patent 5,297,333, March 29, 1994.
8. "Heatspreader for Cavity Down Multi-Chip Module With Flip Chip," U.S. Patent 5,289,337, Feb. 22, 1994.
9. "Process for Flip Chip Connecting a Semiconductor Chip; Mating Solder Bumps on Semiconductor Chip and Corresponding Solder Bumps on Circuit Board, Heating and Breaking away Outer Peripheral Portion," U.S. Patent 5,284,796, Feb. 8, 1994.
10. "Three-Dimensional Memory Card Structure with Internal Direct Chip Attachment," U.S. Patent 5,280,192, Jan. 18, 1994.
11. "Composite Flip Chip Semiconductor Device with an Interposer Having Test Contacts Formed Along Its Periphery," U.S. Patent 5,258,648, Oct. 2, 1993.
12. "Bellows Lid for C4 Flip-Chip Package," U.S. Patent 5,257,162, Oct. 26, 1993.
13. "Semiconductor Device Having Flip Chip Bonding Pads Matched with Pin Photodiodes in a Symmetrical Layout Configuration," U.S. Patent 5,252,852, Oct. 12, 1993.
14. "Flip Chip Encapsulating Compositions and Semiconductor Devices Encapsulated Therewith; Epoxy and Silicone Resins with Imidazole Compounds and Silica," U.S. Patent 5,248,710, Sept. 28, 1993.

15. "Method for Creating Substrate Electrodes for Flip Chip and Others Actions," U.S. Patent 5,246,880, Sept. 21, 1993.
16. "Flip Chip Technology Using Electrically Conductive Polymers and Dielectrics; Bumped Substrate," U.S. Patent 5,237,130, Aug. 17, 1993.
17. "Electrode Bump for Flip Chip Die Attachment," U.S. Patent 5,235,140, Aug. 10, 1993.
18. "Three-Dimensional Memory Card Structure with Internal Direct Chip Attachment; High Density, High Speed Packages Having Reduced Signal Delay," U.S. Patent 5,227,338, July 13, 1993.
19. "Method of Generating a Substrate Electrode for Flip Chip and Other Actions Integrated Circuits," U.S. Patent 5,213,676, May 25, 1993.
20. "Vacuum Infiltration of Underfill Material for Flip-Chip Devices," U.S. Patent 5,203,076, April 20, 1993.
21. "Flip Chip Bonding Method Using Electrically Conductive Polymer Bumps," U.S. Patent 5,196,371, March 23, 1993.
22. "Flexible Attachment Flip-Chip Assembly," U.S. Patent 5,189,505, Feb. 23, 1993.
23. "Method of Bonding Metals, and Method and Apparatus for Producing Semiconductor Integrated Circuit Device Using Said Method of Bonding Metals; Using Flip-Chip System; Radiating, Pressing in Inert Gas; Fluxless; Solder Bumps," U.S. Patent 5,188,280, Feb. 23, 1993.
24. "Method and Apparatus for Isolation of Flux Materials in Flip-Chip Manufacturing," U.S. Patent 5,168,346, Dec. 1, 1992.
25. "Method of Producing Microbump Circuits for Flip Chip Mounting; Depositing Refractory Metals on Either Side of a Layer of Aluminum to Promote Adhesion; Lifting off Photoresist Mask," U.S. Patent 5,118,584, June 2, 1992.
26. "Single Substrate Microwave Radar Transceiver Including Flip-Chip Integrated Circuits," U.S. Patent 5,115,245, May 19, 1992.
27. "Apparatus for Isolation of Flux Materials in "Flip-Chip" Manufacturing," U.S. Patent 5,111,279, May 5, 1992.
28. "Flip Chip Solder Bond Structure for Devices with Gold Based Metallization; Barrier Metallization Layer between Metallization Layer that Would React with Solder," U.S. Patent 5,108,027, April 28, 1992.
29. "Three-Dimensional Memory Card Structure with Internal Direct Chip Attachment," U.S. Patent 5,099,309, March 24, 1992.
30. "Flip-Chip MMIC Oscillator Assembly with Off-Chip Coplanar Waveguide Resonant Inductor," U.S. Patent 5,087,896, Feb. 11, 1992.
31. "Strain Relief Flip-Chip Integrated Circuit Assembly with Test Fixturing," U.S. Patent 5,077,598, Dec. 31, 1991.
32. "Flip Chip Technology Using Electrically Conductive Polymers and Dielectrics; Connecting with Bond Pads," U.S. Patent 5,074,947, Dec. 24, 1991.
33. "Flip-Chip Test Socket Adaptor and Method for Providing Electrical Connection to Integrated Circuit Die," U.S. Patent 5,073,117, Dec. 17, 1991.
34. "Flip Chip Type Semiconductor Device," U.S. Patent 5,046,161, Sept. 3, 1991.
35. "Vernier Structure for Flip Chip Bonded Devices," U.S. Patent 5,022,580, June 11, 1991.
36. "Flip-Chip Package for Integrated Circuits," U.S. Patent 5,019,673, May 28, 1991.
37. "Flip-Chip Test Socket Adaptor and Method," U.S. Patent 5,006,792, April 9, 1991.
38. "Insulated Substrate for Flip-Chip Integrated Circuit Device; Metallized Circuits Supported by an Insulating Layer which Has Been Deposited on a Layer of Silicon Carbide," U.S. Patent 4,979,015, Dec. 18, 1990.
39. "Repairable Flip-Chip Bumping," U.S. Patent 4,817,850, April 4, 1989.
40. "Flip Chip Module," U.S. Patent 4,805,007, Feb. 14, 1989.
41. "Heatsink Package for Flip-Chip IC," U.S. Patent 4,803,546, Feb. 7, 1989.
42. "Electrical Monitoring of Flip-Chip Hybridization," U.S. Patent 4,803,422, Feb. 7, 1989.
43. "Flip-Chip Pressure Transducer," U.S. Patent 4,763,098, Aug. 9, 1988.
44. "Heatsink Package for Flip-Chip IC," U.S. Patent 4,698,663, Oct. 6, 1987.
45. "Method of Batch-Fabricating Flip-Chip Bonded Dual Integrated Circuit Arrays," U.S. Patent 4,416,054, Nov. 22, 1983.

46. "Self-Aligned, Flip-Chip Focal Plane Array Configuration," U.S. Patent 4,369,458, Jan. 18, 1983.
47. "Flip Chip Mounted Diode," U.S. Patent 4,250,520, Feb. 10, 1981.
48. "Method for Making a Plurality of Solid Electrolyte Capacitors and Capacitors Made Thereby; Manganese Dioxide Layers over Oxidized, Porous Valve Metal Pads, Flip-Chip," U.S. Patent 4,188,706, Feb. 19, 1980.
49. "Self Biasing of a Field Effect Transistor Mounted in a Flip-Chip Carrier," U.S. Patent 4,183,041, Jan. 8, 1980.
50. "Process and Device for Unsoldering Semiconductor Modules in the Flip-Chip Technique," U.S. Patent 4,066,204, Jan. 3, 1978.
51. "Flip Chip Structure Including a Silicon Semiconductor Element Bonded to an Si_3N_4 Base Substrate," U.S. Patent 3,955,008, Feb. 1, 1977.
52. "Ultrasonic Thermal Compression Beam Lead, Flip Chip Bonder," U.S. Patent 3,938,722, Feb. 17, 1976.
53. "Flip Chip Cartridge Loader," U.S. Patent 3,937,386, Feb. 10, 1976.
54. "Plastic Power Semiconductor Flip Chip Package," U.S. Patent 3,922,712, Nov. 25, 1975.
55. "Flip Chip Module with Non-uniform Connector Joints," U.S. Patent 3,871,015, March 11, 1975.
56. "Flip Chip Module with Non-uniform Solder Wettable Areas on the Substrate," U.S. Patent 3,871,014, March 11, 1975.
57. "High Heat Dissipation Solder-Reflow Flip Chip Transistor," U.S. Patent 3,823,469, July 16, 1974.
58. "High Heat Dissipation Solder-Reflow Flip Chip Transistor," U.S. Patent 3,772,575, Nov. 13, 1973.
59. "Semiconductor Flip-Chip Soldering Method," U.S. Patent 3,665,590, Jan. 25, 1972.
60. "Flip Chip Integrated Circuit and Method Therefor," U.S. Patent 3,636,619, May 30, 1972.
61. "Method of Providing Flip-Chip Devices with Solderable Connections," U.S. Patent 3,622,385, Nov. 23, 1971.
62. "Semiconductor Devices with Lines and Electrodes which Contain 2 to 3 Percent Silicon with the Remainder Aluminum," U.S. Patent 3,609,470, Sept. 28, 1971.
63. "Method for Bonding the Flip-Chip to a Carrier Substrate," U.S. Patent 3,559,279, Feb. 2, 1971.
64. "Flip Chip Thick Film Device," U.S. Patent 3,539,882, Nov. 10, 1970.
65. "Flip Chip Structure," U.S. Patent 3,517,278, June 23, 1970.
66. "Method of Forming a Flip-Chip Integrated Circuit," U.S. Patent 3,442,012, May 6, 1969.
67. "Method of Joining a Component to a Substrate," U.S. Patent 3,429,040, Feb. 25, 1969.
68. "Method of Rendering Noble Metal Conductive Composition Non-wettable by Solder," U.S. Patent 3,401,126, Sept. 10, 1968.
69. "Method and Apparatus for Reverse Sputtering Selected Electrically Exposed Areas of a Cathodically Biased Workpiece," U.S. Patent 3,410,774, Nov. 12, 1968.
70. "Vapor Depositing Solder," U.S. Patent 3,401,055, Sept. 10, 1968.
71. "Method for Providing Electrical Connections to Semiconductor Devices," U.S. Patent 3,382,568, May 14, 1968.
72. "Terminals for Microminiaturized Devices and Methods of Connecting Same to Circuit Panels," U.S. Patent 3,303,393, Feb. 7, 1967.
73. "Silver Conductors," U.S. Patent 2,961,416, Nov. 22, 1960.

1.7.2　Some non-U.S. flip chip patents

1. WPI Acc. 94-306243/38, "Flip Chip Integrated Circuit Mounting Structure."
2. WPI Acc. 94-302445/37, "Apparatus for Electrically Connecting Flip Chips to Flexible Printed Circuit Substrate."
3. WPI Acc. 94-297751/37, "Multi-chip Module for LSI Circuits."
4. WPI Acc. 94-295505/37, "Packaged Semiconductor Device Especially with Flip-Chip Bonded Integrated Circuit for Microstrip Line Mounting."

5. WPI Acc. 94-293621/36, "Flip-chip Structure with Interposed Preformed Planar Structure Using Embedded Conductors."
6. WPI Acc. 94-288840/36, "Flip Chip Bonding Device Allowing High Density Chip Mounting on Substrate."
7. WPI Acc. 94-288836/36, "Semiconductor IC Device Manufacture Providing Improved Flip Chip Bump Connection."
8. WPI Acc. 94-288834/36, "Flip-chip Bonding Method for Mounting Semiconductor Chip on Wafer."
9. WPI Acc. 94-282413/35, "Solder Cream Printing Method for Flip Chip Bonding of Semiconductor Element on Printed Circuit Substrate."
10. WPI Acc. 94-282377/35, "Bump Manufacturing Flip Chip Packaging."
11. WPI Acc. 94-282176/35, "Resin Sealing of Semiconductor Wafer in Device, as in Flip-Chip Manufacture."
12. WPI Acc. 94-280028/34, "Thermally Conductive Flip Chip IC Package."
13. WPI Acc. 94-279011/34, "Chip Carrier Apparatus for Attaching Semiconductor Die Pads to Printed Circuit Boards."
14. WPI Acc. 94-276563/34, "Semiconductor Device with Flip-Chip System."
15. WPI Acc. 94-274595/34, "Semiconductor Device with Flip-Chip Mounting for High Speed Operation."
16. WPI Acc. 94-274594/34, "Semiconductor Device with Flip Chip Mounting Used for High Speed Operation."
17. WPI Acc. 94-274593/34, "Semiconductor Device Package by Flip Chip Mounting for High Speed Operation."
18. WPI Acc. 94-267698/33, "Alloy Wire for Bump Electrode Formation in Semiconductor Element for Flip Chip Bonding."
19. WPI Acc. 94-258418/32, 'Semiconductor Device for High Speed Switching Operation."
20. WPI Acc. 94-258353/32, "Semiconductor IC Device Manufacturing Method Using Flip Chip System."
21. WPI Acc. 94-244088/30, "Flip Chip Junction Structure."
22. WPI Acc. 94-232084/28, "Encapsulating Flip Chip IC's on Ceramic Substrates."
23. WPI Acc. 94-231474/28, "Flip Chip Bonding for Semiconductor Device Package."
24. WPI Acc. 94-213018/26, "Semiconductor Device with Flip Chip Mounting System."
25. WPI Acc. 94-213015/26, "Electrode Wafer Structure for Flip Chip Mounting with Accurate Positioning of Connector Pad."
26. WPI Acc. 94-213011/26, "Flip Chip Manufacturing."
27. WPI Acc. 94-203896/25, "Flip Chip Bonding Device."
28. WPI Acc. 94-203892/25, "Flip Chip Bonding."
29. WPI Acc. 94-203833/25, "Structure of Flip Chip Bump."
30. WPI Acc. 94-201981/25, "Device Including IC Flip Chip and Heat-Sink."
31. WPI Acc. 94-195140/24, "High Reliability Semiconductor Device and Clock Signal Feed Device."
32. WPI Acc. 94-192920/24, "PCB with Increased Wiring Density in Regions and with Conical Holes."
33. WPI Acc. 94-179857/22, "Image Sensor Mounting Unit."
34. WPI Acc. 94-179633/22, "Flip Chip Mounted on Substrate."
35. WPI Acc. 94-179632/22, "Flip Chip Mounted on PCB."
36. WPI Acc. 94-179629/22, "Bump Electrode of Flip Chip Bonded onto PCB."
37. WPI Acc. 94-171589/21, "Semiconductor Flip Chip and Mount for Transmitting Electrical Signals."
38. WPI Acc. 94-171504/21, "Mounting of Semiconductor Element with Flip Chip."
39. WPI Acc. 94-161965/20, "Sealing of Flip Chip-Type Semiconductor Device Eliminating Air Bubbles between Chip and Wafer."
40. WPI Acc. 94-154997/19, "Printed Wiring Board for Flip Chip."
41. WPI Acc. 94-154672/19, "Semiconductor Wafer with Flexible Circuit Wiring Pattern and Suitable for Flip-Chip Mounting of Circuit Parts."
42. WPI Acc. 94-154669/19, "Reusable Semiconductor Device Flip Chip Mounting Arrangement."
43. WPI Acc. 94-154620/19, "Solder Bump Manufacturing for Flip-Chip Technology."

44. WPI Acc. 94-154618/19, "Half Ball Shape Solder Bump Manufacturing for Flip Chip Technology."
45. WPI Acc. 94-139252/17, "Resin Mold Package Semiconductor Device."
46. WPI Acc. 94-139119/17, "Flip Chip Type Semiconductor Chip Mounting Structure."
47. WPI Acc. 94-139037/17, "Face-Down Bonding Electrode for Flip Chip Semiconductor."
48. WPI Acc. 94-130028/16, "Flip Chip Mounting for IC Device."
49. WPI Acc. 94-113136/14, "Flip-Chip Mounting System Semiconductor IC to Prevent Disconnection Trouble."
50. WPI Acc. 94-113098/14, "Semiconductor to Seal Face Down Flip-Chip Element into Ceramic-Package."
51. WPI Acc. 94-113084/14, "Flip-Chip Type Semiconductor to Couple Chip-Electrode with Substrate-Electrode by CCB-Method."
52. WPI Acc. 94-113033/14, "Flip-Chip Bonding Method to Circuit Board."
53. WPI Acc. 94-113032/14, "Flip-Chip Integrated Circuit Bonding Method to Circuit Board."
54. WPI Acc. 94-113030/14, "Flip-Chip Integrated Circuit Mounting Method."
55. WPI Acc. 94-104855/13, "Semiconductor Device Mounted by Flip Chip Bonding."
56. WPI Acc. 94-103329/13, "Flip-Chip Mounting Process for Substrate with Conductive Paths."
57. WPI Acc. 94-096228/12, "Mounting Semiconductor Element on Substrate by Flip Chip Bonding."
58. WPI Acc. 94-096040/12, "Hybrid Integrated Circuit."
59. WPI Acc. 94-095951/12, "Wiring Substrate Connection to Directly Connect Flip-Chip through Bump."
60. WPI Acc. 94-095950/12, "Semiconductor Package with Particular Flip-Chip Connection to Eliminate Cross-talk."
61. WPI Acc. 94-094174/12, "Semiconductor Device Contact Especially Multiple Ball Contact for Flip Chip."
62. WPI Acc. 94-087268/11, "Flip-Chip Junction State Inspecting Device."
63. WPI Acc. 94-078384/10, "Photocoupler Comprising Sealed Light Receiving and Light Emitting Elements."
64. WPI Acc. 94-078334/10, "Arithmetic Function Attached Photodiode."
65. WPI Acc. 94-069668/09, "Mounting Flip-Chip on Integrated Circuit."
66. WPI Acc. 94-069667/09, "Flip Chip Mounting Method for Semiconductor Device."
67. WPI Acc. 94-069640/09, "Semiconductor Device Manufacturing."
68. WPI Acc. 94-060510/08, "Flip Chip Bonding Method."
69. WPI Acc. 94-060509/08, "Flip Chip Bonding of Integrated Circuit Chip to Circuit Board."
70. WPI Acc. 94-060508/08, "Flip Chip Bonding IC Chip to Circuit Board."
71. WPI Acc. 94-052129/07, "Semiconductor Flip Chip Mounting Structure with Excellent Heat Radiation."
72. WPI Acc. 94-052000/07, "Method of Forming Bumps for Flip Chip Mounting."
73. WPI Acc. 94-051378/07, "Picture Forming Device."
74. WPI Acc. 94-038448/05, "Flip Chip Loading Structure for Electronic Part."
75. WPI Acc. 94-036429/05, "Multi-chip Module."
76. WPI Acc. 94-029838/04, "Multilayer Ceramic Substrate for Electronic Apparatus."
77. WPI Acc. 94-029502/04, "IC Chip Heater Mechanism for TAB and Flip-Chip Bonder."
78. WPI Acc. 94-029436/04, "Manufacturing Flip Chip IC Module."
79. WPI Acc. 94-009718/02, "X-Ray Imager Especially Tomograph for Mammography and Dentistry."
80. WPI Acc. 94-003495/01, "Semiconductor Flip-Chip Mounting Method to Eliminate Parallelism Adjusting Process."
81. WPI Acc. 93-390312/49, "Chip Mounting Method to Mount Chip in Flip-Chip."
82. WPI Acc. 93-390307/49, "Semiconductor Provided Face Down Bonding Flip Chip."
83. WPI Acc. 93-390232/49, "Production of Bump Is Electrode of Flip Chip."
84. WPI Acc. 93-381039/48, "Electrode Pads of High Frequency IC Device."

85. WPI Acc. 93-380970/48, "Mounting Flip Chip Using Face-Down Bonding."
86. WPI Acc. 93-380968/48, "Flip Chip Mounting Method."
87. WPI Acc. 93-380964/48, "Flip Chip Semiconductor Device."
88. WPI Acc. 93-377111/47, "Bonding Flip Chips to a Flexible PCB."
89. WPI Acc. 93-372432/47, "Semiconductor with High-Power IC-Chip, Substrate and Radiation Plate."
90. WPI Acc. 93-372349/47, "Mounting Flip Chip on Printed Wiring Board."
91. WPI Acc. 93-363060/46, "Semiconductor Manufacturing for Fixing Chip to Substrate by Flip-Chip System."
92. WPI Acc. 93-363056/46, "Flip-Chip Structure Semiconductor to Connect with Electrodes by CCB Method."
93. WPI Acc. 93-362992/46, "Forming Bumps for Flip Chip Bonding IC Chip."
94. WPI Acc. 93-355052/45, "Flip-Chip Bonding Semiconductor Device."
95. WPI Acc. 93-355000/45, "Inspection of Flip Chip Bump."
96. WPI Acc. 93-353133/45, "Semiconductor Chip Mounting Method on Board Especially for Flip Chip with Holder."
97. WPI Acc. 93-352335/45, "Semiconductor Device with IC and Hermetic Sealing Cap Especially for Flip Chip Mounting."
98. WPI Acc. 93-344320/43, "Optical Card Fabricated Using Planar Optical Substrate."
99. WPI Acc. 93-341613/43, "Superconductive Flip-Chip."
100. WPI Acc. 93-341480/43, "Flip-Chip System IC Chip Mounting Structure."
101. WPI Acc. 93-341398/43, "Flip-Chip Bonding Circuit Board."
102. WPI Acc. 93-341395/43, "Intermediate Electrode Bump Board for Flip-Chip Bonding."
103. WPI Acc. 93-341390/43, "Inter-substrate Connecting Terminal Frame for Flip-Chip and Circuit Board."
104. WPI Acc. 93-333046/42, "Integrated Circuit Chip Face-Down Mounting Wiring Substrate."
105. WPI Acc. 93-332966/42, "Flip Chip Bonded Circuit Board."
106. WPI Acc. 93-332965/42, "Connecting Structure between Multilayer Circuit Pattern Board and Flip Chip Integrated Circuit."
107. WPI Acc. 93-324131/41, "Flip-Chip Mounting Method Directly Soldered Naked LSI-Chip to Circuit Substrate."
108. WPI Acc. 93-315297/40, "Flip-Chip Mounting."
109. WPI Acc. 93-306907/39, "Multilayer Wiring Substrate for Mounting Chip by Flip-Chip Bonding Method."
110. WPI Acc. 93-306834/39, "IC Pellet Aligning Structure for Flip Chip Bonding Device."
111. WPI Acc. 93-306832/39, "Flip Chip Bonding Device."
112. WPI Acc. 93-306830/39, "Flip Chip Bonding Structure."
113. WPI Acc. 93-306829/39, "Flip Chip Bonding Method."
114. WPI Acc. 93-306828/39, "Flip Chip Bonding Method."
115. WPI Acc. 93-306827/39, "Flip Chip Bonding Method."
116. WPI Acc. 93-306826/39, "Flip Chip Bonding Structure."
117. WPI Acc. 93-306820/39, "IC Chip Device—Has Flip Chip Bump Bonded to Flexible Circuit."
118. WPI Acc. 93-306747/39, "Flip Chip Semiconductor Device."
119. WPI Acc. 93-304950/39, "Chip Carrier for Protecting Optical Device."
120. WPI Acc. 93-298496/38, "Multi-Chip Package for Mounting Flip-Chip Semiconductor IC(s)."
121. WPI Acc. 93-293289/37, "Light Erasable Flip Chip Packaging Method."
122. WPI Acc. 93-291512/37, "Flip-Chip Junction Method Enabling Low Temperature Processing."
123. WPI Acc. 93-283675/36, "Flip-Chip Integrated Circuit Mounting Device."
124. WPI Acc. 93-276181/35, "Semiconductor Having Chip Mounted by Flip-Chip Method."
125. WPI Acc. 93-276161/35, "Flip-Chip Junction Method for Mounting Chip on Substrate."

126. WPI Acc. 93-276122/35, "Flip-Chip Type Semiconductor Mounted on Electric Circuit Wiring Substrate."
127. WPI Acc. 93-276120/35, "Flip-Chip Mounting Semiconductor Device."
128. WPI Acc. 93-274737/35, "Integrated Circuit Connection Method to Conductor in External Circuit Especially Flip Chip Connection."
129. WPI Acc. 93-267889/34, "Wiring Board for Mounting Integrated Circuit Chip by Flip-Chip Bonding."
130. WPI Acc. 93-254496/32, "IC Module Manufacturing—Positioning Circuit Substrate Mounting Flip Chips."
131. WPI Acc. 93-246139/31, "Mounting of Flip-Chip of Semiconductor Device."
132. WPI Acc. 93-230111/29, "Flip Chip Bonding Structure for IC Chip on Circuit Board."
133. WPI Acc. 93-222282/28, "Flip-Chip Component Mounting Structure with Fine Pad Distance."
134. WPI Acc. 93-222227/28, "Flip Chip Bonded Semiconductor Device."
135. WPI Acc. 93-210153/26, "Flip-Chip Mounting Structure with Improved IC Chip Structure."
136. WPI Acc. 93-210137/26, "Electronic Part Mounting Method to Mount Flip-Chip in Circuit Substrate."
137. WPI Acc. 93-206853/26, "Flip-Chip Bonded Defective Resin Encapsulated Semiconductor Die Replacement Method for Direct Chip Attachment Package."
138. WPI Acc. 93-202077/25, "Film Forming Apparatus for Flip Chip Semiconductor Device."
139. WPI Acc. 93-202042/25, "Mounting Structure for Chip Component on Ceramic Substrate."
140. WPI Acc. 93-202040/25, "Semiconductor Device for Flip-Chip Bonding."
141. WPI Acc. 93-178677/22, "Semiconductor Device Having Flip Chip Structure."
142. WPI Acc. 93-178676/22, "IC Flip Chip Mounting."
143. WPI Acc. 93-178675/22, "Coolable Circuit Board Substrate."
144. WPI Acc. 93-173366/21, "Measuring Tilt Angle of Flip-Chips Surface-Mounted Technology Components."
145. WPI Acc. 93-162982/20, "Flip-Chip Type Semiconductor Device."
146. WPI Acc. 93-155718/19, "Pad for Flip Chip Bonding."
147. WPI Acc. 93-147240/18, "Metal Connection Manufacturing for Flip Chip Bonding."
148. WPI Acc. 93-138203/17, "Flip-Chip System IC Device Manufacturing Technology."
149. WPI Acc. 93-122911/15, "Flip Chip Semiconductor Device and Associated Mounting."
150. WPI Acc. 93-122910/15, "Flip Chip with Circular Bumps."
151. WPI Acc. 93-114469/14, "Flip Chip Bonding Semiconductor IC Device."
152. WPI Acc. 93-106122/13, "High Density Semiconductor Package."
153. WPI Acc. 93-105989/13, "Bonding Head for Flip Chip."
154. WPI Acc. 93-097560/12, "Flip Chip Marking Method with Non-removable Marks."
155. WPI Acc. 93-090396/11, "Light Receiver Element with High Speed Response for Flip-Chip Bonding."
156. WPI Acc. 93-090395/11, "Flip-Chip Bonded Light Receiver with Good Optical Connection with Optical Fibre."
157. WPI Acc. 93-082059/10, "Multi-chip Module Flip-Chip Bonding Method."
158. WPI Acc. 93-082058/10, "Semiconductor Element Flip-Chip Connecting Structure."
159. WPI Acc. 93-078322/10, "Semiconductor Chip for Flip-Chip Mounting on Package Substrate."
160. WPI Acc. 93-072463/09, "Flip-Chip Bonding Method for Electrically and Mechanically Connecting Element Chip with Substrate."
161. WPI Acc. 93-072462/09, "IC Package Chip Mounting Method on Wiring Substrate."
162. WPI Acc. 93-055876/07, "Semiconductor Device with Substrate Connected to Chip via Flip Chip System."
163. WPI Acc. 93-055875/07, "Flip Chip Bonding Method on LSI Circuit Substrate."

164. WPI Acc. 93-047824/06, "Multi-flip Chip Module."
165. WPI Acc. 93-047822/06, "Multi-flip Chip Package."
166. WPI Acc. 93-030260/04, "Flip Chip Mounting."
167. WPI Acc. 93-026376/03, "Integrated Circuit Flip-Chip Bonding Method."
168. WPI Acc. 93-013087/02, "Mounted Circuits Device on Wired Substrate by Flip-Chip Connection."
169. WPI Acc. 93-010317/02, "Sealed Flip Chip Device."
170. WPI Acc. 93-001266/01, "Integrated Socket-Type Package for Flip Chip Semiconductor Devices and Circuits."
171. WPI Acc. 92-409516/50, "Semiconductor IC Package Using Flip Chip Method."
172. WPI Acc. 92-408561/50, "Flip Device Bonding Machine."
173. WPI Acc. 92-402897/49, "Semiconductor Bump Connection Method in Mounting Flip Chip."
174. WPI Acc. 92-385545/47, "Flip-Chip Semiconductor Device."
175. WPI Acc. 92-375542/46, "Formulation of Solder Bumps for Use in Flip-Chip Bonding."
176. WPI Acc. 92-369190/45, "Resin Mold Packaged Stacked LSI Chips."
177. WPI Acc. 92-360171/44, "Flip-Chip Type Light Acceptance Device Property Evaluation Nozzle."
178. WPI Acc. 92-358384/44, "Flip-Chip Bonding Process Using Face-Down Method."
179. WPI Acc. 92-340545/41, "Laser Diode Carrier with Semiconductor Cooler."
180. WPI Acc. 92-340255/41, "Indium Bump Contacts for Flip-Chip Bonding."
181. WPI Acc. 92-337238/41, "Flip Chip Joining Apparatus."
182. WPI Acc. 92-336647/41, "Semiconductor Using Air Tight Solder Bump in Inactive Solution."
183. WPI Acc. 92-296984/36, "Registration of Flip Chip Bonding."
184. WPI Acc. 92-289423/35, "Flip Chip Bonder Device."
185. WPI Acc. 92-271735/33, "Hybrid Integrated Circuit."
186. WPI Acc. 92-256734/31, "Flip Chip Method of Semiconductor Device."
187. WPI Acc. 92-253259/31, "Multilayered Glass-Ceramic Substrate."
188. WPI Acc. 92-252872/31, "Parallel Press Mechanism for Flip-Chip Mount."
189. WPI Acc. 92-249607/30, "Adhesive for Reflow Soldering and Encapsulation of Flip Chip IC(s)."
190. WPI Acc. 92-236926/29, "Solder Bump Manufacturing Method for Flip Chip Packaging."
191. WPI Acc. 92-211098/26, "Mounting Flip Chip of Semiconductor Element."
192. WPI Acc. 92-210816/26, "Flip-Chip Mounting on Substrate."
193. WPI Acc. 92-210805/26, "Flip-Chip Bump-Lead Shaping."
194. WPI Acc. 92-210616/26, "Mounting Method of Semiconductor Device Using Flip Chip Bonding."
195. WPI Acc. 92-210615/26, "Flip Chip Bonding Method."
196. WPI Acc. 92-210604/26, "Semiconductor Integrated Circuit Device for Flip-Chip Bonding."
197. WPI Acc. 92-208371/25, "Gallium-arsenic Monolithic Waveguide Switch."
198. WPI Acc. 92-205925/25, "Sealing of Semiconductor with Flip Chip Type Connection."
199. WPI Acc. 92-194639/24, "Semiconductor Integrated Circuit Flip-Chip Packaging System."
200. WPI Acc. 92-188301/23, "Mounting of Semiconductor Chip."
201. WPI Acc. 92-179731/22, "Mounting Structure of Semiconductor Device."
202. WPI Acc. 92-172606/21, "Solder Vapour Depositing Apparatus for Forming Bump Contact in Flip Chip Joint."
203. WPI Acc. 92-163351/20, "IC Chip Circuit Mounting Assembly by Flip-Chip Method."
204. WPI Acc. 92-146465/18, "Flip-Chip Bonding Device for Electrode and Bump."
205. WPI Acc. 92-136963/17, "DIP IC Module—Has Rivets for Fixing Flip Chips-Bonded Wiring Substrate to Tab."
206. WPI Acc. 92-133153/17, "Flip-Chip Assembly."
207. WPI Acc. 92-121326/15, "Flip Chip Socket for Test and Burn-in."

208. WPI Acc. 92-121302/15, "Organo-silicon Polymer Liquid Encapsulant for Flip Chip Encapsulant."
209. WPI Acc. 92-119749/15, "Packaging Flip-Chip Preventing Crushing of Bump."
210. WPI Acc. 92-093422/12, "Bonding Device for Flip Chip."
211. WPI Acc. 92-086551/11, "Soldering Process for DCA Using Solder Foils."
212. WPI Acc. 92-085587/11, "Print Substrate for Packaging Flip Chip of Array Sensor."
213. WPI Acc. 92-085106/11, "Installation Method of Semiconductor Chip."
214. WPI Acc. 92-085090/11, "Flip Chip Mounting Direction for Hybrid IC."
215. WPI Acc. 92-084019/11, "Multi-layered Circuit Substrate."
216. WPI Acc. 92-074749/10, "Flip Chip Bonding Head Apparatus."
217. WPI Acc. 92-074319/10, "Method for Testing Semiconductor Device."
218. WPI Acc. 92-059922/08, "Flip Chip Solder Bonding Apparatus."
219. WPI Acc. 92-058982/08, "Electrical Connection Apparatus for Semiconductor Chips."
220. WPI Acc. 92-052892/07, "Semiconductor with ICs on Printed Substrate."
221. WPI Acc. 92-050625/07, "Film Mounting of Integrated Circuit."
222. WPI Acc. 92-043002/06, "Reliable Flip Chip Semiconductor Device."
223. WPI Acc. 92-036830/05, "Manufacturing Hybrid IC Loading Flip Chip."
224. WPI Acc. 92-027705/04, "Semiconductor Integrated Circuit Module."
225. WPI Acc. 92-027688/04, "Hybrid Integrated Circuit Device."
226. WPI Acc. 92-027687/04, "Flip Chip Bonding Type Hybrid Substrate Integrated Circuit."
227. WPI Acc. 92-019023/03, "Flip Chip Type Semiconductor Device Manufacturing."
228. WPI Acc. 92-013440/02, "Laser Soldering Device for Substrate."
229. WPI Acc. 92-007715/01, "Microbumps for Flip-Chip Mounting."
230. WPI Acc. 92-002897/01, "Hybrid IC Mounted Doubled IC Chips—Uses Flip Chip Connection on One Side and Wire Bonding Connection on Other Side."
231. WPI Acc. 92-002881/01, "Hybrid IC—Mounts Flip Chip IC on Aluminum Double Layer Wiring Silicon Substrates."
232. WPI Acc. 91-328468/45, "Semiconductor Device Mounting Method Using Flip-Chip Bonding."
233. WPI Acc. 91-298319/41, "Flip Chip Bonding—Coating Solder Flux on Chip."
234. WPI Acc. 91-280686/38, "Manufacturing of Page-Width Array."
235. WPI Acc. 91-264485/36, "Bonder for Mounting IC Flip Chip on Circuit Board."
236. WPI Acc. 91-260108/35, "Semiconductor Module Densely Packing IC Chips."
237. WPI Acc. 91-255885/35, "Soldering Jig and Method for Flip Chip Joining."
238. WPI Acc. 91-243051/33, "Bonding Machine for Connection of Flip-Chip to Circuit Board."
239. WPI Acc. 91-219490/30, "Bonding Semiconductor Flip Chip to Control Board."
240. WPI Acc. 91-218638/30, "Solder Bump Structure in Flip Chip Method."
241. WPI Acc. 91-218178/30, "Flip-Chip Type Semiconductor IC."
242. WPI Acc. 91-218177/30, "Flip Chip Type Semiconductor IC Device Manufacturing."
243. WPI Acc. 91-196711/27, "Flip Chip Bonding Device."
244. WPI Acc. 91-189429/26, "Semiconductor IC Device Formed by Using Flip Chip System."
245. WPI Acc. 91-189338/26, "Flip Chip IC."
246. WPI Acc. 91-169445/23, "Direct Chip Attach Thin Film Package for Fibre-Optic Transceiver."
247. WPI Acc. 91-129385/18, "Semiconductor Device."
248. WPI Acc. 91-128609/18, "High Sensitivity Semiconductor Acceleration Sensor for IC."
249. WPI Acc. 91-114799/16, "Semiconductor Device."
250. WPI Acc. 91-106112/15, "Flip Chip Element Packaging."
251. WPI Acc. 91-099372/14, "Semiconductor Device Bonding to Circuit Board."
252. WPI Acc. 91-084073/12, "Flip-Chip Semiconductor Bonding Method."
253. WPI Acc. 91-083587/12, "Flip Chip Mounting Device."
254. WPI Acc. 91-069435/10, "Chip-Mounted Circuit Device."

255. WPI Acc. 91-062880/09, "Forming Bump Electrode on Semiconductor IC Chip."
256. WPI Acc. 91-062272/09, "Semiconductor-Chip-Mounted Circuit Board Unit."
257. WPI Acc. 91-062114/09, "Electrode Connecting Structure of IC Chip and Circuit Board."
258. WPI Acc. 91-040221/06, "Wiring Improving Liquid Crystal Display Method."
259. WPI Acc. 91-025027/04, "Semiconductor Device with Short Circuit Prevention."
260. WPI Acc. 91-017521/03, "Flip Chip Type Semiconductor Device."
261. WPI Acc. 91-011339/02, "Wiring Substrate for Flat or Flip Chip Package."
262. WPI Acc. 91-002763/01, "Image Sensor for CCD Camera."
263. WPI Acc. 91-002648/01, "Flip-Chip Joining Semiconductor Chip to Circuit Board."
264. WPI Acc. 90-380294/51, "Production of Ceramic Circuit Board for Mounting of Semiconductor Chip."
265. WPI Acc. 90-372743/50, "Bonding Semiconductor Chip to Circuit Board."
266. WPI Acc. 90-357681/48, "Flip-Chip Bonding Method."
267. WPI Acc. 90-344821/46, "Flip Chip Delaminating Heat Tool."
268. WPI Acc. 90-344819/46, "Mask for Printing Solder Paste on Hybrid IC."
269. WPI Acc. 90-334457/44, "Multilayer Bonding and Cooling of Integrated Circuit Devices."
270. WPI Acc. 90-330984/44, "Flip Chip Bonding Using Thick Super-elastic Metal Film."
271. WPI Acc. 90-330983/44, "Flip Chip Bonding."
272. WPI Acc. 90-330982/44, "Flip Chip Bonding Method."
273. WPI Acc. 90-325843/43, "Coupling Method of Circuit Wafer with Reduced Manufacturing Cost."
274. WPI Acc. 90-324822/43, "Solder Bump-Pad Junction Composition by Flip Chip Mounting."
275. WPI Acc. 90-301577/40, "Mounting IC Chip on Circuit Board."
276. WPI Acc. 90-293759/39, "IC Flip Chip Mounting Method."
277. WPI Acc. 90-280571/37, "Fabricating Thick Film Solder Attach Pads for Fine Pitch Flip Chips."
278. WPI Acc. 90-264599/35, "Flip Chip Bonding Substrate, e.g., for Semiconductor Element."
279. WPI Acc. 90-236398/31, "Solid-State Image Sensor."
280. WPI Acc. 90-227151/30, "Flip-Chip Bonding Method."
281. WPI Acc. 90-190059/25, "Hybrid IC Device Production."
282. WPI Acc. 90-175590/23, "Hybrid IC Device—Comprises Flip Chips Mounted on Silicon Circuit."
283. WPI Acc. 90-168677/22, "IC Chip Bump Formation Performing Flip-Chip Bonding."
284. WPI Acc. 90-166383/22, "Flip Chip Mounting Circuit Substrate."
285. WPI Acc. 90-166265/22, "Flip Chip Element for Memory Element."
286. WPI Acc. 90-104737/14, "Flip-Chip Component for Installing onto PCB."
287. WPI Acc. 90-104736/14, "Flip-Chip Component for Installing onto PCB."
288. WPI Acc. 90-104735/14, "Flip-Chip Component for Installing onto PCB."
289. WPI Acc. 90-093729/13, "Flip Chip Mounting for Thermal Printing Head."
290. WPI Acc. 90-062507/09, "Flip Chip Mounting Structure Enabling Reliable Chip Replacement."
291. WPI Acc. 90-055856/08, "Flip-Chip Mounter for Installing Semiconductor on Circuit Board."
292. WPI Acc. 90-037595/06, "Flip-Chip Mounting Process."
293. WPI Acc. 90-011278/02, "Mounting Semiconductor Flip Chip on Printed Circuit Board."
294. WPI Acc. 89-377807/51, "Plating Frame for Producing Bumps on Surface of Semiconductor Wafer."
295. WPI Acc. 89-374309/51, "Machine in Which Flip-Chip Mounted on LSI Chip."
296. WPI Acc. 89-373121/51, "Flip Chip Type Mounting for Semiconductor Device."
297. WPI Acc. 89-343989/47, "Soldering Flip Chip to Bump in Test Board."
298. WPI Acc. 89-329039/45, "Flip Chip Mounting Circuit Substrate."
299. WPI Acc. 89-328985/45, "Flip-Chip Type Semiconductor Device."
300. WPI Acc. 89-328148/45, "Flip Chip Having Gourd-Shaped Electrode."

301. WPI Acc. 89-318813/44, "Package Structure for Flip Chip Type Semiconductor Device."
302. WPI Acc. 89-313829/43, "Bonding Flip Chips onto Circuitry Base."
303. WPI Acc. 89-312119/43, "Flip Chip Bonding Process."
304. WPI Acc. 89-305387/42, "Integrated Circuit Carrier, Placing IC Package in Socket."
305. WPI Acc. 89-296121/41, "Manufacturing High Density Mounting Flip Chip."
306. WPI Acc. 89-296120/41, "Manufacturing High Density Mounting Flip Chip."
307. WPI Acc. 89-296116/41, "Flip Chip with Solder Bumps Arranged in Matrix."
308. WPI Acc. 89-281732/39, "Flip Chip Bump Electrode for Bonding onto Circuitry Base."
309. WPI Acc. 89-281724/39, "Flip Chip Mounted Circuitry Base."
310. WPI Acc. 89-268473/37, "Flip-Chip-Mounted Semiconductor Device."
311. WPI Acc. 89-267191/37, "Flip Chip Manufacturing."
312. WPI Acc. 89-267190/37, "Manufacturing Flip Chip with Stress Relief Bump Electrode."
313. WPI Acc. 89-266378/37, "Wiring Structure of Bistable Semiconductor Laser."
314. WPI Acc. 89-266322/37, "Flip Chip Type IC Mounting for Improving Bonding Reliability."
315. WPI Acc. 89-261160/36, "Semiconductor Hybrid IC Device."
316. WPI Acc. 89-259413/36, "Semiconductor Device Used in Flip Chip System Assembling."
317. WPI Acc. 89-253641/35, "Flip-Chip Semiconductor Device."
318. WPI Acc. 89-245736/34, "Mounting Microchip on Printed Circuit Base."
319. WPI Acc. 89-245668/34, "Flip Chip Bonding Method."
320. WPI Acc. 89-224881/31, "Manufacturing Flip-Chip Semiconductor IC Device."
321. WPI Acc. 89-216055/30, "Semiconductor Incorporating Flip Chip Type Chip."
322. WPI Acc. 89-209529/29, "Substrate for Mounting Flip Chip."
323. WPI Acc. 89-209528/29, "Flip Chip Bump Electrodes Bonded onto Circuit Substrate."
324. WPI Acc. 89-209526/29, "Automatic Flip Chip Bonding Equipment."
325. WPI Acc. 89-209485/29, "LCD Circuit Substrate with Electrode Excitation Pellet."
326. WPI Acc. 89-183731/25, "Electrode Structure of Flip-Chip Transistor."
327. WPI Acc. 89-169714/23, "Semiconductor Chip Carrier—Connects Chips in Surface of Flip-Chip."
328. WPI Acc. 89-147795/20, "Liquid Crystal Display Element with Mounted Flip Chip."
329. WPI Acc. 89-135169/18, "Fixture Board for Testing or Aging Semiconductor Flip Chips."
330. WPI Acc. 89-133296/18, "Bi-facial Flip-Chip Bonding Multilayer Circuitry Board."
331. WPI Acc. 89-133295/18, "Fault-Finding Facilitation Circuitry Board Mounting Flip-Chips."
332. WPI Acc. 89-133292/18, "Flip Chip Bonded Package Circuitry Manufacturing."
333. WPI Acc. 89-125480/17, "Flip Chip Position Photodetector for Circuitry Manufacture."
334. WPI Acc. 89-125393/17, "Flip-Chip with Buffer Electrode Connection."
335. WPI Acc. 89-123459/17, "Damage Preventive Flip Chip Auto-bonding Carrier Tape."
336. WPI Acc. 89-116940/16, "Semiconductor Flip-Chip Mounter."
337. WPI Acc. 89-096455/13, "Flip Chip Bonding Electrode for Hybrid IC."
338. WPI Acc. 89-089557/12, "Circuit Board Designed to Mount Flip Chips without Shorting."
339. WPI Acc. 89-088323/12, "Hybrid Integrated Circuit."
340. WPI Acc. 89-080891/11, "Improved Exothermal Escape Performance IC Pellet Carrier."
341. WPI Acc. 89-072955/10, "Bonding Semiconductor Flip Chip to Circuit Board."
342. WPI Acc. 89-056555/08, "Semiconductor Device Used for Film Carrier or Flip Chip System."
343. WPI Acc. 89-041657/06, "Flip Chip Semiconductor Device."

344. WPI Acc. 89-011996/02, "Flip Chip Bonding Method."
345. WPI Acc. 89-010852/02, "Flip-chip Semiconductor Device."
346. WPI Acc. 89-003220/01, "Durably Mounting Pellet IC on Printed Circuit Board."
347. WPI Acc. 88-350555/49, "Semiconductor Memory Device."
348. WPI Acc. 88-341188/48, "Flip Chip Type Semiconductor Device."
349. WPI Acc. 88-327269/46, "Manufacturing Flip Chip Junction LSI Reinforced by Resin."
350. WPI Acc. 88-317885/45, "Production of Hybrid Package LSI."
351. WPI Acc. 88-317749/45, "Computer Assisted Flip Chip Auto-mounting Step Management."
352. WPI Acc. 88-311227/44, "Forming Bump Electrode Connecting to Flip-Chip IC."
353. WPI Acc. 88-311078/44, "Flip-Chip Bonding."
354. WPI Acc. 88-310328/44, "Gallium Arsenide MESFET with Flip Chip Structure."
355. WPI Acc. 88-303407/43, "Flip Chip Bump Electrode Production without Delaminating."
356. WPI Acc. 88-281053/40, "Multiple Flip Chip Mounting Structure for Circuit Board."
357. WPI Acc. 88-247236/35, "Bonding Method for Flip Chip."
358. WPI Acc. 88-231045/33, "Conical Bump Contact Manufacturing for Flip-Chip IC."
359. WPI Acc. 88-185821/27, "Probe for Flip Chip with Dense Terminal Array."
360. WPI Acc. 88-170463/25, "Cooling Circuit Board Having Flip Chip."
361. WPI Acc. 88-162715/24, "Head with LED Assembly for Electrophotographic Printer."
362. WPI Acc. 88-152878/22, "Mounting Structure for LSI Chip."
363. WPI Acc. 88-145349/21, "Package for Semiconductor IC Flip Chip."
364. WPI Acc. 88-143091/21, "High Density Mounting Structure for Hybrid Integrated Circuit."
365. WPI Acc. 88-122982/18, "Forming Solder Bump in Wireless Bonding Flip Chip."
366. WPI Acc. 88-122382/18, "Hybrid IC Chip Structure without Bonding Wire."
367. WPI Acc. 88-115927/17, "Soldering Flip Chip onto Circuit Board."
368. WPI Acc. 88-115867/17, "Dustproof Flip Chip Bonding Method."
369. WPI Acc. 88-094152/14, "Circuit Substrate for Mounting Flip Chip."
370. WPI Acc. 88-074792/11, "Semiconductor Device."
371. WPI Acc. 88-058752/09, "Flip-Chip Bonded Hybrid Integrated-Circuit Device."
372. WPI Acc. 88-047682/07, "Forming of Solder Bump for Flip Chip."
373. WPI Acc. 88-038629/06, "Connecting Flip Chip to ITO Pattern of Circuit Substrate."
374. WPI Acc. 88-033334/05, "Connection Structure for Flip Chip."
375. WPI Acc. 88-032613/05, "Installing Flip-Chip Semiconductor Element on Circuit Board."
376. WPI Acc. 88-032295/05, "Spherical Bump Electrode Formation Method."
377. WPI Acc. 88-018985/03, "Bump for Bonding Flip Chip."
378. WPI Acc. 88-018444/03, "Semiconductor Device with Eutectic Alloy between Terminal and Substrate."
379. WPI Acc. 88-012015/02, "Flip Chip System Semiconductor Unit."
380. WPI Acc. 87-345936/49, "Flip-Chip Joint Method for Manufacturing of Hybrid IC."
381. WPI Acc. 87-345935/49, "Semiconductor Hybrid IC Device."
382. WPI Acc. 87-345934/49, "Method for Bonding Semiconductor Flip Chip."
383. WPI Acc. 87-339339/48, "Semiconductor Chip."
384. WPI Acc. 87-310474/44, "Optical Printer Head for Electronic Photograph Printer Light Source."
385. WPI Acc. 87-304688/43, "Economic Production of Flip Chip Bonding Bump Electrode."
386. WPI Acc. 87-304687/43, "Robust Flip Chip Bonding Bump."
387. WPI Acc. 87-275097/39, "Flip-Chip Semiconductor Device."
388. WPI Acc. 87-253665/36, "Production Method for Flip-Chip Semiconductor Device."
389. WPI Acc. 87-240363/34, "Cooled Flip Chip LSI Package Module."
390. WPI Acc. 87-238318/34, "Bonding System Soldering Flip-Chip Elements to Substrate."

391. WPI Acc. 87-223843/32, "Manufacturing Flip Chip."
392. WPI Acc. 87-211149/30, "Flip Chip Mounting Construction."
393. WPI Acc. 87-210008/30, "Semiconductor Package with Flip-Chip Connection and Rear Side Cooling."
394. WPI Acc. 87-199336/29, "Contacting Semiconductor Device Using Flip-Chip Method."
395. WPI Acc. 87-188798/27, "Stress-Absorbing IC Package Module."
396. WPI Acc. 87-188795/27, "Flip Chip-Mounted Board with Reinforced Cohesive Strength."
397. WPI Acc. 87-102858/15, "Mounting Flip-Chip onto Substrate with Coolant Channels."
398. WPI Acc. 87-091181/13, "Positioning Flip Chip in Mounting Part of Substrate."
399. WPI Acc. 87-079792/11, "Solder Bumps Flip-Chip Bonding Process for Optical Device."
400. WPI Acc. 87-069975/10, "Flip Chip Bonder."
401. WPI Acc. 87-040472/06, "Flip Chip Pattern for Connection on PCB."
402. WPI Acc. 87-040452/06, "Flip Chip-Bonded Electrode Bumps for Semiconductor."
403. WPI Acc. 87-033114/05, "Solder Bump Formation for Semiconductor Flip-Chip Bonding."
404. WPI Acc. 86-329731/50, "Manufacturing of Flip-Chip Semiconductor Device."
405. WPI Acc. 86-328818/50, "Thermal Head with Flip-Chip Bonding."
406. WPI Acc. 86-276807/42, "Flip-Chip Bonding for Optical Coupler."
407. WPI Acc. 86-249305/38, "Multilayer Ceramic Substrate is Mounted with Flip Chip Comprising Semiconductor."
408. WPI Acc. 86-249304/38, "Multilayer Ceramic Substrate Mounted with Flip Chip."
409. WPI Acc. 86-242867/37, "Thermal Head with Flip-Chip Bonding."
410. WPI Acc. 86-236902/36, "Wire Bonder for Flip Chip Semiconductor Device."
411. WPI Acc. 86-236901/36, "Flip Chip Type Semiconductor Device Repairer."
412. WPI Acc. 86-230109/35, "Quality-Examinable Flip-Chip-Mounted Device."
413. WPI Acc. 86-229346/35, "Sticking IC Chip of Flip-Chip-Type on Lamina."
414. WPI Acc. 86-216951/33, "Flip-Chip Integrated Circuit Package Module."
415. WPI Acc. 86-216409/33, "Semiconductor Flip Chip System."
416. WPI Acc. 86-201512/31, "Flip Chip for Photocoupler Applications."
417. WPI Acc. 86-193048/30, "Semiconductor Apparatus with Efficient Heat Transfer Loads Semiconductor Chips Using Flip-Chip Method."
418. WPI Acc. 86-171727/27, "Hermetically Sealing Flip Chip."
419. WPI Acc. 86-140357/22, "Flip-Chip Style Semiconductor Device Has Dummy Project Electrode for Heat Radiation between Chip and Substrate."
420. WPI Acc. 86-135273/21, "Flip Chip on Substrate Mounting Method."
421. WPI Acc. 86-128738/20, "Mounting Flip Chip and Bonding Wire with Resin."
422. WPI Acc. 86-128736/20, "Mounting Flip Chip on Substrate and Covering with Intermediate Protection Material."
423. WPI Acc. 86-127262/20, "Flip-Chip Packaging Using Leadless Chip Carrier."
424. WPI Acc. 86-103348/16, "Semiconductor-Chip Mounted on Substrate in Flip-Chip-System Has Closely Metallic Cap."
425. WPI Acc. 86-072798/11, "Flip-Chip Semiconductor Device Has Wiring for Probe on Circuit Board Loading Chip, and Coupling by Face-Down Bonding."
426. WPI Acc. 86-052046/08, "Flip-Chip Bonded Low-Noise LSI Manufacturing."
427. WPI Acc. 86-017710/03, "Flip-Chip Semiconductor Device Has Heat Sinking Thermally Conducting Body between Back Surface of Chip and Cap."
428. WPI Acc. 86-002685/01, "Semiconductor Wafer with Flip-Chip Face-Down Bonding Forms Hole for Superposing Bump Electrode on Chip by Matching with Mark."
429. WPI Acc. 85-286542/46, "Flip Chip Type Semiconductor Device Mounting Method."
430. WPI Acc. 85-280849/45, "Semiconductor Device Manufacturing for Flip-Chip."
431. WPI Acc. 85-279913/45, "Circuit Substrate Mounted with Flip Chip."
432. WPI Acc. 85-267116/43, "Flip Chip Bonding Type Semiconductor Device Has Two Insulator Layers with Coaxial Openings between Wiring Layer and Pump."
433. WPI Acc. 85-259911/42, "Flip Chip Bonding as Positioning Chip so that Edges of Dyed Portion Coincide with Those of Substrate."

434. WPI Acc. 85-239117/39, "Semiconductor Device with Wire Bonding Pads Comprises Semiconductor Element with Flip Chip Bonding Bump Having Different Pitch from Wire Bonding Pad Pitch."
435. WPI Acc. 85-157095/26, "Flip Chip Integrated Circuit Manufacturing by Forming Bump on Electrode Pad of Semiconductor Pellet by Wire Bonding Method."
436. WPI Acc. 85-148785/25, "Flip-Chip Bonding on Card."
437. WPI Acc. 85-089775/15, "Flip-Chip Type Semiconductor Device Has Low Melting Point Metal Bump Formed in Direction of Stress."
438. WPI Acc. 85-023595/04, "Bonding Flip Chip to Wiring Conductor on IC Through Solder Layers on Conductor and Bump Electrodes on Chip."
439. WPI Acc. 85-022620/04, "Flip-Chip Bonding Method for Semiconductor Pellet Using Bump Whose Tip Is Formed with Low-Melting-Point Metal."
440. WPI Acc. 85-009739/02, "High Loading Density Semiconductor Device Has Flip Chip Bonding Electrode and Wire Bonding Electrode Formed on Same Substrate."
441. WPI Acc. 85-009642/02, "Formation Method of Bump for Flip Chip Bonding."
442. WPI Acc. 84-303928/49, "Burn-in Apparatus for Flip Chip."
443. WPI Acc. 84-297634/48, "Semiconductor Device Having Elements with Bumps Uses Pressing Device for Flip Chip Bonding."
444. WPI Acc. 84-285957/46, "Flip Chip Loading Method."
445. WPI Acc. 84-253868/41, "Flip Chip IC Testing Method."
446. WPI Acc. 84-240438/39, "Covering Semiconductor with Flip-Chip Electrode with Protective Resin by Mounting Element on Circuit Substrate, and Covering Element with Light Curing Resin."
447. WPI Acc. 84-210016/34, "Component to Lead Frame Joining Method Using Flip-Chip Bonding."
448. WPI Acc. 84-174284/28, "Semiconductor Device with Vertically Super Imposed Dielectric Plate Has at Least One Device Contacting Plate Flip Chip Type."
449. WPI Acc. 84-168402/27, "Flip Chip Bonding Device."
450. WPI Acc. 84-155011/25, "Hybrid Device Mounted with Flip-Chip System."
451. WPI Acc. 84-137759/22, "Flip Chip Bonding Device."
452. WPI Acc. 84-123030/20, "Manufacturing Bump for Flip Chip Applications."
453. WPI Acc. 84-110961/18, "Semiconductor Device for Flip Chip Has Reinforcing Metal Film on Other Main Surface of Semiconductor Plate."
454. WPI Acc. 84-098516/16, "Semiconductor Device Taking Flip Chip Structure by Wireless Bonding."
455. WPI Acc. 84-097189/16, "Electronic Circuit Module Connects and Mounts Flip Chip Type Semiconductor Chip on Multilayer Substrate by Low Melting Point Metal."
456. WPI Acc. 84-097184/16, "Semiconductor Module Device Has Module Substrate Forming Wiring Conductor Film, Flange and Flip Chip."
457. WPI Acc. 84-078921/13, "Semiconductor Device for Flip Chip Bonding."
458. WPI Acc. 84-066525/11, "Semiconductor Device for Flip Chip Fixes Silicon Plate on Ceramic Base."
459. WPI Acc. 84-045407/08, "Hybrid Device for Flip Chip Bonding Process."
460. WPI Acc. 83-840892/50, "Insulating Substrate for Bonding Flip Chip Semiconductor Element."
461. WPI Acc. 83-830675/48, "Semiconductor Device that Can Perform Flip-Chip Bonding."
462. WPI Acc. 83-823797/47, "Semiconductor Device with Flip-Chip Has Chip Reliably Bonded on Package or Carrier."
463. WPI Acc. 83-768541/38, "Flip-Chip Bonding Machine Precisely Joints Semiconductor Element on Die Pad by IR Ray Irradiation."
464. WPI Acc. 83-762153/37, "Mounting Flip Chips on Substrate."
465. WPI Acc. 83-743147/34, "Internal Matching Type Semiconductor Device with Flip Chip Structure."
466. WPI Acc. 83-743132/34, "Manufacturing Flip Chip Mounting Structure for Integrated Circuit Absorbs Horizontal Stress with Elasticity of Solder Bump."
467. WPI Acc. 83-707586/28, "Flip Chip Type Semiconductor Device."
468. WPI Acc. 83-11329K/05, "Resin-Sealed Flip-Chip Semiconductor Device."
469. WPI Acc. 83-H3066K/22, "Flip Chip Bonding Method Soldering Bump Electrode to Flat Pad Formed on Semiconductor Substrate."

470. WPI Acc. 83-F2864K/16, "Flip-Chip Bonding Device Connects Semiconductor Chip to Board and Has Solder-Plated Back Surface."
471. WPI Acc. 83-E2462K/13, "Substrate for Flip Chip Bonding in Integrated Circuit."
472. WPI Acc. 83-C6379K/08, "Package Structure for Semiconductor Device has Flip-Chip Semiconductor Dice Bonded onto Package."
473. WPI Acc. 83-B8829K/06, "Hybrid Integrated Circuits Has Soldered Flip-Chip Capacitors and Resistors."
474. WPI Acc. 83-B5248K/05, "Alignment Apparatus for Flip Chip Applications."
475. WPI Acc. 83-B1771K/04, "Bump Electrode for Flip Chip Device."
476. WPI Acc. 83-A8294K/03, "Semiconductor Flip Chip Element with Improved Bump Electrodes."
477. WPI Acc. 83-A8288K/03, "Semiconductor Flip Chip Element Has Substrate with Bump Electrode."
478. WPI Acc. 82-86585E/41, "Manufacturing Solder Bump Electrodes for Face-Down-Bonding of Flip-Chip Semiconductor Devices."
479. WPI Acc. 82-84750E/40, "Forming Solder-Bump Electrode Used for Making Face Down-Bonding on Flip-Chip Semiconductor Device."
480. WPI Acc. 82-84749E/40, "Forming Solder-Bump Electrode Is Used for Making Face Down-Bonding on Flip-Chip Semiconductor Device."
481. WPI Acc. 82-16650E/09, "Gallium Arsenide FET with Flip Chip Structure."
482. WPI Acc. 82-16649E/09, "Gallium Arsenide FET with Flip Chip Structure."
483. WPI Acc. 82-J6345E/29, "Light Emitting Device Comprises Flip Chip Type Light Emitting Diode Element Mounted to Stem with Lead Having Reflective Surface."
484. WPI Acc. 82-E5032E/16, "Integrated Circuit Package Contains Elastic Rod Conductors for Low Heat Resistance and Sealed with Flip Chip System."
485. WPI Acc. 82-E5031E/16, "Integrated Circuit Package Contains Elastic Rod Conductors for Low Heat Resistance and Sealed with Flip Chip System."
486. WPI Acc. 82-E2251E/15, "LSI Circuit with Upper and Lower Surfaces with Flip Chip Bonding Connection."
487. WPI Acc. 82-C9481E/11, "Flip-Chip Type Monolithic Gallium Arsenide Integrated Circuit."
488. WPI Acc. 82-C6162E/10, "Flip Chip Type Field Effect Transistor."
489. WPI Acc. 82-A7503J/49, "Flip Chip Connection System Joins Semiconductor Integrated Circuit Chips to Substrate Using Solder Mounds."
490. WPI Acc. 79-67257B/37, "Forming Bump Contacts on a Flip Chip."
491. WPI Acc. 79-A8870B/04, "Reverse Channel Gallium Arsenide FET Oscillator."
492. WPI Acc. 77-80149Y/45, "Electrodes Used for Outer Lead Connections."
493. WPI Acc. 77-80148Y/45, "Electrodes Used for Outer Lead Connections."
494. WPI Acc. 75-76636W/46, "Flip Chip Transistor Manufacture."
495. WPI Acc. 75-74182W/45, "Flip Chip Bonding to Substrate."
496. WPI Acc. 75-07137W/04, "Multiple Flip-Chip Array Bonding."
497. WPI Acc. 74-K9170V/47, "Flip Chip Mounted Transistor."
498. WPI Acc. 74-G7470V/33, "Connecting Semiconductor Chip to Substrate by Flip-Chip Method."
499. WPI Acc. 70-94930R/51, "Flip-Chip—Schottky Cascade Diodes."
500. WPI Acc. 70-15388R/10, "Galvanically Reinforced Metal Micro Structures."

Acknowledgments

The author would like to thank his eminent colleagues throughout the industry for their stimulating discussions in electronic packaging and interconnection problems. He learned a lot from them.

References

1. Lau, J. H., *Chip on Board Technologies for Multichip Modules,* Van Nostrand Reinhold, New York, 1994.

2. Lau, J. H., *Ball Grid Array Technology,* McGraw-Hill, New York, 1995.
3. Frear, D., H. Morgan, S. Burchett, and J. H. Lau, *The Mechanics of Solder Alloy Interconnects,* Van Nostrand Reinhold, New York, 1993.
4. Lau, J. H., *Handbook of Fine Pitch Surface Mount Technology,* Van Nostrand Reinhold, New York, 1993.
5. Lau, J. H., *Thermal Stress and Strain in Microelectronics Packaging,* Van Nostrand Reinhold, New York, 1993.
6. Lau, J. H., *Handbook of Tape Automated Bonding,* Van Nostrand Reinhold, New York, 1992.
7. Lau, J. H., *Solder Joint Reliability: Theory and Applications,* Van Nostrand Reinhold, New York, 1991.
8. Wong, C. P., *Polymers for Electronic and Photonic Applications,* Academic Press, San Diego, 1993.
9. Manzione, L. T., *Plastic Packaging of Microelectronic Devices,* Van Nostrand Reinhold, New York, 1990.
10. Hwang, J. S., *Solder Paste in Electronics Packaging,* Van Nostrand Reinhold, New York, 1989.
11. Hymes, L., *Cleaning Printed Wiring Assemblies in Today's Environment,* Van Nostrand Reinhold, New York, 1991.
12. Gilleo, K., *Handbook of Flexible Circuits,* Van Nostrand Reinhold, New York, 1991.
13. Engel, P. A., *Structural Analysis of Printed Circuit Board Systems,* Springer-Verlag, New York, 1993.
14. Suhir, E., *Structural Analysis in Microelectronic and Fiber Optics Systems,* Van Nostrand Reinhold, New York, 1991.
15. Seraphim, D. P., R. Lasky, and C. Y. Li, *Principles of Electronic Packaging,* McGraw-Hill, New York, 1989.
16. Tummala, R. R., and E. Rymaszewski, *Microelectronics Packaging Handbook,* Van Nostrand Reinhold, New York, 1989.
17. Vardaman, J., *Surface Mount Technology, Recent Japanese Developments,* IEEE Press, New York, 1992.
18. Johnson, R. W., R. K. Teng, and J. W. Balde, *Multichip Modules: System Advantages, Major Construction, and Materials Technologies,* IEEE Press, New York, 1991.
19. Senthinathan, R., and J. L. Prince, *Simultaneous Switching Noise of CMOS Devices and Systems,* Kluwer Academic Publishers, New York, 1994.
20. Sandborn, P. A., and H. Moreno, *Conceptual Design of Multichip Modules and Systems,* Kluwer Academic Publishers, New York, 1994.
21. Nash, F. R., *Estimating Device Reliability: Assessment of Credibility,* Kluwer Academic Publishers, New York, 1993.
22. Gyvez, J. P., *Integrated Circuit Defect-Sensitivity: Theory and Computational Models,* Kluwer Academic Publishers, New York, 1993.
23. Doane, D. A., and P. D. Franzon, *Multichip Module Technologies and Alternatives,* Van Nostrand Reinhold, New York, 1992.
24. Messuer, G., I. Turlik, J. Balde, and P. Garrou, *Thin Film Multichip Modules,* International Society for Hybrid Microelectronics, Silver Spring, Md., 1992.
25. Matisoff, B. S., *Handbook of Electronic Packaging Design and Engineering,* Van Nostrand Reinhold, New York, 1989.
26. Prasad, R. P., *Surface Mount Technology,* Van Nostrand Reinhold, New York, 1989.
27. Manko, H. H., *Soldering Handbook for Printed Circuits and Surface Mounting,* Van Nostrand Reinhold, New York, 1986.
28. Morris, J. E., *Electronics Packaging Forum,* vol. 1, Van Nostrand Reinhold, New York, 1990.
29. Morris, J. E., *Electronics Packaging Forum,* vol. 2, Van Nostrand Reinhold, New York, 1991.
30. ASM International, *Electronic Materials Handbook,* vol. 1, *Packaging,* Metals Park, Ohio, 1989.
31. Hollomon, J. K., Jr., *Surface-Mount Technology,* Sams, Indianapolis, 1989.
32. Solberg, V., *Design Guidelines for SMT,* TAB Professional and Reference Books, New York, 1990.
33. Hutchins, C., *SMT: How to Get Started,* Hutchins, Raleigh, N.C., 1990.

34. Bar-Cohen, A., and A. D. Kraus, *Advances in Thermal Modeling of Electronic Components and Systems,* vol. 1, Hemisphere, New York, 1988.
35. Bar-Cohen, A., and A. D. Kraus, *Advances in Thermal Modeling of Electronic Components and Systems,* vol. 2, ASME Press, New York, 1990.
36. Kraus, A. D., and A. Bar-Cohen, *Thermal Analysis and Control of Electronic Equipment,* Hemisphere, New York, 1983.
37. Harper, C. A., *Handbook of Microelectronics Packaging,* McGraw-Hill, New York, 1991.
38. Pecht, M., *Handbook of Electronic Package Design,* Marcel Dekker, New York, 1991.
39. Pecht, M., *Placement and Routing of Electronic Modules,* Marcel Dekker, New York, 1993.
40. Pecht, M., *Integrated Circuit, Hybrid, and Multichip Module Package Design Guidelines,* Wiley, New York, 1994.
41. Pecht, M., A. Dasgupta, J. W. Evans, and J. Y. Evans, *Quality Conformance and Qualification of Microelectronic Packages and Interconnects,* Wiley, New York, 1994.
42. Hannemann, R., A. Kraus, and M. Pecht, *Physical Architecture of VLSI Systems,* Wiley, New York, 1994.
43. Pecht, M., *Soldering Processes and Equipment,* Wiley, New York, 1993.
44. Harman, G., *Wire Bonding in Microelectronics,* International Society for Hybrid Microelectronics, Reston, Va., 1989.
45. Lea, C., *A Scientific Guide to Surface Mount Technology,* Electrochemical Publications, Ayr, Scotland (U.K.), 1988.
46. Lea, C., *After CFCs? Options for Cleaning Electronics Assemblies,* Electrochemical Publications Ltd., IOM, U.K., 1992.
47. Wassink, R. J. K., *Soldering in Electronics,* Electrochemical Publications, Ayr, Scotland (U.K.), 1989.
48. Pawling, J. F., *Surface Mounted Assemblies,* Electrochemical Publications, Ayr, Scotland (U.K.), 1987.
49. Ellis, B. N., *Cleaning and Contamination of Electronics Components and Assemblies,* Electrochemical Publications, Ayr, Scotland (U.K.), 1986.
50. Sinnadurai, F. N., *Handbook of Microelectronics Packaging and Interconnection Technologies,* Electrochemical Publications, Ayr, Scotland (U.K.), 1985.
51. Motorola, *C4 Product Design Manual,* vol. 1: *Chip and Wafer Design,* 1993.
52. Murphy, B. T., "Cost-Size Optima of Monolithic Integrated Circuits," *Proc. IEEE,* **52:**1537–1545, Dec. 1964.
53. Seeds, R. B., "Yield, Economic, and Logistic Models for Complex Digital Arrays," *IEEE Internatl. Convention Records,* **6:**61–66, April 1967.
54. Stapper, C. H., F. M. Armstrong, and K. Saji, "Integrated Circuit Yield Statistics," *Proc. IEEE,* **71:**453–470, April 1983.
55. Stapper, C. H., "LSI Yield Modeling and Process Monitoring," *IBM J. Research Devel.,* **20:**228–234, May 1976.
56. Stapper, C. H., "Defect Density Distribution for LSI Yield Calculations," *IEEE Transact. Electron Devices,* **ED-20:**655–657, July 1973.
57. Stapper, C. H., "On a Composite Model to the IC Yield Problem," *IEEE J. Solid-State Circuits,* **SC-10:**537–539, Dec. 1975.
58. Okabe, T., M. Nagata, and S. Shimada, "Analysis on Yield of Integrated Circuits," *Electronic Eng. Jpn.,* **92:**135–141, Dec. 1972.
59. Gulett, M. R., "A Practical Method of Predicting IC Yields," *Semiconductor Internatl.,* **4:**87–94, Feb. 1981.
60. Stapper, C. H., "On Murphy's Yield Integral," *IEEE Transact. Semiconductor Manufact.,* **4:**294–297, Nov. 1991.
61. Cunningham, J. A., "The Use and Evaluation of Yield Models in Integrated Circuit Manufacturing," *IEEE Transact. Semiconductor Manufact.,* **3:**60–71, May 1990.
62. Michalka, T. L., R. C. Varshney, and J. D. Meindl, "A Discussion of Yield Modeling with Defect Clustering, Circuit Repair, and Circuit Redundancy," *IEEE Transact. Semiconductor Manufact.,* **3:**116–127, Aug. 1990.
63. Kooperberg, C., "Circuit Layout and Yield," *IEEE J. Solid-State Circuits,* **23:**887–892, Aug. 1988.
64. Lau, J. H., and S. Erasmus, "Review of Packaging Methods to Complement IC Performance," *Electron. Packag. Product.,* 51–56, June 1993.

65. Matsui, N., S. Sasaki, and T. Ohaski, "VLSI Chip Interconnection Technology Using Stacked Solder Bumps," *Proceedings of the 37th IEEE Electronic Components Conference,* pp. 573–578, 1987.

66. Moresco, L., "Electronic System Packaging: The Search for Manufacturing the Optimum in a Sea of Constraints," *IEEE Transact. Components, Hybrids, Manufact. Technol.,* **13**(3):494–508, Sept. 1990.

67. Lau, J. H., "On Murphy's Integrated Circuit Yield Integral," *ASME Transact., J. Electron. Packag.,***117**(2):159–164, June 1995.

68. Lau, J. H., "Temperature and Stress Time History Responses in Electronic Packaging," *Proceedings of the 45th IEEE Electronic Components & Technology Conference,* pp. 952–958, 1995.

69. Hagge, J. K., and R. J. Wagner, "High-Yield Assembly of Multichip Modules through Known-Good ICs and Effective Test Strategies," *Proc. IEEE,* **80**(12):1965–1994, Dec. 1992.

70. Radke, C. E., L. S. Su, Y. M. Ting, and J. Vanhorn, "Known Good Die and Its Evolution—Bipolar and CMOS," *Proceedings of the 2nd International Conference and Exhibition on Multichip Modules,* pp. 152–159, April 1993.

71. Bracken, R. C., B. P. Kraemer, R. Paradiso, and A. Jensen, "MULTICHIP-MODULES, Die and MCM Test Strategy: The Key to MCM Manufacturability," *Proceedings of the 1st International Conference and Exhibition on Multichip Modules,* pp. 456–460, April 1992.

72. Trent, J. R., "Test Philosophy for Multichip Modules," *Proceedings of the 1st International Conference and Exhibition on Multichip Modules,* pp. 444–452, April 1992.

73. Smitherman, C. D., and J. Rates, "Methods for Processing Known Good Die," *Proceedings of the 1st International Conference and Exhibition on Multichip Modules,* pp. 436–443, April 1992.

74. Martin, S., D. Gage, T. Powell, and B. Slay, "A Practical Approach to Producing Known-Good Die," *Proceedings of the 2nd International Conference and Exhibition on Multichip Modules,* pp. 139–151, April 1993.

75. Corbett, T., "A Process Qualification Plan for KGD," *Proceedings of the 2nd International Conference and Exhibition on Multichip Modules,* pp. 166–171, April 1993.

76. Begay, M., "Bare-Die Test Strategies for the MCM Market," *Solid State Technol.,* 65–75, June 1994.

77. Harr, R., "Easing the Availability of Bare Die: Die Information Exchange (DIE) Format," *Proceedings of the ISHM,* pp. 78–83, Boston, Nov. 1994.

78. Charles, H., and M. Uy, "Design Aid for Multichip Modules," *Proceedings of the ISHM,* pp. 42–48, Boston, Nov. 1994.

79. Genin, D. J., and M. M. Wurster, "Probing Considerations in C-4 Testing of IC Wafers," *Proceedings of the 2nd International Conference and Exhibition on Multichip Modules,* pp. 166–171, April 1993.

80. Bauer, C. E., K. W. Posse, and K. T. Wilson, "Good Enough Die? A Cost Model for KGD Test Decisions," *Proceedings of the 8th International Microelectronics Conference,* pp. 266–270, Omiya, Japan, 1994.

81. Kelly, M., and J. H. Lau, "Low Cost Solder Bumped Flip Chip MCM-L Demonstration," *Proceedings of the 16th IEEE/CPMT IEMTS,* pp. 147–153, Sept. 1994.

82. Davis, E., W. Harding, R. Schwartz, and J. Corning, "Solid Logic Technology: Versatile, High Performance Microelectronics," *IBM J. Research Devel.,* 102–114, April 1964.

83. Suryanarayana, D., and D. S. Farquhar, "Underfill Encapsulation for Flip Chip Applications," in *Chip on Board Technologies for Multichip Modules,* Lau, J. H., ed., Van Nostrand Reinhold, New York, pp. 504–531, 1994.

84. Totta, P. A., and R. P. Sopher, "SLT Device Metallurgy and Its Monolithic Extension," *IBM J. Research Devel.,* pp. 226–238, May 1969.

85. Goldmann, L. S., and P. A. Totta, "Chip Level Interconnect: Solder Bumped Flip Chip," in *Chip on Board Technologies for Multichip Modules,* Lau, J. H., ed., Van Nostrand Reinhold, New York, pp. 228–250, 1994.

86. Goldmann, L. S., R. J. Herdzik, N. G. Koopman, and V. C. Marcotte, "Lead Indium for Controlled Collapse Chip Joining," *Proceedings of the IEEE Electronic Components Conference,* pp. 25–29, 1977.

87. Totta, P., "Flip Chip Solder Terminals," *Proceedings of the IEEE Electronic Components Conference,* pp. 275–284, 1971.

88. Goldmann, L. S., "Geometric Optimization of Controlled Collapse Interconnections," *IBM J. Research Devel.,* 251–265, May 1969.

89. Goldmann, L. S., "Optimizing Cycle Fatigue Life of Controlled Collapse Chip Joints," *Proceedings of the 19th IEEE Electronic Components and Technology Conference,* pp. 404–423, 1969.

90. Goldmann, L. S., "Self Alignment Capability of Controlled Collapse Chip Joining," *Proceedings of the 22nd IEEE Electronic Components and Technology Conference,* pp. 332–339, 1972.

91. Shad, H. J., and J. H. Kelly, "Effect of Dwell Time on Thermal Cycling of the Flip Chip Joint," *Proceedings of the ISHM,* pp. 3.4.1–3.4.6, 1970.

92. Hymes, I., R. Sopher, and P. Totta, "Terminals for Microminiaturized Devices and Methods of Connecting Same to Circuit Panels," U.S. Patent 3,303,393, 1967.

93. Karan, C., J. Langdon, R. Pecararo, and P. Totta, "Vapor Depositing Solder," U.S. Patent 3,401,055, 1968.

94. Seraphim, D. P., and J. Feinberg, "Electronic Packaging Evolution," *IBM J. Research Devel.,* 617–629, May 1981.

95. Tsukada, Y., Y. Mashimoto, and N. Watanuki, "A Novel Chip Replacement Method for Encapsulated Flip Chip Bonding," *Proceedings of the 43rd IEEE/EIA Electronic Components & Technology Conference,* pp. 199–204, June 1993.

96. Tsukada, Y., Y. Maeda, and K. Yamanaka, "A Novel Solution for MCM-L Utilizing Surface Laminar Circuit and Flip Chip Attach Technology," *Proceedings of the 2nd International Conference and Exhibition on Multichip Modules,* pp. 252–259, April 1993.

97. Tsukada, Y., S. Tsuchida, and Y. Mashimoto, "Surface Laminar Circuit Packaging," *Proceedings of the 42nd IEEE Electronic Components and Technology Conference,* pp. 22–27, May 1992.

98. Tsukada, Y., "Solder Bumped Flip Chip Attach on SLC Board and Multichip Module," in *Chip on Board Technologies for Multichip Modules,* Lau, J. H., ed., Van Nostrand Reinhold, New York, pp. 410–443, 1994.

99. Miller, L. F., "Controlled Collapse Reflow Chip Joining," *IBM J. Research Devel.,* 239–250, May 1969.

100. Miller, L. F., "A Survey of Chip Joining Techniques," *Proceedings of the 19th IEEE Electronic Components and Technology Conference,* pp. 60–76, 1969.

101. Miller. L. F., "Joining Semiconductor Devices with Ductile Pads," *Proceedings of ISHM,* pp. 333–342, 1968.

102. Norris, K. C., and A. H. Landzberg, "Reliability of Controlled Collapse Interconnections," *IBM J. Research Devel.,* 266–271, May 1969.

103. Oktay, S., "Parametric Study of Temperature Profiles in Chips Joined by Controlled Collapse Technique," *IBM J. Research Devel.,* 272–285, May 1969.

104. Bendz, D. J., R. W. Gedney, and J. Rasile, "Cost/Performance Single Chip Module," *IBM J. Research Devel.,* 278–285, 1982.

105. Blodgett, A. J., Jr., "A Multilayer Ceramic Multichip Module," *IEEE Transactions on Components, Hybrids, and Manufacturing Technology,* pp. 634–637, 1980.

106. Fried, L. J., J. Havas, J. Lechaton, J. Logan, G. Paal, and P. Totta, "A VLSI Bipolar Metallization Design with Three-Level Wiring and Area Array Solder Connections," *IBM J. Research Devel.,* 362–371, 1982.

107. Clark, B. T., and Y. M. Hill "IBM Multichip Multilayer Ceramic Modules for LSI Chips—Designed for Performance Density," *IEEE Transactions on Components, Hybrids, and Manufacturing Technology,* pp. 89–93, 1980.

108. Blodgett, A. J., and Dr. R. Barbour, "Thermal Conduction Module: A High-Performance Multilayer Ceramic Package," *IBM J. Research Devel.,* 30–36, 1982.

109. Dansky, A. H., "Bipolar Circuit Design for a 5000-Circuit VLSI Gate Array," *IBM J. Research Devel.,* 116–125, 1981.

110. Oktay, S., and H. C. Kammer, "A Conduction-Cooled Module for High-Performance LSI Devices," *IBM J. Research Devel.,* 55–66, 1982.

111. Chu, R. C., U. P. Hwang, and R. E. Simons, "Conduction Cooling for an LSI Package: A One-Dimensional Approach," *IBM J. Research Devel.,* 45–54, 1982.
112. Howard, R. T., "Packaging Reliability and How to Define and Measure It," *Proceedings of the IEEE Electronic Components Conference,* pp. 376–384, 1982.
113. Howard, R. T., "Optimization of Indium-Lead Alloys for Controlled Collapse Chip Connection Application," *IBM J. Research Devel.,* 372–389, 1982.
114. Kamei, T., and M. Nakamura, "Hybrid IC Structures Using Solder Reflow Technology," *Proceedings of the IEEE Electronic Components Conference,* pp. 172–182, 1978.
115. Liu, J., "Reliability of Surface-Mounted Anisotropically Conductive Adhesive Joints," *Circuit World,* 4–11, 1993.
116. Greer, S. E., "Low Expansivity Organic Substrate for Flip-Chip Bonding," *Proceedings of the IEEE Electronic Components Conference,* pp. 166–171, 1978.
117. Wong, C. P., J. M. Segelken, and C. N. Robinson, "Chip on Board Encapsulation," in *Chip on Board Technologies for Multichip Modules,* Lau, J. H., ed., Van Nostrand Reinhold, New York, pp. 470–503, 1994.
118. Rai, A., Y. Dotta, H. Tsukamoto, T. Fujiwara, H. Ishii, T. Nukii, and H. Matsui, "COB (Chip on Board) Technology: Flip Chip Bonding onto Ceramic Substrates and PWB (Printed Wiring Boards)," *ISHM Proceedings,* pp. 474–481, 1990.
119. Rai, A., Y. Dotta, T. Nukii, and T. Ohnishi, "Flip-Chip COB Technology on PWB," *Proceedings of IMC,* pp. 144–149, June 1992.
120. Lowe, H., "No-Clean Flip Chip Attach Process," *International TAB/Advance Packaging and Flip Chip Proceedings,* pp. 17–24, Feb. 1994.
121. Giesler, J., S. Machuga, G. O'Malley, and M. Williams, "Reliability of Flip Chip on Board Assemblies," *International TAB/Advance Packaging and Flip Chip Proceedings,* pp. 127–135, Feb. 1994.
122. Murphy, C. F., L. Gilg, and C. Spooner, "Mechanical, Electrical and Thermal Evaluation of Known Good Die Carriers," *Proceedings of ICEMCM,* pp. 428–433, 1995.
123. Lau, J. H., T. Krulevitch, W. Schar, M. Heydinger, S. Erasmus, and J. Gleason, "Experimental and Analytical Studies of Encapsulated Flip Chip Solder Bumps on Surface Laminar Circuit Boards," *Circuit World,* 19(3):18–24, March 1993.
124. Koopman, N., S. Bobbio, S. Nangalia, J. Bousaba, and B. Peikarski, "Fluxless Soldering in Air and Nitrogen," *Proceedings of the IEEE Electronic Components & Technology Conference,* pp. 376–384, 1993.
125. Koopman, N., and S. Nangalia, "Fluxless Flip Chip Solder Joining," *Proceedings of NEPCON West,* pp. 919–931, 1995.
126. Zakel, E., J. Gwiasda, J. Kloeser, J. Eldring, J. Engelmann, and H. Reichl, "Fluxless Flip Chip Assembly on Rigid and Flexible Polymer Substrates Using the Au-Sn Metallurgy," *Proceedings of IEEE/CPMT IEMTS,* pp. 177–184, 1994.
127. Zakel, E., *Untersuchung von Cu-Sn-Au, Cu-Au, und Cu-Sn Metallisierungssystemen fur die TAB-Technologie,* Ph.D. thesis, Technical University, Berlin, 1994.
128. Zakel, E., R. Aschenbrenner, J. Gwiasda, G. Azdasht, A. Ostmann, J. Eldring, and H. Reichl, "Fluxless Flip Chip Bonding on Flexible Substrates," *Proceedings of NEPCON West,* pp. 909–919, 1995.
129. Han, B., Y. Guo, T. Chung, and D. Liu, "Reliability Assessment of Flip Chip Package with Encapsulation," *Proceedings of NEPCON West,* pp. 600–602, 1995.
130. Chung, T., D. Carey, and B. Gardner, "Development of Large High I/O Flip-Chip Technology," *Proceedings of NEPCON West,* pp. 1527–1536, 1994.
131. Degani, Y., and T. D. Dudderar, "AT&T Lead-Free Solder Paste for the Cost Effective Manufacture of Flip-Chip Silicon-On-Silicon MCMs," *Proceedings of IEEE/CPMT IEMTS,* pp. 20–24, 1994.
132. Dudderar, T. D., T. Degani, J. Spadafora, K. Tai, and R. Frye, "AT&T μSurface Mount Assembly: A New Technology for the Large Volume Fabrication of Cost Effective Flip-Chip MCMs," *Proceedings of the International Conference on Multichip Modules,* pp. 266–272, 1994.
133. Scharr, T., R. Lee, W. Lytle, R. Subrahmanyan, and R. Sharma, "Gold Wire Bump Process Development for Flip-Chip Application," *Proceedings of the 8th Electronic Materials and Processing Conference,* pp. 97–102, 1994.

134. Subrahmanyan, R., "Micromechanical and Accelerated Testing of Flip Chip Interconnect Systems," *Proceedings of MRS, Electronic Packaging in Materials Science VII*, pp. 395–406, 1994.

135. Estes, R., F. Kulesza, D. Buczek, and G. Riley, "Environmental and Reliability Testing of Conductive Polymer Flip Chip Assemblies," *Proceedings of IEPS*, pp. 328–342, 1993.

136. Lee, C. H., and K. Loh, "Fine Pitch COG Interconnection Using Anisotropically Conductive Adhesives," *Proceedings of the IEEE Electronic Components & Technology Conference*, pp. 121–125, 1995.

137. Keswick, K., R. German, M. Breen, and R. Nolan, "Compliant Bumps for Adhesive Flip Chip Assembly," *Proceedings of the IEEE Electronic Components & Technology Conference*, pp. 7–15, 1994.

138. DiStefano, T., "The μBGA as a Chip Size Package," *Proceedings of NEPCON West*, pp. 327–333, 1995.

139. Matthew, L. C., and T. H. DiStefano, "Future Directions in TAB: The TCC/MCM Interconnect," *Proceedings of ITAP and Flip-Chip Symposium*, pp. 228–231, Feb. 1994.

140. Martinez, M., D. Gibson, L. Matthew, T. DiStefano, and J. Cofield, "The TCC/MCM: μBGA on a Laminated Substrate," *Proceedings of the International Conference and Exhibition on Multichip Modules*, pp. 161–166, April 1994.

141. Matthew, L., Z. Kovac, G. Karavakis, and T. DiStefano, "Beyond the Barriers of Known Good Die," *Proceedings of NEPCON East*, pp. 153–156, June 1994.

142. Loo, M., and K. B. Gilleo, "Area Array Chip Carrier: SMT Package for Known Good Die," *Proceedings of ISHM*, pp. 318–323, Nov. 1993.

143. Baba, S., W. Carlomagno, D. Cummings, and F. Guerrero, "Bonded Interconnect Pin (BIP) Technology, a New Bare Chip Assembly Technique for Multichip Modules," *Proceedings of Japan IEEE/CHMT IEMTS*, pp. 156–139, 1991.

144. Moresco, L., D. Love, B. Chou, and V. Holalkere, "A New Flip Chip to Substrate Connection Method," *Proceedings of the ASME INTERpack*, p. 65, 1995.

145. Adema, G., C. Berry, N. Koopman, G. Rinne, E. Yung, and I. Turlik, "Flip Chip Technology: A Method for Providing Known Good Die with High Density Interconnections," *Proceedings of the 3rd International Conference & Exhibition on Multichip Modules*, pp. 41–49, 1994.

146. Kusagaya, T., H. Kira, and K. Tsunoi, "Flip Chip Mounting Using Stud Bumps and Adhesives for Encapsulation," *Proceedings of the 2nd International Conference & Exhibition on Multichip Modules*, pp. 238–245, 1993.

147. Lau, J. H., "Thermoelastic Problems for Electronic Packaging," *J. Hybrid Circuit*, 11–15, May 1991.

148. Date, H., Y. Hozumi, H. Tokuhira, M. Usui, E. Horikoshi, and T. Sato, "Anisotropic Conductive Adhesive for Fine Pitch Interconnections," *Proceedings of ISHM*, pp. 570–575, 1994.

149. Yung, E. K., and I. Turlik, "Electroplated Solder Joints for Flip-Chip Technology," *IEEE Transact. CHMT*, 549–559, 1991.

150. Yu, K., and F. Tung, "Solder Bump Fabrication by Electroplating for Flip-Chip Applications," *Proceedings of IEEE/CHMT IEMTS*, pp. 277–281, 1993.

151. Imler, B., K. Scholz, M. Cobarruviaz, R. Haitz, V. Nagesh, and C. Chao, "Precision Flip-Chip Solder Bump Interconnects for Optical Packaging," *Proceedings of the IEEE Electronic Components Conference*, pp. 508–512, 1992.

152. Yost, F. G., "Aspects of Lead-Indium Solder Technology," *Proceedings of ISHM*, pp. 61–65, 1976.

153. Warrior, M., "Reliability Improvements in Solder Bump Processing for Flip Chips," *Proceedings of the IEEE Electronic Components & Technology Conference*, pp. 460–469, 1990.

154. Takenaka, T., F. Kobayashi, T. Netsu, and H. Imada, "Reliability of Flip-Chip Interconnections," *Proceedings of ISHM*, pp. 419–423, 1984.

155. Guo, Y., W. T. Chen, and K. C. Lim, "Experimental Determinations of Thermal Strains in Semiconductor Packaging Using Moire Interferometry," *Proceedings of the 1st ASME/JSME Advances in Electronic Packaging Conference*, pp. 779–784, April 1992.

156. Lau, J. H., "Thermal Fatigue Life Prediction of Encapsulated Flip Chip Solder Joints for Surface Laminar Circuit Packaging," ASME paper no. 92-WA/EEP-34, ASME Winter Annual Meeting, Anaheim, Calif., Nov. 8–13, 1992.

157. Lau, J. H., M. Heydinger, J. Glazer, and D. Uno, "Design and Procurement of Eutectic Sn/Pb Solder-Bumped Flip Chip Test Die and Organic Substrate," *Circuit World,* **21**(3):20–24, March 1995.

158. Hirosawa, H., and T. Ohta, "Double-Layer Anisotropic Conductive Adhesive Films," *Proceedings of Display Manufacturing Technology Conference,* pp. 17–18, 1995.

159. Shiozawa, N., I. Tsukagoshi, A. Nakajima, and T. Itoh, "Anisotropic Conductive Adhesive Film Anisolm AC-2052 for Connecting the One Metal Electrode to Another," *Hitachi Chemical Technical Report,* no. 23, pp. 23–26, 1994.

160. Suryanarayana, D., R. Hsiao, T. Gall, and J. McCreary, "Flip-Chip Solder Bump Fatigue Life Enhanced by Polymer Encapsulation," *Proceedings of the 40th IEEE Electronic Components and Technology Conference,* pp. 338–344, May 1990.

161. Wun, B., and J. H. Lau, "Characterization and Evaluation of the Underfill Encapsulants for Flip Chip Assembly," *Circuit World,* **24**(3):25–27, March 1995.

162. Satoh, R. K., "A Prediction of the Thermal Fatigue Life of Solder Joints Using Crack Propagation Rate and Equivalent Strain Range," in *Thermal Stress and Strain in Microelectronics Packaging,* Lau, J. H., ed., Van Nostrand Reinhold, New York, pp. 500–531, 1993.

163. Suhir, E., and J. M. Segelken, "Mechanical Behavior of Flip-Chip Encapsulants," *ASME J. Electron. Packag.,* **112**:327–332, Dec. 1990.

164. Lau, J. H., L. Powers, J. Baker, D. Rice, and W. Shaw, "Solder Joint Reliability of Fine Pitch Surface Mount Technology Assemblies," *IEEE Transact. Components, Hybrids, Manufact. Technol.,* **13**(3):534–544, Sept. 1990.

165. Lau, J. H., G. Dody, W. Chen, M. McShane, D. Rice, S. Erasmus, and W. Adamjee, "Experimental and Analytical Studies of 208-Pin Fine Pitch Quad Flat Pack Solder-Joint Reliability," *Internatl. J. Interconnection Technol. (Circuit World),* **18**(2):13–19, Jan. 1992.

166. Lau, J. H., "Thermal Stress Analysis of SMT PQFP Packages and Interconnections," *J. Electron. Packag., Transact. ASME,* **111**:2–8, March 1989.

167. Lau, J. H., "Thermal Fatigue Life Prediction of Flip Chip Solder Joints by Fracture Mechanics Method," *Internatl. J. Eng. Fracture Mech.,* **45**(5):643–654, July 1993.

168. Solomon, H. D., "Fatigue of 60/40 Solder," *IEEE Transact. Components, Hybrids, Manufact. Technol.,* **CHMT-9:**423–432, Dec. 1986.

169. Suhir, S., "Interfacial Stresses in Bimetal Thermostats," *ASME J. Appl. Mech.,* **55**:595–600, 1989.

170. Lau, J. H., "A Note on the Calculation of Thermal Stresses in Electronic Packaging by Finite Element Methods," *ASME J. Electron. Packag.,* **111**:313–320, Dec. 1989.

171. Logsdon, W. A., P. K. Liaw, and M. A. Burke, "Fracture Behavior of 63Sn-37Pb," *Eng. Fracture Mechanics,* **36**(2):183–218, 1990.

172. Lau, J. H., and D. W. Rice, "Solder Joint Fatigue in Surface Mount Technology: State of the Art," *Solid State Technol.,* 91–104, Oct. 1985.

173. Abramowitz, M., and I. A. Stegun, *Handbook of Mathematical Functions,* Dover Publications, New York, 1965.

174. Manson, S. S., *Thermal Stress and Low Cycle Fatigue,* McGraw-Hill, New York, 1966.

175. Kanninen, M. F., and C. H. Popelar, *Advanced Fracture Mechanics,* Oxford University Press, London, 1985.

176. Knott, J. F., *Fundamentals of Fracture,* Wiley, New York, 1973.

Fluxless Flip Chip Solder Joining

Nick Koopman and S. Nangalia

2.1 Introduction

Fluxless soldering has long been a dream of many researchers and has gained importance in recent years because of increasing concern for the environment. The use of liquid chemicals in the processing of electrical components, printed wiring boards, and packages is degrading our air, water, and land. Soldering is the single major electronic assembly process using liquid chemicals. Soldering uses liquid fluxes to chemically dissolve metal oxides that inhibit the flow of solder. Solvents or other liquids are used to clean these flux/metal reaction products (usually known as *flux residue*) left on the components, as it affects the reliability of the product. Chlorofluorocarbons (CFCs) and other solvents used to clean flux residues from soldered assemblies are believed to be damaging the earth's protective ozone layer, and their use has been banned by the Montreal protocol[1] of November 25, 1992.

Because of the reliability problems associated with the use of highly activated fluxes (corrosion from residues of acidic activators), most soldering applications in the microelectronics industry use a mildly activated or nonactivated fluxes. Even so, we are just swapping one chemical for another and reducing, not eliminating, the problem. The use of these mildly activated or nonactivated fluxes has created problems of its own—namely, higher defect levels and high rework costs. An estimate suggests that 700 million dollars per year could be saved on military microelectronics alone if a fluxless process which had the effectiveness of an activated flux[2] was to be adopted.

Flux residues are always present on microelectronics components assembled using flux. Flux residues, besides being reliability risks due to corrosion,[3] can also have a negative effect on upcoming optoelectronic technology. Flux residues can interfere with laser signals in optoelectronic assemblies using free-space signal paths. Laser signals can be deflected or attenuated by flux residues and impact the performance of emitter or receiver components.

Flux residues can also seriously affect the reliability and performance of surface mount and flip chip assemblies. The shrinking size of all electronic components and bonding pads has made postreflow cleaning of flux residues increasingly difficult.[4] The small gaps between assembled parts, and solidification cavities in mixed solder joints[5] are very resistive to penetration by cleaning liquids. Entrapped flux in solder joints can cause gross solder voids and premature failure of the solder connection. Flip chip and direct chip attach to printed wiring boards are compromised by flux residues interfering with the flow of underfill encapsulants and encapsulant adhesion.[6] As interconnection density increases and the pitch decreases, the gaps reduce in size, promising to worsen the postreflow cleaning problem in the future.

What we need is a technique that will eliminate both flux and post-solder cleaning. In addition, it should be widely applicable to the many types of assemblies and processes that exist in the soldering world, not just flip chip. Such a process would not only be of major benefit to the environment but also reduce costs and energy consumption, and have a positive impact on performance. After summarizing several of the fluxless approaches, we will concentrate the rest of the chapter on one technique which appears to have solved the aforementioned problems, and is on its way to starting a new soldering technology.

2.2   Approaches to Fluxless Soldering

Many approaches to fluxless soldering are reported in the literature. Most of these processes try to reduce the oxide on the solder surface back to pure metal. Others try to physically remove the oxide covering the solder surface to expose fresh solder. Some processes try to modify the composition of the surface oxide to change its chemical and physical properties. The underlying theme of all these techniques is to address the problem of the surface oxides covering the soldering surfaces that prevent joining from taking place.

Gaseous reduction.   The use of reducing gases, especially hydrogen, to reduce metal oxide back to pure metal has occasionally been used at 350°C with high-lead solder.[7] This is a very clean process; however, it requires the reflow to take place in a highly flammable gas. Further, hydrogen is extremely inefficient and slow, especially at lower temperatures typical of eutectic PbSn processing (235 to 250°C). This has led numerous investigators to pursue ways to make the hydrogen more active. These include such as reflowing in hydrogen plasmas,[8] activating the surfaces with palladium[9,10] or a platinum catalyst.[11] These platinum group metals (PGMs) adsorb hydrogen and cause the molecules to dissociate into adsorbed hydrogen atoms. The six stems in hydrogen reduction are listed below, where H, O, and M represent hydrogen, oxygen and metal, respectively:[12]

$$H_2 \text{ (gas)} = H_2 \text{ (ad)}$$

$$H_2 \text{ (ad)} = 2H \text{ (ad)}$$

$$M_xO_y = xM_m + yO^- \text{ (ad)}$$

$$H \text{ (ad)} + O^- \text{ (ad)} = HO \text{ (ad)} + e^-$$

$$HO \text{ (ad)} + H \text{ (ad)} = H_2O \text{ (ad)}$$

$$H_2O \text{ (ad)} = H_2O \text{ (gas)}$$

Here e^- and "ad" represent electron and adsorption, respectively, and M_m is the metal ion on normal lattice site in the metal oxide. These equations indicate that the adsorbed hydrogen atoms and the HO are involved in the reaction. The hydrogen plasma approach[8] eliminates the requirement of the catalyst. Here the active atomic hydrogen species is generated in a microwave plasma system with the samples downstream from the plasma. Again, reflow must be in situ as any exposure of the metal surfaces to air after the reduction step would regrow oxide, even at room temperature. This regrowth of oxide applies to many other techniques which have been tried to remove oxides, such as argon sputtering,[13] ion beam cleaning,[14] laser radiation, or other reducing gases such as carbon monoxide or silane.[15] Sputtering away the oxide layer just prior to reflow allowed lower hydrogen contents to be used (forming gas), and even very pure nitrogen (<2 ppm oxygen) for a gold tin solder alloy.

Formic acid.   In contrast to hydrogen, gaseous formic acid is much more effective in reducing the surface oxide at lower temperatures. The reaction can be described as follows:[16]

$$MeO + 2HCOOH \rightarrow Me(COOH)_2 + H_2O \qquad T > 150°C$$

$$Me(COOH)_2 \rightarrow Me + CO_2 + H_2 \qquad T > 200°C$$

Reflow is carried out in a controlled atmosphere chamber where various gases at different flow rates can be introduced. Deshmukh et al. have described an optoelectronic application where an optical device is passively attached to silicon surmounts using flip chip bonding with a gaseous formic acid reflow.[17] Surmounts are placed on a graphite strip heater capable of rapid heating rates (150–200°C/min). The subassembly can be viewed through a glass window of the sample chamber allowing observation of the self-alignment, with an optical microscope.

The reflow chamber is continuously flushed with dry nitrogen. Gaseous formic acid is introduced through a different port, and the concentration of the acid is controlled by regulating the flow rate. Once the eutectic solder melts, the subassembly takes about 30 s to self-align.

Mechanical scrubbing. Mechanical scrubbing has been used to break up oxides and assist flip chip bonding. Some success[18,19] has been achieved using ultrasonics in traditional thermosonic bonding mode. Loading is required; therefore, its application appears limited to stud bump structures and applications not requiring self-alignment. Full solder connections would short-circuit without the stud.

Wet chemical dips have also been reported to remove oxides. An example is *reduced-oxide soldering activation* (ROSA), which relies on reduction of the surface oxide to a metallic surface in an aqueous solution containing highly reduced vanadous ions.[20] Because it is a wet chemical dip, it requires a postprocess cleaning, and compatibility of chemicals used with all the components. As such, it truly does not fit our definition of a fluxless, no-clean process. The other techniques look promising, but some of them have shown severe restrictions in one or more of the following areas: flammability hazards, line-of-sight requirements, low throughput, and requiring application at the point of solder reflow and joining (in situ) to effectively prohibit reoxidation of the solder.

While it is possible to do the reflow-join operation in the sample chamber, or in the plasma itself, the real power of any viable technique is being able to move the samples after fluxless treatment to an existing soldering tool for reflow. This may sound simple in concept and almost self-evident, but presents an interesting challenge for researchers. No one can deny the fact that tin forms oxides in the presence of oxygen even at room temperature. Therefore, any technique that removes the constraint of in situ reflow has to deal with the reoxidation issue. This is a backbreaking task for processes that rely on oxide reduction to achieve fluxless joining. It is an easy task for the new PADS process described next.

2.3 Plasma-Assisted Dry Soldering (PADS)

2.3.1 Introduction

The new fluxless, no-clean process, called *plasma-assisted dry soldering* (PADS), neither cleans nor reduces the surface oxides. PADS is a dry pretreatment that *converts* the surface oxides to oxyfluorides. *Convert* is a key term here, as it defines the PADS process. This conversion film passivates the solder surface and has the unique property of breaking up when the solder melts, exposing free solder and allowing reflow and joining to occur in inert (nitrogen) or even oxidizing (air) ambients, depending on the rate of heating of the sample. Further, the pretreatment does not have to be coupled to the reflow process, so existing tools can be used. This is one process that does not have the limitations discussed above, and offers the promise of broad industry acceptance. The rest of the chapter will be devoted to this technique.

2.3.2 PADS process background

The ground-breaking work for PADS was initiated in 1988, and the first patent was issued in 1990.[21] This method used an RF-generated plasma to disassociate an innocuous fluorine containing source gas such as CF_4 or SF_6 and produce a very reactive species, atomic fluorine, to remove the oxide. Various experiments with the original plasma system—a split-cathode magnetron system,[22] and with an IBM flexible diode system,[23] gave promising but sporadic results.[24] Both systems were designed as in situ, high-rate reactive-ion etching (RIE) systems for IC manufacturing. Temperature control was a problem and sample damage (melting and sputtering) occasionally occurred. A modified downstream system was needed to alleviate these problems. So, a prototype system was designed and built to explore the downstream configuration. The pretreatment tool used is a microwave plasma chamber upstream from a sample chamber. The microwave plasma system also temporarily dissociates an inert, nonflammable, fluorine-containing gas (SF_6) into its constituent elements, including atomic fluorine. The fluorine lasts long enough to diffuse downstream to the sample chamber and convert the oxides to oxyfluorides. Being downstream, the samples are not exposed to the high temperatures of the plasma, or the intense sputtering which could damage sensitive microelectronic devices. The conversion film passivates the solder and has a longer shelf life than does the in situ generated film. The unique property of breaking up when the solder melts remains, thus exposing free solder and allowing reflow and joining to occur in inert (nitrogen) or even oxidizing (air) ambient. Since the pretreatment does not have to be coupled to the reflow process, existing soldering tools can be used. One simply eliminates the flux and flux cleaning, and adds a pretreatment tool prior to the soldering step.

Simple consumer microwave ovens have been used to produce inexpensive prototype tools for evaluation and demonstration purposes. The first-generation tool used a single microwave and small sample chamber to generate the proof-of-concept data.[25] This study quantified the soldering metrics in a general fashion using pellets, wires, and preforms. These metrics will be described in detail later as they form the base of understanding of the PADS process. The following reaction occurs in the sample chamber:

$$SnO_x + yF\text{—}SnO_xF_y$$

Without the microwave plasma system turned on, the SF_6 is ineffective and surface conversion does not take place. For scaling up to real product applications, a second-generation tool was designed and built.[33] A picture and schematic of this Phase 2 pretreatment tool is shown in Fig. 2.1. Two (or three) microwave ovens provide a higher F atom density and better uniformity than the original single-oven de-

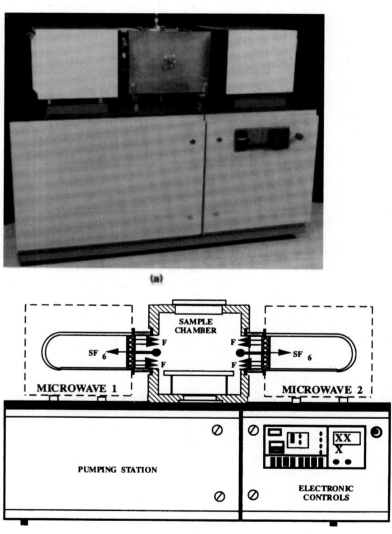

(a)

(b)

Figure 2.1 Phase 2 PADS tool.[33]

sign. The chamber can accommodate up to 12-in^2 assemblies, in single or multiple tiers. The surface conversion process, typically taking minutes to perform, is done here as a batch. Once converted, the assemblies can be stored for up to a week in air, or 2 weeks in nitrogen before the reflow or joining operations. Numerous demonstrations of fluxless soldering of real product formfactors have been performed using this tool. These will be described in detail in later sections.[33]

2.4 Fluxless Soldering Optimization

Optimization and characterization of the PADS process was performed by using simple soldering tests such as balling and spreading to quantify the soldering behavior.[25] The process windows and dependences are determined for specific process variables, such as exposure time and temperatures. Once the process windows were determined, a simple "optimal" cycle was chosen to further characterize the systems performance. The reflow tests called metrics are described in the following section, along with the performance measured for the "optimal" cycle. These tests can be useful in evaluating or predicting compatibility and performance expected with different solder alloys and/or terminal metallurgies for flip chip.

Several types of reflow metrics are described in the following subsections.

2.4.1 Balling of solder preforms

Solder balling is one of the simplest soldering processes.[25] A flat solder disk is melted to see if it reflows into a spherical ball. The surface tension of the clean liquid solder must be sufficient to overcome gravity and force the molten solder into the lowest free-energy configuration—a perfect sphere. Figure 2.2 shows the metric used to assess the efficiency of flux or fluxless processes in their ability to reduce or break up surface oxide so that the solder can reflow; d_0 is calculated from the volume of the solder preform. Figure 2.3 shows an SEM (scanning electron microscopic) view of a typical 1-mm solder disk preform re-

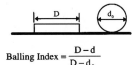

$$\text{Balling Index} = \frac{D - d}{D - d_0}$$

D = Original Diameter
d = Final Diameter
d_0 = Minimum Possible Diameter

Figure 2.2   Balling metric.[25]

Figure 2.3 Comparison between as-received and reflowed disk without any treatment.[25]

flowed in air (bottom disk), without any fluxless treatment. It also shows an as-received preform, resting on top of the reflowed disk for comparison. The bottom disk has clearly melted, but the tenacity of the surface oxide forms a bag preventing solder from flowing.

An SEM view after a successful fluxless treatment and reflow should resemble the ball shown in Fig. 2.4. Reflow is carried out on a polished silicon surface (nonwettable by the solder). These samples can be rapid-reflowed in air by dropping the preform onto a preheated stage, such that the time to melt is less than 2 s, or slow-reflowed by placing the disk on a stage and heating at the appropriate rate with the stage controller. Samples of eutectic SnPb show 100 percent balling when successfully treated. By contrast, untreated preforms exhibit near-zero balling.

2.4.2 Balling of flip chip solder bumps[25]

Flip chip solder bumps[26–29,34,36] are typically very small compared to the solder disk preforms used above. To verify that solder ball size

Figure 2.4 Good fluxless reflow.[25]

does not influence the reflow behavior, flip chips with 40-μm-diameter bumps were fabricated by an electroplating process at MCNC[29] with the same composition as the disk preforms (63/37 SnPb eutectic). As received, the solder bumps were in the as-reflowed condition (from a previous flux reflow and clean). The bumps were flattened using a Research Devices Aligner Bonder to 5 μm thickness (Fig. 2.5).

Reflow behavior of these mini–solder disks was identical to the large 1-mm solder disk preforms, with 100 percent balling observed on treated chips (Fig. 2.6), and zero balling observed on untreated chips (see Fig. 2.7). The fluxless reflowed solder is not quite as smooth and shiny as the flux reflowed solder (Fig. 2.8). This has the potential to impact inspection criteria for joints but is not expected to impact their functional properties.

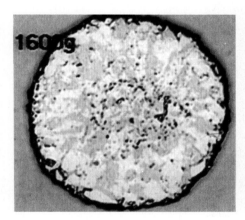

Figure 2.5 As-flattened structure.[25]

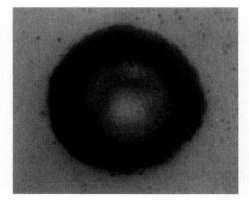

Figure 2.6 A good fluxless reflow.[25]

Figure 2.7 Reflow without any treatment.[25]

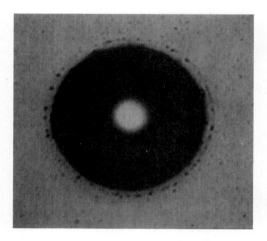

Figure 2.8 Flux reflow.[25]

2.4.3 Joining of solder preforms (solder–solder interaction)[25]

The most visually dramatic indicator of soldering performance is the actual joining of two solder masses. The use of preforms allows an unobstructed view of the solder pieces melting and agglomerating into a single mass. Using the same solder for both preforms simplifies the joining process and cuts down on the number of variables. It is directly applicable to a solder-coated component joining to a solder-tipped pad on a substrate. Actual joining of solder disks has been captured on videotape in a hot-stage microscope.[25] Figure 2.9 shows the sequence during joining.

As with balling, the final equilibrium shape if two disks are joined as a sphere is shown in Fig. 2.10. The measurement metric is similar

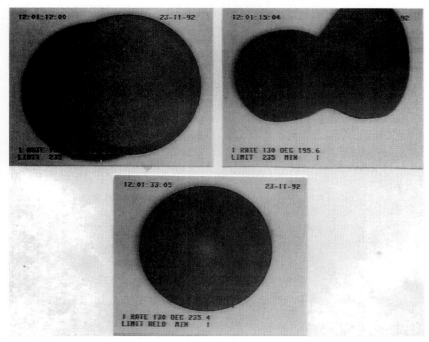

Figure 2.9 Joining sequence.[25]

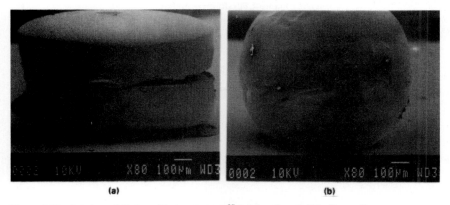

Figure 2.10 Joining of disks with treatment.[25] (*a*) As placed; (*b*) after reflow.

to that used for balling (Fig. 2.11). The glue showing in the as-placed view is just aquadag (graphite-water suspension) used for holding the samples while taking the SEM picture and is not present during actual reflow. The slight dimples seen in the as-reflowed ball are due to solidification shrinkage. Figure 2.12 shows two disks reflowed without first being pretreated. They have obviously melted; however, no mixing of the two solder masses has occurred. The wrinkling of the oxide skin is clearly visible. With pretreated disks, successful joining (100 percent ideal shape) has been demonstrated with reflow in nitrogen and in air (dried) with standard 2-min ramp heating time.

An interesting side experiment was performed with only one of the two disks pretreated with the PADS; 100 percent joining was observed regardless of whether the treated disk is initially on top or on the bottom of the stack. This opened the door for processes in which only one of the joining components is treated; and, in the extreme, use of a pretreated preform to join two components that are not pretreated. It is expected that an air ambient would be acceptable only for very rapid

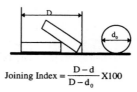

Joining Index $= \dfrac{D-d}{D-d_0} \times 100$

D = Original Diameter

d = Final Diameter

d_0 = Minimum Possible Diameter

Figure 2.11   Joining metric.[25]

Figure 2.12   Reflow of disks without treatment.[25]

processes, and that nitrogen would be required for slow processes to protect the untreated parts from excessive oxidation.

2.4.4 Joining with mixed solders (DCA)[25]

There are applications where mixed solder systems are required. For example, IBM has joined high-melting solder flip chips to printed-circuit boards (PCBs) at low temperature by having eutectic solder on the board.[30] The high-melting-point solder mass does not melt during the low-temperature reflow. Instead, the low-melting-point solder, which does melt, "wets" the high-melting-point solder, flowing up the sides to form the joint with a reaction zone intermediate in composition to the two initial solder masses.

We have simulated this application by using solder disk preforms, one with eutectic composition 63/37 SnPb, the other with high-melting 97/3 PbSn.[25] Figure 2.13 shows the successful mixed joint obtained with pretreated disks in nitrogen reflowed at 150°C. By comparison, Fig. 2.14 shows the control with no pretreatment of the solder disks. When treated, the low-melt solder can break through the conversion film on the unmelted disk, and allow wetting and spreading to occur on that surface. Joining of two components with high-lead solder termination with a eutectic solder preform in between is simulated in Fig. 2.15. Joining was in nitrogen at 250°C with pretreated disks. Precise self-alignment of the three disks has occurred similar to the self-alignment of flip chips. This illustrates the potential for joining of pad grid array and ball grid array substrates to PCBs.

97/3 PbSn   M.P. 310°C

Figure 2.13 Successful mixed joint obtained with pretreated disks.[25]

63/37 SnPb   M.P. 183°C

97/3 PbSn, M.P. 310°C

63/37 SnPb, M.P. 183°C

Figure 2.14 Reflow of disks without treatment.[25]

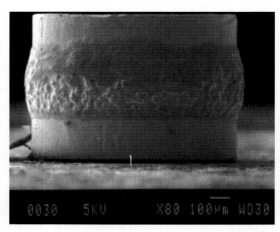

Figure 2.15 Joining of two treated high-lead solder disks with a eutectic solder preform.[25]

2.4.5 Dip soldering with wetting balance[25]

A wetting balance measures the force exerted by liquid solder on a metal sample when it is inserted into a solder pot. A wire or sheet of the same material as the bond pad can be used. The Multicore unit used in these tests is automated and records the force versus time as the solder wets the sample and climbs by surface tension forces. One measure of the wettability index is the force relative to the ideal force governed by the surface tension and meniscus formed by the solder as shown in Fig. 2.16.

Pretinned copper resistor leads were used for the wetting balance evaluations. A typical curve for the resistor leads dipped in air ambient is shown in Fig. 2.17. Full 100 percent of ideal force is attained with PADS treated samples, while near-zero force is measured with untreated samples. Wetting time to two-thirds maximum force is 0.76 s. By comparison, wetting forces measured with RMA flux typically reaches 85 to 90 percent (see Fig. 2.18). Activated flux reaches the full 100 percent as shown in Fig. 2.19. The time to reach zero wetting force, usually considered a measure of the heatup time for the wire to reach solder melting temperature, is slightly faster for the flux-coated samples. This is believed to be due to the presence of the flux liquid aiding heat transfer. With this exception, the fluxless process is equivalent to activated flux. While this alone does not predict equivalent yields in a real manufacturing process, it strongly indicates that the desired result may be attainable.

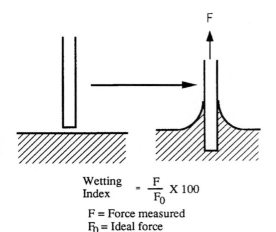

$$\text{Wetting Index} = \frac{F}{F_0} \times 100$$

F = Force measured
F_0 = Ideal force

Figure 2.16 Dip solder wetting index.[25]

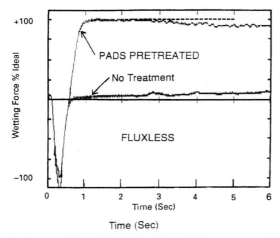

Figure 2.17   Wettability balance in air.[25]

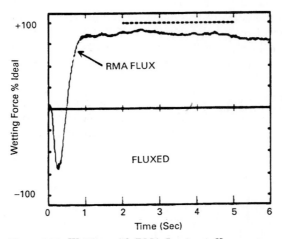

Figure 2.18   Wetting with RMA flux in air.[25]

2.4.6 Spreading[25]

Spreading of solder on a wettable metal surface is the most complex of soldering processes. It involves not only the requirement of solder flow but also the condition of the mating metallic surface and the reaction between them and the ambient. This is the least understood phenomenon—as it applies to the PADS process. The metric is simple and involves the spreading diameter and height relative to the initial volume as shown in Fig. 2.20. Spreading is the only metric where 100 percent is not attainable, as this would require an infinitely large

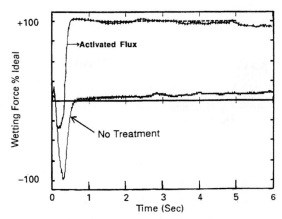

Figure 2.19 Wetting with activated flux.[25]

$$\text{Spreading Index} = \frac{D-h}{D} \times 100$$

Figure 2.20 Spreading index.[25]

substrate. Measurements have been made on numerous metals, with the data on bare copper reported in Fig. 2.21. The comparative data shown in Fig. 2.21 indicates spreading on bare copper with PADS can be *better than with activated flux.*[41]

2.4.7 Summary of fluxless joining metrics

This section described the laboratory experiments to measure solderability metrics with many different but standard accepted techniques. In general, for the PADS process, wettability was equivalent to, or better than, activated flux.[25] Most importantly, this process appeared applicable to a wide variety of soldering processes since these were not changed except to eliminate flux and cleaning. This was made possible by a pretreatment that converted the oxides to oxyfluorides, and promoted subsequent solder wetting in inert, even oxidizing ambients. Joining, balling, and wetting have been shown to be successful with pretreatment, unsuccessful without pretreatment. With the optimal pretreatment cycle, ideal behavior is realized. There are, however, more items to consider in attaining a fully characterized process. These include rate of heating and ambient impurity levels. This will be discussed in a later section on reflow and joining variables.

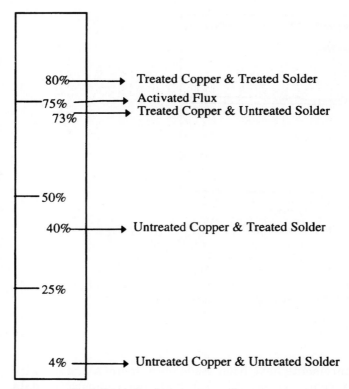

Figure 2.21 Spreading index for bare copper.[41]

2.5 The Five Keys[25]

If one looks at the overall timeline (Fig. 2.22) for the fluxless process, five distinct periods can be recognized. Each of these areas has a different set of concerns. Some comments are in order on each of these time periods.

2.5.1 Preplasma[25]

The primary concern of the preplasma period is the sample history, especially as it relates to the oxidation state of the surface, and contamination of the surface by organics, etc. In the course of this work, we have successfully processed samples with the following histories.

- Acid-etched
- Aged in air

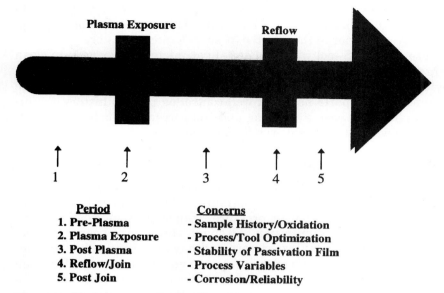

Figure 2.22 Fluxless process timeline.[25]

- Aged in nitrogen
- Air-oxidized to 160°C
- Air-reflowed (183°C)
- Steam-aged ($\leq$ 4 h)
- Double-cycled (previously fluxless reflowed)
- Previously flux-reflowed and cleaned

This attests to the broad process window regarding the sample condition. However, it is not unlimited. Two notable exceptions were encountered. Samples steam-aged over 4 h exhibit degraded wettability as determined by the wetting balance. To take care of the much thicker oxide on these samples, one would either have to extend the "optimal" plasma treatment cycle time to allow further solid-state diffusion to take place (reattain the proper fluorine/oxygen range), or remove the oxide before the plasma fluoridation step.

The other exception has been parts contaminated with organic residues from prior processing or storage. While the oxide variations can be severe on incoming parts, organics cannot. They inhibit the oxyfluoride conversion and need to be removed. While solvents have been used to do this, it is preferable to use dry processes. Air or oxygen plasmas have been used successfully in conjunction with PADS to restore solderability to "dirty" hardware. This has been done in a separate system prior to the PADS exposure. Experiments are under way to use the PADS tool with sequential gas inlets of air/oxygen to clean organics, then SF_6 to convert the oxides. It is not known at this time which configuration is optimal—separate chambers or the same chamber. With parts having a well-controlled process history, the organics have not been a problem, and 100 percent yields have been attainable without a special organic removal step.[33]

2.5.2 Plasma exposure[25]

The conditions used in the plasma chamber and sample chamber of the PADS machine need to be optimized. The variables to be understood are time, temperature, pressure, sample position (relative to the plasma), and gas composition. Although these have been explored, most of the details are proprietary and have not been published, but are available to licensees of the process. The conditions used for small samples in small chambers are not necessarily the same ones to be used for high-volume manufacturing in a large chamber. Attention to process uniformity (sample to sample) and process speed are very important. At the time of this writing MCNC has joined with a number of companies to build and qualify a very high-rate, automated, manufacturing tool and qualify it in a high-volume manufacturing product line. MCNC should be contacted for details.

2.5.3 Postplasma[25]

In the postplasma period, the primary concern is the stability of the passivation film. With the original in situ RF plasma process, the working time was only 3 h before the joining-reflow step needed to be accomplished. If the delay exceeded this time, performance in reflow was degraded. Thus, a degradation mechanism exists which somehow modifies the passivation film. The lifetime of the conversion film has been greatly extended with the use of the latest PADS process, which incorporates the use of downstream microwave systems, coupled with defining an optimized cycle. Wetting balance measurements indicate a 1-week lifetime for samples stored in air and a 2-week lifetime if samples are stored in a nitrogen box (Fig. 2.23). At the time these experiments were done, the nitrogen box was in constant use—numerous times a day opening and closing the door as samples were either put in or retrieved. The lifetime in a "sealed" container could possibly be significantly longer. The mechanism for the degradation has not yet been explored, but it is believed to be due to reaction with atmospheric moisture. Thus, any contact of the PADS-treated parts with water prior to reflow is to be avoided.

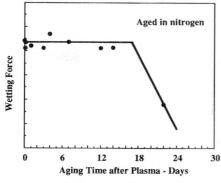
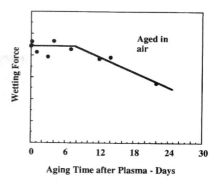

Figure 2.23 Aging effects.[25]

2.5.4 Reflow-join

Rate of heating.[25,31] The rate of heating is a key variable for all soldering operations, especially for fluxless processes. When the conversion film breaks up and exposes fresh solder, there is nothing to prevent it from reoxidizing. Therefore, the working time depends on the soldering atmosphere. It was found that for quick reflow, with heatup time of < 5 s, the soldering atmosphere was not critical. But, for slow-ramp profiles such as a 2-min belt furnace, strict ambient control was necessary. Good reflow characteristics were also obtained for ramp rates of ≤ 5.25°C (45-min ramp) in a pure dry nitrogen atmosphere.

The reflow of PADS treated samples is not very sensitive to the oxygen content if the water vapor is below a certain level. To achieve good joining, it is critical to control the water vapor content of the reflow environment. Good soldering performance was obtained in any ambient with rapid heating rates (< 5 s). For slow processes (ramp times ≤ 2 min), control of ambient purity was necessary. Good reflow could be obtained in air (21% oxygen) if the water vapor content was maintained below 22 ppm in the reflow cover gas. Good reflow characteristics were obtained when the water vapor levels in nitrogen were below 130 ppm for joining and in the range of 20 to 80 ppm for balling. High-purity nitrogen can extend the ramp heating time to 45 min with good reflow results. The cause of poor fluxless performance in a belt furnace was traced to high water vapor content. This was corrected by ensuring sufficiently long purge times after furnace turnon to flush the water vapor down to acceptable levels, as shown in the next section.

Trace impurity measurement and control.[31] The atmospheric control and measurement of water vapor and oxygen is one of the most important factors to consider outside the fluxless pretreatment itself. An understanding of the samples' interactions with the reflow ambient is absolutely critical. Water vapor and oxygen are the main active species present in the reflow atmosphere that have to be studied. This section describes a method to do the same. Evaluation of O_2 and N_2 were carried out using Airco's atmospheric-pressure in situ analysis system. This system is capable of measuring water vapor, oxygen, and total hydrocarbon levels in environments at or above atmospheric pressure. Detection limits for each impurity are

$$\text{Water vapor} = 50 \text{ ppb} \le 2\%$$

$$\text{Oxygen} = 50 \text{ ppb} \le 25\%$$

$$\text{THC} = 100 \text{ ppb} \le 2\%$$

The analysis system also contains an on-board calibration gas supply to supply zero gas (< 10 ppb of listed impurities). Trace oxygen is measured by a Teledyne analytical model 3160. The 3160 uses a microfuel cell to measure the oxygen concentration in the gas stream. The *microfuel cell* is an electrochemical transducer which produces a linear output with oxygen concentrations from 0 to 25%. The 3160 is generally used for oxygen concentrations below 1%. High-level oxygen measurements are taken using a Panametrics TMO_2 thermoparamagnetic oxygen analyzer. Measurement range for the TMO_2 is 0 to 25%. Water vapor measurements were taken using a Panametrics MIS hygrometer. The MIS uses an aluminum oxide sensor to detect water vapor. The capacitance of the aluminum oxide sensor is a function of the water vapor concentration in the sample gas stream.

Gas sampling of a belt furnace and hot-stage microscope was accomplished by using the stainless-steel sampling line attached to the sample inlet of the in situ analysis system. The analysis system has switching manifolds which permit the operator to change flow to the analyzers from the sample gas stream to the on-board calibration/zero-gas stream. Flow is drawn through the water vapor and oxygen analyzers by a downstream vacuum pump. During the sampling procedure the analyzers receive flow from the zero gas stream until the hot stage is closed and ready for sampling. At the start of sampling the analyzer flow is switched to the sampling manifold. Once sampling is complete, the analyzers are switched back to zero gas to maintain stability.

Belt furnace results.[31]   Early fluxless reflow experiments were performed in a hot-stage microscope under a nitrogen purge. The success of these experiments was not repeated when scaled up to reflow in an infrared belt furnace. This was initially puzzling since the samples, pretreatment, and the nitrogen gas source were common to both. Characterization of the ambient in the belt furnace, coupled with am-

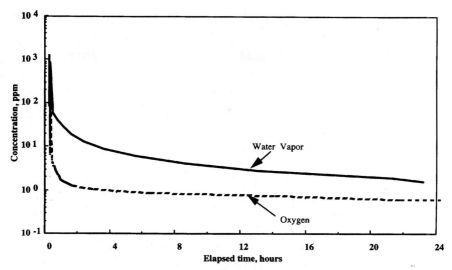

Figure 2.24   Oxygen and moisture concentration in the furnace as a function of hours of operation.[31]

bient effects on reflow, led to a simple explanation and solution. The initial water vapor level prior to furnace heating was approximately 400 ppm. When the furnace heat was turned on, the water vapor level increased to approximately 1000 ppm and then fell to 2 ppm following a 24-h purge at temperature. The furnace oxygen level fell to 2 ppm after 1 h and then stabilized around 1 ppm. There was no apparent affect of standby conditions on oxygen level. Good reflow was obtained when the water vapor and oxygen levels were kept below 10 and 2 ppm, respectively. Thus the oxygen and water vapor levels in the furnace were tracked until they were below those values. The change in the oxygen and water vapor levels versus time are compared in Fig. 2.24. The oxygen level decreases from >1000 to less than 2 ppm in the very first hour of operation, while the water vapor drops from 1400 to only 100 ppm. It took the furnace at least 5 h of operation before the water vapor level would reduce to below 18 ppm, where good reflow could start to take place. The level of oxygen remained fairly constant at around 2 ppm. It was only after 24 h of continuous operation that the water vapor level decreased to a level (3 ppm) where good reflow could consistently be obtained. This could be due to the fact that the water vapor adsorbed in the ceramic insulator of the furnace is very slow to drive off.

Reflow dependences.[31]   Reflow characteristics of solder preform under nitrogen purge was measured, and the results are shown below. Experiments include balling and joining of eutectic disks. Most of the

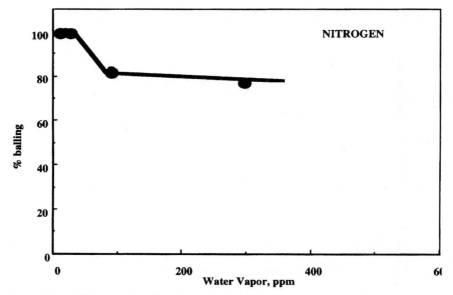

Figure 2.25 Balling as a function of water vapor in purge gas (nitrogen).[31]

reflow was done using a heatup ramp rate of 130°/min. Figure 2.25 plots balling as a function of water vapor content in the purge gas (nitrogen). As seen from the figure, the balling effectiveness decreased from 100 to 83 percent for an increase in water vapor level from 20 to 90 ppm. Once the water vapor level reached about 90 ppm, the curve flattens. The same information for joining is shown in Fig. 2.26. The joining was eutectic to eutectic. Fairly good joining could be achieved with water vapor levels close to 130 ppm. One important point to be noted is that the degradation of the joining efficiency is much slower as compared to balling. The sharp drop seen in Fig. 2.25 is not present in Fig. 2.26. In joining, two disks melt and join at the interface and coalesce into one single ball. The driving force for this is larger than simple balling of a disk (Fig. 2.27).

In comparison, joining in air showed a marked decline in the 20- to 30-ppm moisture range (Fig. 2.24). Thus, the furnace problems were traced to transients in the water vapor and oxygen levels caused by the intermittent use of the furnace and water-oxygen pickup during furnace off periods. Proper reflow conditions are now maintained by lowering the temperature and nitrogen flow to a standby level instead of turning the furnace off. This prevents water and oxygen pickup by the furnace walls and permits a very rapid response of the furnace to use conditions.

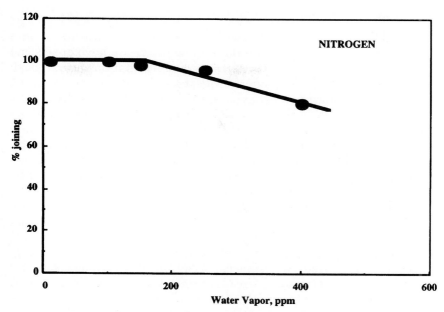

Figure 2.26   Joining as a function of water vapor in purge gas (nitrogen).[31]

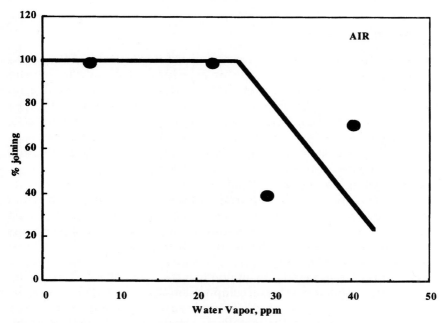

Figure 2.27   Joining as a function of water vapor in purge gas (air).[31]

2.5.5 Postjoin reliability

A full qualification needs to be performed for any application using a new process. This has not been done yet. However, we have done some preliminary experiments on the primary concern that has been raised: corrosion. Specifically, do residues from the PADS pretreatment affect the corrosion or migration behavior of the interconnections? Two studies indicate that the reliability risks of samples soldered with the PADS process are low. Surface insulation resistance (SIR) testing and ionizable contaminant assessment by solvent extraction conductivity (SEC) have been performed on FR-4 PCB coupons with a standard comb pattern.[32] Using both bare copper and pretinned coupons, SIR and SEC levels were well within specification for "clean" PCBs. Readings remained stable through 130 h of temperature and humidity testing. In a separate study,[25] we took samples processed with our optimal conversion process and exposed them for up to 334 h at $T = 85°C$ and 85% RH. Preforms, flip chip solder bumps, and solder-dipped copper coupons were used. No visible corrosion was observed on the solder either by optical microscope or SEM. Surface oxide was measured with the dispersive x-ray detector in the SEM and indicated the occurrence of only a simple oxidation, with flux-treated and plasma-treated samples oxidizing at the same rate. Thus, the oxyfluoride residues are not acting as a catalyst to accelerate the rate of corrosion.

2.6 Applications[33]

In this section we will be describing several applications of the PADS process. All have bearing on flip chip attachment by solder reflow and illustrate the application as well as the materials and structures compatibility with the PADS process. All are experimental, in that no products are in volume manufacturing at the present time using PADS processes. However, many of the Phase 2 tools, as well as homemade systems, are in use in laboratories around the world, and application papers are being reported regularly.

Before going into details, a brief discussion of tacking requirements for flip chip is in order. During traditional flip chip joining with liquid fluxes, the flux acts like a glue holding the chip in position after chip placement. Thus, subsequent handling and mass reflow processes such as belt furnace reflow can occur without yield losses due to loose chips being shaken or knocked out of position prior to joining. Fluxless processes require a tacking to accomplish the same effect. During chip placement, a temporary thermocompression bond is made between the solder ball and the bond pad.[39] The pressure, time, and temperature requirements depend on the metallurgies involved and the throughput required from the chip placement machine. Higher placement speeds require shortened tack times. This is facilitated by increasing temperature and pressure during tacking. It is not necessary for all bumps to make contact, nor for precise bump-to-pad alignment. One is not trying to make a "permanent" tack connection. The subsequent melt cycle and solder flow will result in "controlled collapse," self-alignment, and interconnection to all contact sites. Experiments are underway with metallurgy and process variations to reduce tack times to about 0.1 s at room temperature.

2.6.1 Flip chip joining examples

The first several examples presented here are flip chip interconnections.[33–40] Figure 2.28 shows a schematic of the chip and substrate structures for a direct-chip-attach (DCA) demonstration. The chip is silicon with 1679 high-lead solder bumps. It is to be joined to an FR-4 organic board substrate that has copper traces covered with a thin layer of eutectic tin lead solder on the bond pads. The high-lead bumps were electroplated in the MCNC Flip Chip Technology Center, and reflowed in a nitrogen inerted belt furnace at 355°C. The eutectic cap on the copper of the substrate was fabricated by a simple dipping of the substrate into a eutectic solder pot at 235°C.[33]

Both chip and substrate were pretreated with PADS before joining. Placement and tacking were done in a Research Devices Aligner Bonder (model M8). The joining was performed in the same nitrogen inerted belt furnace used earlier, this time with a peak temperature of 233°C. At this temperature the high-lead bump does not melt; only the eutectic tin lead does. The eutectic wets the high-lead bump and

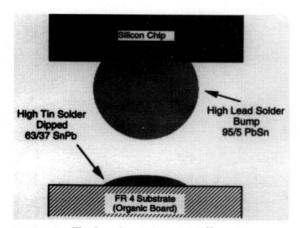

Figure 2.28 Fluxless direct chip attach.[33]

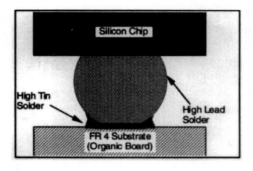

Figure 2.29 Direct-chip-attach joint.[33]

achieves the fully joined configuration as shown schematically in Fig. 2.29. The fillet is shown small on the FR-4 side because of the very low solder volume of eutectic deposited with the dip solder method. While this is not optimum for reliability, there is sufficient volume for demonstrating the joining of the high I/O density chip. Pulling the assembly apart separates the interconnections at the FR-4/copper interface as shown schematically in Fig. 2.30. This allows for scanning electron microscope pictures to be taken of the structures. SEM photomicrographs of the actual parts taken before and after joining are shown in Fig. 2.31. All joints were made in the array.

The cross sections shown in Fig. 2.32 clearly delineate the high-tin and the high-lead zones in the microstructure. Figure 2.33 shows the joining of chips with square bumps. The configuration was chip to chip, with both the chips having eutectic solder bumps (Fig. 2.34). An alternative low-temperature joining option to an organic board is to have the chip bumped with a low-melting-point solder. In this case the true controlled-collapse chip connection (C4) is made, which relies on solder surface tension of the completely molten bump to support

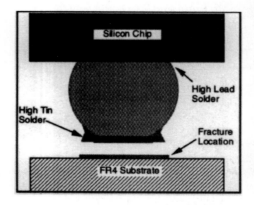

Figure 2.30 Tensile pulled fluxless DCA joint.[33]

Figure 2.31   SEMs of fluxless flip chip DCA[33]: (*a*) chip solder bump; (*b, c*) chip after pull; (*d*) substrate pad; (*e*) substrate pad after pull; (*f*) substrate after pull.

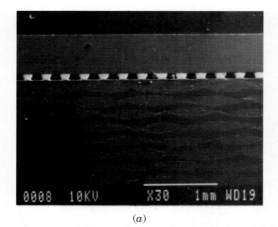

(a)

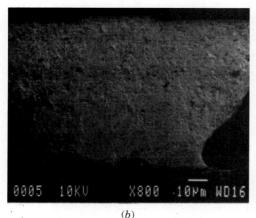

(b)

Figure 2.32 Cross sections of DCA solder joints on FR-4.[40]

the weight of the chip. We have demonstrated this structure by joining eutectic tin lead solder-bumped chips to other chips, with and
without solder bumps (Fig. 2.35). The bare chip is most significant
and is shown in Fig. 2.36 where the eutectic solder is joined to bare
copper without flux. As with the first case, both parts are pretreated
with PADS, the chips were placed in position, and joining and reflow
were performed in a nitrogen inert belt furnace. The fracture surface
is shown in Fig. 2.37, where complete wetting to the bare copper terminals is indicated.

The PADS process has allowed the integration of MEMS and electronics via fluxless flip chip joining.[38] The MEMS structures are extremely delicate and contain movable micromachined elements. One
such resonant structure is shown in Fig. 2.38. Measurements of the
resonant frequency of the vibrating shuttle during actuation indicates

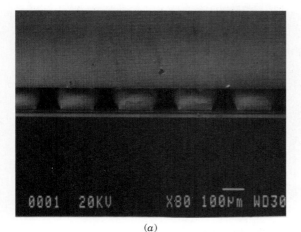

(a)

(b)

Figure 2.33   Square bumps.[40]

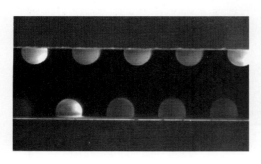

Figure 2.34   Chip-to-chip connection.[40]

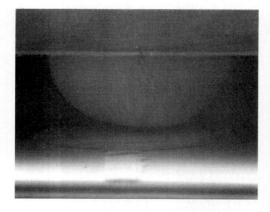

Figure 2.35   Bump-to-pad connection.[33]

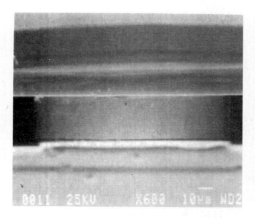

Figure 2.36   Connecting fluxless flip chip to bare copper.[33]

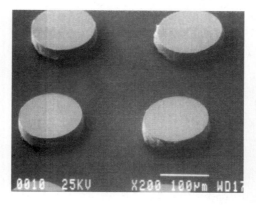

Figure 2.37   Full wetting to copper.[33]

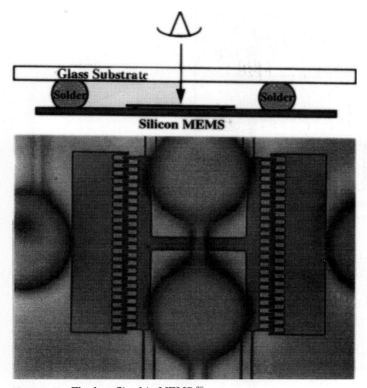

Figure 2.38   Fluxless flip chip MEMS.[38]

zero change due to the fluxless joining. Other MEMS structures such as piezoresistors, double supported beams, and diaphragms are also being fabricated by PADS and their response being measured before and after flip chip joining.[38]

A well known MEMS application is an automobile air bag accelerometer.[4] The full eutectic tin lead structure is shown in Fig. 2.39 where an accelerometer is joined using the PADS process.[34] The terminal metallurgy is TiNiAu on the ASIC to which the solder bumped capacitor sense chip is joined. The hermetic seal is achieved simultaneously with the interconnection as both the ring and the bumps are 60/40 SnPb. In this example, the IEI PADS tool was not employed, but a downstream microwave chamber was set up at LETI in France.[34] 100 percent yields were achieved with the PADS process.

2.6.2 Related applications

Wave soldering of PCBs containing surface mount or pin-in-hole components, or both, represent a high proportion of microelectronics as-

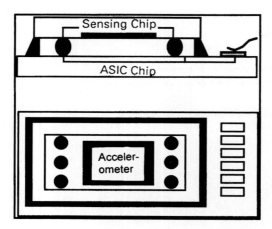

Figure 2.39 Fluxless automobile accelerometer.[34]

semblies in the marketplace. Direct chip attach by flip chip to these boards is a very important application. The next example illustrates the compatibility of the PADS process to the materials and structures commonly encountered on these boards, and the compatibility with wave soldering. One particular PCB was populated with adhesively tacked surface mount components and top-loaded pin-in-hole components at Northern Telecom (NT), pretreated at MCNC, and sent back to NT for wave soldering *without flux or postsolder cleaning.*[33] The fillets were well formed and bright, and the board was pristine clean. Some 100 percent good boards (no opens or shorts) were obtained on the first run in manufacturing. One of these boards is shown in Fig. 2.40. The surface-mount components are adhesively tacked to the board prior to the PADS exposure, so all surfaces are treated as they go into the solder wave. Following a period of process optimizations, the single layer of boards loaded into the PADS tool was expanded to include 3 tiers, and a significantly larger batch size. We have now obtained multiple *zero-defect batches* at NT with this configuration,

Figure 2.40 Fluxless/no-clean printed-circuit board.[33]

each batch containing 7704 solder joints; 36 3 × 4-in cards are in a batch, with 2772 surface-mount and 4932 pin-in-hole solder joints. The board metallurgy in this case is hot-air solder-leveled (HASL), and the wave solder machine is inerted with nitrogen. Of 13 batches fabricated in manufacturing trials between October 1994 and May 1995, 1 defect was found in 100,152 solder joints.[37]

We have also run samples in air and with alternative metallurgies. The joints made in air are also well formed but not as bright and shiny as the ones done in nitrogen. Figures 2.41 and 2.42 show surface-mount and pin-in-hole joints made in an air wave on boards with gold-plated copper traces. Defect levels, both skips and bridges, appear to be higher with air waves than with nitrogen waves. At the time of this writing, controlled experiments are just being done in this area. Dip soldering of FR-4 PCBs has also been demonstrated with bare copper traces. Figure 2.43 shows line traces of copper before and after dip soldering in eutectic tin lead in air at 235°C. The bare

Figure 2.41 Wave solder surface mount.[33]

Figure 2.42 Wave solder pin-in-hole.[33]

Figure 2.43 Fluxless dip soldering of bare copper.[33]

Figure 2.44 Fluxless solder wetting of TAB.[33]

copper board was pretreated with PADS prior to the immersion in the liquid solder. To check the compatibility of TAB and flexible circuits, a sample of polyimide tape with pretinned copper traces was PADS-treated and reflowed with a eutectic tin lead solder disk at 250°C in nitrogen. The disk wetted the TAB traces, elongating the solder of the disk as the solder flowed out on the traces (see Fig. 2.44).

If we could sum up the applications work so far, it would seem that the promise of gold soldering performance of PADS predicted by the proof-of-concept studies is being realized. Some of the results have been spectacular, and additional successes appear to be coming in the future.

2.7   Future Work

We have just scratched the surface of the fluxless soldering potential with PADS. Future work includes an equipment effort to ramp up to a high-rate, high-volume PADS tool that would be fully automated. It will be qualified in high volume for both process yields and product reliability.[37] At the same time, we will continue to evaluate the compatibility of the process with other companies' products, structures, and soldering tools to ensure the broadest impact and widest possible implementation in the marketplace. Numerous companies and organizations are working with us on this effort. Evaluations are being planned with ball grid arrays, die bonds, optoelectronic fiber attachment, hermetic seals, and heatsink attachment, to name only a few.

2.8 Summary

PADS (plasma-assisted dry soldering) enables fluxless, no-clean soldering to be accomplished by means of a pretreatment which enables solder reflow in inert and even oxidizing ambients. Conventional soldering tools can be used, just eliminating the flux dispense and flux cleaning steps, and adding the pretreatment step. The proof-of-concept paper "Fluxless Soldering in Air and Nitrogen," was presented to the Electronics Components and Technology Conference in 1993. Since that time MCNC has had a Phase 2 PADS tool built and has been using it to explore applications with real products which are typically much larger than the original Phase 1 tool could accommodate (such as 12-in PCBs).

A review of the tool and process has been discussed, and highlights of the applications studies presented with an emphasis on experiments which have a bearing on the processing of flip chip and surface-mount assemblies. Both organic PCBs and silicon substrate assemblies are represented. Examples include eutectic solder-bumped flip chips joined in a nitrogen inerted belt furnace at 250°C to bare copper, high-lead flip chips joined to an FR-4 organic board, capacitors adhesively tacked to the bottom of PCBs and wave-soldered in air or nitrogen, and dip soldered in air of insulation resistance comb-pattern test sites in eutectic PbSn solder. A commercial tool is now available for prototype explorations on real product formfactors. Plans are in place to enter Phase 3, where high-volume manufacturing tooling will be built and products qualified.

References

1. Newboe, B., "Warning: Is Your Product Made with CFCs?" *Semiconductor Internatl.,* 50–55, Nov. 1993.
2. Irving, B., "How $1-Billion per Year Can Be Saved in the Soldering of Electronic Components," *Welding J.,* 54–56, 1991.
3. Benson, R. C., et al., "Metal Migration Induced by Solder Flux Residue in Hybrid Microcircuits," *IEEE Transact. Components, Hybrid, Manufact. Technol.,* 2(4): Dec. 1988.
4. Dishon, G., S. Bobbio, N. Koopman, and G. Rinne,"Plasma Assisted Fluxless Soldering," *NEPCON West,* Feb. 23–27, 1992.
5. Koopman, N., "Application of E-SEM to Fluxless Soldering," *Microsc. Research Tech.,* 25 (5,6):493–502, August 1993.
6. Partridge, J., and P. Viswanadham, "Organic Carrier Requirements for Flip Chip Assemblies," *Proceedings: NEPCON,* Reed Exhibition Companies, 11: 1519–1526, 1994.

7. Chance, D., "Fluxless C4 Joints and Reflow without Degradation," *IBM Tech. Disc. Bull.*, **23**(7):2990, Dec. 1980.
8. Pickering, K., P. Southworth, C. Wort, A. Parsons, and D. Pedder, "Hydrogen Plasmas for Flux Free Flip-Chip Solder Bonding," *J. Vacuum Sci. Technol.*, **A8**(3):1503–1508, May/June 1990.
9. Yeh, H., S. Strickman, N. Koopman, C. Chang, J. Roldan, and K. Srivastava, "Palladium Enhanced Fluxless Soldering and Bonding of Semiconductor Device Contacts," U.S. Patent 5,048,744, Sept. 1991.
10. Yeh, H., and S. Strickman, "Palladium Enhanced Dry Soldering Process," *Proceedings of 42nd Electronic Components and Technology Conference*, pp. 492–501, 1992.
11. Moskowitz, P., and H. Yeh, "Thermal Dry Process Soldering," *J. Vacuum Sci. Technol.*, **4**(3): 838, 1986.
12. Chiu, W. F., and A. Rahmel, "The Kinetics of the Reduction of Chrome Oxide by Hydrogen," *Metallurg. Transact.*, **10B**:402, 1979.
13. Nishikawa, T., M. Ijuin, and R. Satoh, "Fluxless Soldering Process Technology," *Proceedings of 44th Electronic Components and Technology Conference*, pp. 268–292, May 1994.
14. Harper, J. M., H. Yeh, and K. Grebe, "Ion Beam Joining Technique," *J. Vacuum Sci. Technol.*, **20**(3): 359, 1982.
15. Howard, R., "Fluxless Soldering Process Using a Silane Atmosphere," U.S. Patent 4,646,958, March, 1987.
16. Trovato, R. A., *SMTCON Technical Proceedings*, Atlantic City, N.J., p. 283, April 1990.
17. Deshmukh, R. D., M. Brady, R. Roll, L. King, J. Shmulovich, and D. Zolonowski, "Active Atmosphere Solder Self Alignment and Bonding of Optical Components," *International Electronic Packaging Conference*, Austin, Tex., p. 1037, Sept. 27–30, 1992.
18. Vollmer, A., "German Firm Unveils Flux-Free Soldering," *Electronics*, p. 3, Nov. 1994.
19. Kloeser, J., E. Zakel, A. Ostmann, J. Eldring, and H. Reichl, "Cost Effective Flip-Chip Interconnections on FR-4 Boards," *Proc. ISHM*, 491–500, 1994.
20. Tench, M., D. Hillman, and G. Lucey, "Environmentally Friendly Closed-Loop Soldering," *1993 International CFC and Halon Alternative Conference*, Washington, D.C., 1993.
21. Dishon, G., and S. Bobbio, "Fluxless Soldering Process," U.S. Patent 4,921,157, 1990.
22. Bobbio, S., and Y. Ho, "Shared Loop and Multiple Field Apparatus and Process for Plasma Processing," U.S. Patent 4,738,761, April 1988.
23. Ephrath, L. M., and R. S. Bennet, *Proc. 1st Symposium on VLSI Science and Technology*, Electrochem. Soc., Pennington, N.J., p. 108, 1982.
24. Dishon, G., S. Bobbio, N. Koopman, and G. Rinne, "Plasma Assisted Fluxless Soldering," *NEPCON West*, Feb. 23–27, 1992.
25. Koopman, N., S. Bobbio, S. Nangalia, J. Bousaba, and B. Peikarski, "Fluxless Soldering in Air and Nitrogen," *Proceedings for the 43rd ECTC*, Orlando, Fla., June 2–4, 1993.
26. Koopman, N., T. Riley, and P. Totta, "Chip-to-Package Interconnections," in *Microelectronics Packaging Handbook*, Tummala, R., and E. Rymaszewski, eds., Van Nostrand Reinhold, New York, pp. 361–453, 1989.
27. Koopman, N., "I/O Options for MCNC Multichip Package," *Proceedings 41st Electronic Components and Technology Conference*, pp. 234–244, 1991.
28. Koopman, N., "Flip Chip Interconnections," *Concise Encyclopedia of Semiconducting Materials and Related Technologies*, Pergamon Press, U.K., 1992.
29. Yung, E., and I. Turlik, "Electroplated Solder Joints for Flip-Chip Applications," *10th International Electronic Packaging Conference*, Marlboro, Mass., Sept. 9–13, 1990.
30. Tsukada, Y., "Surface Lamilar Circuit and Flip Chip Attach Packaging," *Proceedings of 42nd Electronic Components and Technology Conference*, May 1992.
31. Nangalia, S., F. Tapp, and N. Koopman, "Effect of Water Vapor on Fluxless Reflow in Air and Nitrogen," *First International Flip Chip Symposium*, San Jose, Calif., Feb. 1994.
32. Koopman, N., S. Nangalia, T. Shankoff, and D. Culver, "Reliability Studies of 'PADS' Fluxless Soldering Process," in press.

33. Koopman, N., S. Nangalia, C. Lizzul, S. Bobbio, and J. Peterson, "Fluxless Soldering for Chip-on-Board and Surface Mount," *Surface Mount Symposium SMTA 1994,* Research Tiangle Park, N.C., pp. 54–60, Oct. 19, 1994.
34. Caillat, P., and G. Nicolas, "Fluxless Flip-Chip Technology," *First International Flip Chip Symposium,* San Jose, Calif., Feb. 1994.
35. Tsukada, Y., and Y. Mashimoto, "Low Temperature Flip Chip Attach Packaging on Epoxy Based Carrier," *Surface Mount Internatl.,* 1994.
36. Adema, G., C. Berry, N. Koopman, G. Rinne, E. Yung, and I. Turlik, "Flip-Chip Technology: A Method for Providing Known Good Die with High Density Interconnection," *International Conference and Exhibition on Multichip Module,* 1994.
37. Koopman, N., S. Nangalia, V. Rogers, J. Peterson, P. Brinkley, E. Yow, S. Bobbio, and M. Pennington, "Fluxless, No-Clean Solder Processing of Components, Printed Wiring Boards, and Packages in Air and Nitrogen," *Surface Mount International,* San Jose, Calif., Aug. 1995.
38. Markus, K., V. IDhuler, D. Roberson, A. Cowen, M. Berry, S. Nangalia, "Smart MEMS: Flip Chip Integration of MEMS and Electronics," *SPIE Smart Materials Conference,* Feb. 1995.
39. Koopman, N., G. Adena, S. Nangalia, M. Schneider, and V. Saba, "Flip Chip Process Development Techniques using a Modified Laboratory Aligner Bonder," *IEMT,* Oct. 1995.
40. Koopman, N., and S. Nangalia, "Fluxless Flip Chip Solder Joining," *NEPCON West,* pp. 922–931, Feb. 1995.
41. Koopman, N., and S. Nangalia, "Fluxless Soldering of Copper," U.S. Patent 5,407,121, April 1995.

Solder-Bumped Flip Chip Interconnect Technologies: Materials, Processes, Performance, and Reliability

Ravi Sharma and Ravi Subrahmanyan

3.1 Introduction

Flip chip interconnects are being used in the electronics industry primarily because of their high I/O density capability, small profiles, and good electrical performance. Demands on performance, reliability, and cost have resulted in the development of a variety of flip chip technologies using solder,[1–6] conductive epoxy,[7] hard-metal bump (such as gold[8,9]), and anisotropic conductive epoxy[10–12] interconnects. Among these materials, solders have remained a preferred choice as the material forming electrical connections in flip chip assemblies.

Solder flip chip interconnect systems consist of essentially three basic elements (see Fig. 3.1). These include the chip, the solder bump, and the substrate. The bumps are first deposited on a wafer and reflowed. The wafer is then diced into chips. The chips are flipped over, aligned to a substrate, tacked, and reflowed. An underfill is often used to improve the reliability of the interconnects. Each of these elements and the processes used to assemble them together affect the performance and cost of the interconnect system. Therefore, the performance and cost must be compared on the basis of the interconnect system as a whole, and not merely on any single element of the interconnect assembly.

Selection of a solder flip chip interconnect system for any application is influenced by its performance. The solder bumps in a flip chip interconnect system provide three functions. First, the solder joint is the electrical connection between the chip and the substrate. In some instances, the solder joint may also serve as a path for heat dissipation

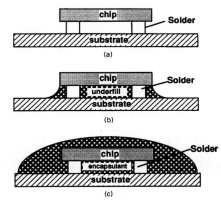

Figure 3.1 Schematic of a solder flip chip interconnect system (*a*) without underfill, (*b*) with underfill, (*c*) encapsulated.

from the chip. Finally the solder joint often provides the structural link between the chip and the substrate. The structural integrity of the solder joint affects both the electrical and thermal performance of the flip chip interconnect system. A degradation in the structural integrity can be a reliability concern.

The materials and processes involved in the manufacture of the flip chip interconnect system determine its performance. For example, the semiconductor device or the chip may be silicon or gallium arsenide. The bond pad metallization on the wafer can be Cr-Cu-Au,[1] TiW-Cu,[13,14] Ti-Cu,[15] or TiW-Au.[16] The bump material can be one of a variety of Pb-based[1–4,13–15] or Pb-free solders.[5,6,16] The substrate could be silicon, alumina, glass, or one of a variety of organic substrates. The substrate metallization can be gold or copper. In addition to the variety of materials available, the process steps used in the manufacture of the interconnect systems are also varied. Several bumping process technologies such as plating,[13–16] evaporation,[1–6] wire bumping,[17,18] dispensing,[19] and printing[20] are available. The reflow process may be performed in air with flux or in a controlled ambient.[21] Flip chip bonding processes include those based on the controlled-collapse chip connection (C4)[1–4] approach or those in which the geometry of the bump is controlled by the bonding equipment.[21]

Damage to the structure of the solder joint can significantly affect the performance of the flip chip interconnect system. For example, during temperature cycles the differential expansion of the members of the solder flip chip interconnect system can produce stresses in the solder joint which result in crack growth. This phenomenon can result in increases in both electrical and thermal resistance and is therefore a reliability concern. The kinetics of such damage can be significantly influenced by the design and materials used in the flip chip interconnect system.

The performance and reliability often influence the selection and design of a solder flip chip interconnect system. The geometry of the interconnect system and the substrate and underfill material influence the electrical performance. The thermal resistance of the interconnect system is governed by the thermal conductivity of the bump and the substrate. The structural integrity and reliability are influenced by the mechanical behavior of the assembly members and their thermal expansion coefficients. These requirements can impose adequate constraints to reduce the number of alternatives. Then, the task of solder flip chip interconnect system selection is reduced to a comparison of the manufacturability of the interconnect system and total cost.

For example, let us assume that we have an application which requires relatively large heat dissipation in addition to electrical performance. The preferred bump material for these two criteria would be gold. We would then have several alternatives for the substrate. We could use a ceramic substrate such as alumina or aluminum nitride. Alternatively, a low-dielectric organic substrate can be used. While the surface planarity and high-temperature resistance of ceramic substrates are favorable, cost factors may dictate the use of organic boards. This choice may be based on manufacturability–cost tradeoffs.

While the details of manufacturability and cost analysis are application-dependent, some generalizations apply. Perhaps one of the most important aspects of manufacturability involves preserving the temperature hierarchy in the package processing. One or more process steps in the fabrication of the solder interconnect system often require temperatures above room temperatures. Materials in the interconnect system must be selected such that temperature cycles in the process flow do not adversely affect the structural integrity of the assembly members. The second important aspect of manufacturability involves the integrity of the interfaces. The process technologies must be capable of producing structurally sound and uniform interfaces between the solder and the adjoining bond pad metallizations. Bump height and volume uniformity, substrate planarity, and compositional control are some of the factors which influence bump interfacial uniformity. A process technology which is robust with respect to these factors can significantly impact the manufacturability of the interconnect system. Environmental friendliness is another important aspect of manufacturability, affecting the choice of the materials used to produce a solder flip chip interconnect system.

The cost of manufacturing a solder flip chip interconnect system is related to the manufacturing process technologies. Some basic elements of a cost model are the materials cost, number of process steps, equipment costs, floor space, and labor. In many instances the materials cost of the interconnect system is dominated by that of the substrate. The substrate material, the design rules (minimum feature size and spacing), number of layers and the via/through-hole design are some factors. The number of process steps has a significant influence on the cost, since they also affect equipment costs, floor space, and labor. A smaller number of process steps invariably results in lower costs.

The objective of this chapter is to provide an overview of solder flip chip technologies. On the basis of a typical process flow, we shall discuss the process technology options available to manufacture the flip chip interconnects. Then, some aspects of the electrical and thermal performance of solder flip chip interconnect systems are discussed. Phenomena which cause degradation of the performance that can affect the reliability of the interconnect system are then discussed. It is apparent that the number of alternative technologies are numerous and their reliability assessment can consume a significant fraction of the development cycle. We have outlined a methodology to develop accelerated test strategies for evaluating new interconnect systems and designs. This methodology is based on a damage integral approach.[22-25] A micromechanical measurement system is used to measure the mechanical behavior of such interconnects. Then, using a damage integral methodology, the micromechanical test methodology is used to describe the thermal cycling behavior of such interconnects. A correlation is shown between the isothermal micromechanical measurement data and thermal fatigue. Then, the significance of isothermal fatigue data on the reliability of some solder flip chip interconnect systems is discussed.

3.2   Materials and Processes

A typical process sequence for producing flip chip bonded modules is shown in Fig. 3.2. Typically, the wafer is obtained with an aluminum alloy metallization and some type of organic (e.g., polyimide-based) or inorganic (e.g., oxide, nitride) passivation. For many bumping processes, a metallization step is first required to form the bond pad or pad limiting metallurgy. This metallization consists of a adhesion/barrier layer (such as Cr[1] or TiW[13-16]), a solderable surface (such as Cu[13-15]) and a protective layer (such as Au[1]). The metallization may be either deposited using evaporation or sputtering, or plated using electrolytic or electroless processes. Following the metallization step, the wafer is bumped. Several methods are used. The bumps are evaporated (e.g., IBM C4 process[1-6]), plated,[13-16] wire-bumped,[17,18] dispensed,[19] or printed.[20] The bumping processes will be discussed in greater detail later. Following the bumping step a reflow process is usually preferred. The reflow temperature–time profiles are dependent on the composition of the solder used. The reflow may be performed in an inert or

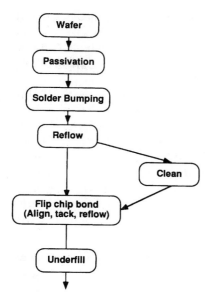

Figure 3.2 Typical process flow for manufacturing solder flip chip interconnect systems.

slightly reducing atmosphere. Alternatively, we may use ambient environment and fluxes. If necessary, the flux residues may be cleaned using a solvent. The wafer is then cut into individual chips or die in a dicing or sawing operation. The chip is aligned to the substrate using either separate alignment marks or the bumps and placed on the substrate. Then the module is reflowed. The reflow may be performed on the bond station or on a separate furnace. After flip chip bonding the module can be encapsulated or underfilled.

3.2.1 Materials for solder flip chip technologies

Bump materials. The melting point of the solder, the bumping process capability, manufacturability, and reliability are some of the major factors which influence the choice of a solder composition. The melting point of the solder is indicative of the maximum process temperatures encountered by the assembly elements. It is desirable that the substrate and the chip be structurally and chemically stable at these processing temperatures. Similarly, in order to preserve the microstructural and geometric stability of the solder joint, the flip chip bonded module must be retained at temperatures well below its melting point. The latter constraint is often referred to as the *temperature hierarchy,* which must be preserved at all times to ensure structural integrity of the solder flip chip interconnect.

Solder for flip chip applications can be either lead-based or lead-free. The most commonly lead based alloy systems for flip chip applications are PbSn[1,2,4,13–15] and PbIn.[3] Lead-free solders include those based on In and its alloys with bismuth, tin, and silver; 97.5Sn2.5Ag alloy is used in wire-bumping applications.[17,18] The composition of the bump materials is influenced by the bumping process capability. For example, plating processes can be used to deposit a variety of metals including lead, tin, silver, indium, and bismuth. But, increasing the number of elements in a plated bump increases the number of process steps. Therefore, it may be desirable to restrict solder compositions to binary systems. Such constraints may not apply to other processes such as printing or wire bumping. Bond pad metallization is another factor which influences bump composition selection. Often indium-based solders are preferred for gold bond pads because of the low dissolution rates of gold in indium.

Substrate materials. Perhaps one of the most important factors influencing the choice of a flip chip interconnect system is the substrate. The substrate influences the maximum allowable process temperature, the reliability of the interconnect, minimum pitch and I/O density, the composition of interconnect and cost. The choice and design of the substrate significantly impact the electrical and thermal performance of the interconnect system. The purpose of this section is to provide a brief overview of the factors influencing substrate selection with particular emphasis on its influence on the solder flip chip processes. Design and routing considerations are outside the scope of this chapter.

The substrates may be broadly classified into organic and ceramic or glass. In some instances, silicon may also be used as a substrate. Organic substrates include those based on FR-4, Kapton, or BT. Ceramic substrates, especially those based on alumina, are common in the electronics industry. The metallization on these substrates are deposited either using thick-film (e.g., screen printing) or thin-film (e.g., evaporation/sputtering) technologies. Although ceramic substrates are more expensive, the advantages of ceramic substrates are numerous. Higher thermal conductivity, lower dielectric constant, smaller expansion coefficient, better planarity, and higher process temperatures can improve the performance and process capability of interconnect systems using ceramic substrates. The thin-film technology is generally more expensive, but it can produce fine pitches and planar substrates.

Underfill materials. Underfills are used primarily to improve the reliability of flip chip interconnect systems. These materials fill the gap between the chip and substrate around the solder joints, reducing the thermal stresses imposed on the solder joint. The cure time and temperature of the underfill constitute a significant factor in the choice of an underfill material for a flip chip interconnect system. Since the underfill process follows the flip chip bonding step, cure temperatures lower than the melting point of the solder joint is desirable. From the manufacturability standpoint, short cure cycles are generally preferred since they can reduce the process times.

Other factors which influence the performance of an underfill are its expansion coefficient and adhesion to the interconnect system materials. The expansion coefficient of the underfill material must match those of the solder joint as closely as possible. Good adhesion of the underfill material to the substrate and die generally improves the reliability of the interconnect system.

3.2.2 Bumping processes

Several processes have been demonstrated to deposit solder bumps on a wafer or substrate. These include plating, wire bumping, evaporation, dispensing, and printing. The choice of one technology over another is influenced primarily by the bump dimensions and pitch, composition, and cost.

Plating. Perhaps one of the most cost-effective solutions to solder bumping involves plating. The plating process involves the following steps. First, the wafer with devices is metallized in such a way as to provide a current path to the individual bond pads. Two approaches are commonly used for this purpose. In the first, metallization traces are patterned between the die to the bond pad. In order to electrically isolate the bumps after the plating process, these traces are cut during the dicing operation. This method of metallization typically provides uniform plating heights with relatively good process control. However, such a method requires an additional patterning step, and it may not be desirable for chip designs with bumps in the interior of the die. The alternative approach is to deposit a blanket metallization on the entire wafer. This metallization is called the *bus* or *seed* metal. After the plating step, the seed metal is etched away to isolate individual bumps.

Following the deposition of the seed metal, the wafer is patterned with photoresist in such a way that the seed metal is exposed in the regions where the bump is desired. The photoresist process often determines the geometry and pitch limitations of plated bump technology. The wafer is then mounted in a cup plater as discussed in Refs. 13 and 16. A static or pulsed current is then applied along with the fluid flow in the cup with the wafer as the cathode. The time for which the current is applied generally determines the thickness of the plated bump. If the bump contains more than one material, the plating process can be repeated for each metal in the bump.

Wire bumping.   Solder wire bumping is a relatively low-cost alternative for low-volume prototyping applications. In this approach, a thermosonic bonder is used to deposit solder wire bumps directly on aluminum[17,18] (see Fig. 3.3). A rapidly solidified SnSbAg wire is subject to an arc discharge in an argon-hydrogen environment forming a solder ball. This ball is bonded directly to an aluminum bond pad at about 150°C in air using thermosonic energy. In this approach, the bond pad metallization step can be eliminated reducing the number of process steps required. Furthermore, this wire bump may be reflowed and processed as any other solder interconnect. The size and pitch ca-

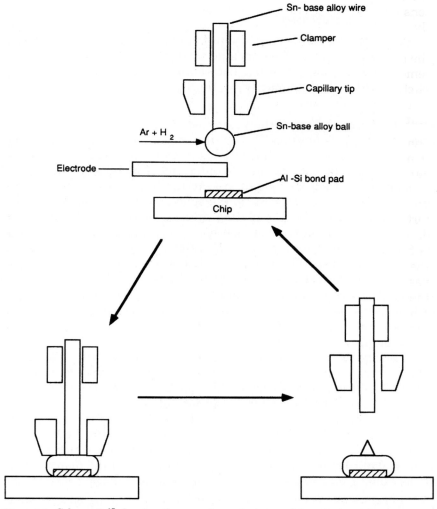

Figure 3.3 Schematic[17] showing the procedures during a solder wire bumping process.

pabilities of the wire-bumping process are limited by the diameter of the wire and the alignment capabilities of the wire-bumping equipment. Ogashiwa et al. demonstrated 100-μm-diameter bumps on a 250-μm pitch using a 44.8-μm-diameter wire.[17]

Evaporated bumps. Evaporation can provide bumps with the best uniformity in composition and volume. The most commonly used evaporated bumps are those based on PbSn (e.g., IBM C4 process[1]) deposited on a wafer through a molybdenum metal mask. The molybdenum metal masks are first aligned to the bond pads on the wafer and clamped. The assembly is then mounted into an evaporator to deposit the bond pads. For interconnect systems such as C4 this metallization consists of chromium, copper, and gold.[1] Following the metallization, the wafers along with the shadow mask assembly are transferred into a solder evaporation system. Here a known composition and volume of the solder is evaporated onto the bond pads. The shadow mask is then removed. For several decades, 100-μm-diameter bumps on a 250-μm pitch have been demonstrated in manufacturing environments. An obvious factor influencing the minimum size and pitch capability of evaporated solder bumps is the metal mask technology.

Printing. Printing of bumps using either stencils and screens has traditionally been used in the surface-mount industry.[20] More recently advances in the stencil technology[26] and pastes have extended this technology to flip chip applications, particularly to coarse-pitch (> 600-μm) applications. Applications of this technology to finer pitches require further refinements in stencil and screen technologies, and solder paste development. Printing processes typically involve three steps. The paste is first pushed into the holes in the stencil or screen by a squeegee. It then makes contact with the bond pads on the substrate. Then, the paste is transferred to the substrate while the stencil retracts. Finally, the paste bump is prebaked in order to retain its shape during handling. The rheology of the paste, the print speed, and the separation between the stencil (or screen) and the substrate are some factors which affect the printed bump geometry and uniformity.

Dispensing. A relatively new technology for depositing solder bumps on substrates is based on the ink-jet technology.[19] Molten Indalloy-158 solder droplets, approximately 100 μm in diameter, were used to form solder balls and bumps on copper. Such a technology may be useful as a low-cost flexible bump process for solder-bump deposition.

3.2.3 Reflow process

In general, solder bumps are reflowed twice during the process flow. The first reflow step occurs immediately following the bumping process. The second reflow occurs after the die placement during the flip chip bonding process. During the reflow process the solder bump is heated to a temperature above the melting point of the solder. When the solder melts, it forms a metallurgical bond with the bond pad metallization. For lead tin solders, this operation can be performed in a reducing atmosphere[21] such as hydrogen or nitrogen-hydrogen. For other solders, such as those based on indium, a flux may be required. In bumping operations such as plating or evaporation, the surface of the bump is often oxidized. The reflow operation reduces the oxide from the surface. In addition, the reflow processes promote uniformity in the solder composition and heights.

The reflow temperature profile can influence the kinetics of the reaction of the solder composition with the metallization. Typically a peak temperature between 10 and 50°C above the melting point is used. At low temperatures the rate of melting of the solder can be slow, resulting in poor ball formation and height nonuniformity. At very high temperature, the solder might react extensively with the bond pad metallization, resulting in intermetallics and even dewetting.[21] Extensive intermetallic formation can be deleterious to the reliability of the interconnect.

The choice of the environment or flux for bump reflow is determined by the composition of the solder bump. Fluxes commonly contain three constituents: a solvent (e.g., alcohol), a vehicle (e.g., a high-boiling-point solvent such as aliphatic alcohols), and an activator (e.g., carboxylic acids). The solvent facilitates uniform spreading of the flux on the bond pads. The reflow process usually consists of preheat step where the solvent is vaporized. This promotes a uniform coating of the flux on the solder and bond pad metallization. The flux also becomes more viscous and tacky. Further increase in temperature causes the vehicle to flow along with the activator. The activator reduces the oxides, while both the vehicle and the activator volatilize. Flux residues commonly contain residues from the carrier, the wetting agent, and reaction by-products of the reduction reaction. These can be cleaned using a variety of organic and inorganic solvents. Environmental concerns and cost factors has resulted in the development of some no-clean fluxes in which the cleaning step may be eliminated.

3.2.4 Assembly processes

The assembly of flip chip interconnect systems involves two tasks: (1) flip chip bonding and (2) encapsulation or underfill. This section addresses some key issues affecting assembly of solder flip chip interconnect systems.

During flip chip bonding, the bumped die is first aligned and attached to the bond pads on the substrate using a tacky flux. Then the module is heated so that the solder melts and forms a metallurgical bond with the bond pad. The shape and geometry of the solder inter-connect is determined by the process conditions. Two methods are commonly used to control bump shape and size. In one method the surface tension of the solder is used to control the bump shape. This method, the controlled-collapse chip connection (or C4) approach,[1] is commonly used to attach silicon die bumped with a lead-rich, lead-tin solder to a ceramic substrate. The number of bumps can be so de-signed to accommodate the weight of the silicon chip by the surface tension forces of the molten solder. The bumped die is first attached to the substrate typically using a tacky flux. Then the module is re-flowed in a belt furnace. During the reflow operation, the bumps wet the bond pads on the substrate. This wetting operation tends to pro-mote self-alignment of the die to the substrate. The height of the bonded module is determined by the balance of the surface tension forces on the solder and the weight of the chip. The calculation of restoring forces during reflow is discussed elsewhere.[27,28]

In the second method of flip chip bonding, the reflow is performed while the solder joint is physically deformed in its molten state. Al-though this process can be slower than the previous method, it can pro-duce interconnects which are more reliable than those produced using the controlled-collapse technique described above.[21] The sequence of tasks is as follows. The chip, typically at a temperature below the melt-ing point, is aligned to the substrate and placed on the bond pad. The temperature is then raised to the peak temperature. The bumps melt and wet the substrate bond pads. The chip is then pulled away from the substrate while the solder is molten, and then the entire system is cooled. Thus, it is possible to obtain larger standoff gaps between the die and the substrate. Such solder joints have been produced using cen-trifugal reflow[21] or self-stretching technology,[29] and by stretching a molten solder joint in a high-accuracy flip chip bonder.[30]

Flip chip bonder requirements are usually determined on the basis of the throughput, alignment accuracy and performance require-ments. Higher throughput can be achieved if accuracy requirements are not very stringent. In such cases, we rely on the self-alignment capability during the reflow operations. Typically the alignment accu-racy for self-alignment is specified as less than one-quarter of the bond pad dimensions.[27,28] That is, if the chip is misaligned to the sub-strate by less than 25 percent of the bond pad dimensions, the reflow process is capable of self-aligning the chip to the substrate. If greater reliability is needed or self-alignment is not possible, then a bonder with better accuracy is needed.

Following the flip chip bonding process the flux residues are cleaned. Then the module is underfilled or encapsulated. In this step, the gap between the chip and the substrate is filled with an epoxy. Most underfills require at least a 25-μm gap between the chip and the substrate. Smaller gaps can result in segregation of fillers and nonuniform chip coverage. Underfill or encapsulation provides two functions. First, it protects the chip and the interconnects during subsequent processes; but second and more importantly, it improves the reliability of the interconnect system. In fact, a factor of 10 improvement in reliability has been reported in the literature by underfilling a flip chip interconnect system.[21]

Given the multitude of materials and processes used in solder flip chip interconnect systems, it is important to be able to evaluate the performance of the interconnect system. We are concerned essentially with three aspects of the solder flip chip interconnect system performance. The electrical behavior of the interconnect must be well understood. For cases where power dissipation is significant, the thermal resistance of the interconnect system must be evaluated. Finally, the reliability of the module must be established in terms of the interconnect performance.

3.3   Performance

The emphasis in this section will be on the electrical and thermal performance of a solder flip chip interconnect system. Furthermore, only those aspects of performance which are significantly impacted by the assembly materials and processes are discussed.

3.3.1   Test vehicle design

The best method for evaluating any flip chip interconnect system is in an actual application. This, however, may not always be practical, especially during process development and optimization. Test vehicles with relatively simple structures may be desirable. Not only would this vehicle help in the electrical characterization of the flip chip interconnect system, but such a vehicle may also be useful for reliability assessment.

Flexibility, functionality, and ease of processing are some considerations affecting test chip design for benchmarking solder flip chip interconnect systems. The test vehicle design must be such that it can be processed using all the technologies of interest. In addition, the wafer layout, such that the user has flexibility in selecting an appropriate chip size for the application, is sometimes desirable to evaluate the technology capability. The test structure must be able to measure the total bump resistance. Continuity patterns can be used in reliability testing to determine behavior of a series of bumps during, for example, reliability testing. Serpentine and comb patterns measure the leakage in the circuit lines while the chip is biased at a predetermined temperature and humidity. Some additional structures relevant to performance evaluation are diodes for sensing temperature and piezoresistive element to estimate stresses. Given the increased complexity of processing such wafers and associated costs, a careful analysis may be required to establish the marginal benefit of test structures over fully functional chips.

Figure 3.4a shows a flip chip accelerated testing (FCAT) vehicle designed to rapidly evaluate the electrical performance and reliability of flip chip interconnect systems as a function of process, materials, and design. The structure essentially consists of three elements: a bump contact resistance pattern, a continuity pattern, and a serpentine pattern. Given that this test vehicle would be fabricated using a variety of bumping methods and substrates, the design rules for the test vehicle was 150-μm lines and 150-μm spaces. The minimum die cell size is 2.7 mm. This cell is stepped and repeated over the entire wafer. By appropriately dicing the wafer, other die sizes can be obtained. An example of a substrate design for a single layer board for a 2.7-mm-square die is shown in Fig. 3.4b.

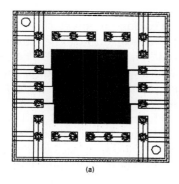

(a)

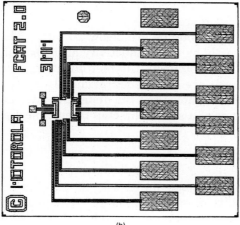

(b)

Figure 3.4 An example of a test structure to evaluate thermomechanical behavior and reliability of flip chip interconnects: (a) die cell; (b) substrate.

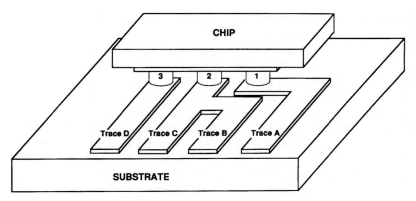

Figure 3.5 A schematic illustration showing the measurement of bump resistance of a flip chip interconnect.

3.3.2 Electrical performance

In order to measure the bump resistance, one approach is to use a structure as shown in Fig. 3.5. This structure involves three bumps. We measure the bump resistance of the center bump. The traces used as voltage sense and current sense could be interchanged. The principle of operation of this structure is as follows. Let us assume that the current flows through trace *A,* through bump 1, over the trace on the chip, down bump 2 and out of trace *B.* Note that traces *C* and *D* have no current flowing through them. Furthermore, bump 3 and trace *D* are at the same potential as the top of bump 2. And trace *C* is at the same potential at the bottom of bump 2. Therefore, any potential difference measured across traces *C* and *D* would represent the potential drop across bump 2. Given the current flowing through bump 2 and the measured potential difference across traces *C* and *D,* the bump resistance may be readily estimated.

Typical resistance values for solder bumps are below 20 mΩ.[31] The bump resistance is influenced mostly by the conductivity of the solder, and by the contact resistance of the bump to the metallization. The bump inductance and capacitance are influenced primarily by the height and diameter of the bump, the pitch, and the underfill material. Typical inductance and capacitance of a solder flip chip interconnect is 0.06 nH and 0.17 pF, respectively.[31]

3.3.3 Thermal performance

It is well recognized that one of the major limitations of flip chip interconnect systems is its capability of extracting heat.[32] The primary path for heat conduction in most solder flip chip interconnect systems is through the solder joint. The resistivity of the solder bumps is typically about an order of magnitude higher than those typically seen in die attach or backbonded chips.[21,33] However, the overall thermal conductivity of solder flip chip interconnect systems is only about half of a comparable wire bond–die attach module.[33,34]

Calculations of the thermal performance of solder flip chip interconnect systems often require numerical analysis.[32,35] The thermal resistance of the module is influenced by the location of the chip on the substrate; the number, size, and pitch of the connections; the proximity of the chip to the substrate; the metallizations; and the thermal conductance of the substrate.

Several approaches can be used to enhance the thermal conductivity of solder flip chip interconnect systems. Given the high-density capability of flip chip interconnects, some solder bumps can be used to primarily dissipate heat. High-thermal-conductivity underfills can also be used to extract heat from the chip surface. Liquid immersion cooling[36] provides heat dissipation rates as high as 20 W/cm^2. Thermal vias or materials such as AlN[37] or SiC[38] can reduce the thermal resistance of substrates. Finally, since the backside of the chip is unused, heat can be extracted from the backside of the chip.

3.4 Reliability

Another consideration in the selection of a solder flip chip interconnection system is its reliability. Both the thermal fatigue and the corrosion of the solder joint can significantly affect the performance of a solder-based interconnect system during service. Some solders such as those based on indium are typically thermal fatigue-resistant but are not reliable in high-humidity environments. The relative kinetics of damage accumulation due to these phenomena is influenced by the solder composition, the substrate and underfill materials, and the severity of the environment. We shall examine some aspects affecting reliability of a solder interconnect system below.

3.4.1 Thermomechanical behavior and reliability

Thermomechanical behavior is most significantly influenced by the composition of solder joint. For example, 95/5PbSn solder interconnects exhibit three times the fatigue life of eutectic PbSn solder.[39] Pure In, used in optoelectronic devices,[6] shows about 20 times the thermal life of 95/5PbSn solder joints.[3] Increasing Pb content in the indium results in reduced fatigue life. 5/95InPb and 50/50InPb alloys show a factor of 6 and 10 times reduction in fatigue life compared to pure In solder.[3] Similarly, 50/50InSn (used for chip on glass applications[5]) shows better fatigue resistance at temperatures near room temperature. The selection of the solder composition for optimum thermomechanical integrity is most often empirical. This is because there is no direct correlation between the composition and the melting point of the solder and its fatigue resistance.

During service, ambient-temperature fluctuations result in differential thermal expansion of the various members of a flip chip interconnect assembly. The expansion mismatch between the members of the assembly impose displacements which produce mechanical stresses in the solder joint. These stresses are the driving force[22–25] for damage mechanisms such as crack growth. The magnitude of the stresses are determined by the assembly stiffness and inelastic deformation properties of the solder joint. The rate of the damage process is influenced by the stress, temperature, and the environment. The damage process can result in structural failures in the solder joint. Consequent electrical failures is potential reliability concern.

Traditional accelerated test methodologies reduce the time required for reliability assessment by increasing the rate of the damage process. For example, during accelerated temperature cycling experiments, the temperature profiles are prescribed such that they are more severe than those commonly encountered in service. If a correlation between the fatigue life under accelerated test conditions and service life can be experimentally established, such a method can provide an effective means of solder joint life estimation.

Test structures such as those shown in Fig. 3.4 can be used to assess the reliability of flip chip interconnect systems. Typically, the total resistance of a continuity or daisy-chain pattern is monitored as the flip chip bonded module is subjected to temperature cycles. The temperature range and rate of heating or cooling is determined by the application. Failure criterion is often specified in terms of percent resistance increase or opens.

Isothermal mechanical test methods have been used to assess the reliability of solder flip chip interconnects.[24,40] A method based on the Coffin-Manson correlation[41] was used by Norris and Landzberg[42] to develop life correlation for a 95/5PbSn solder flip chip interconnect system. More recently, Subrahmanyan[24] has used isothermal measurements to describe thermal fatigue behavior of Pb-based solder interconnects. We shall discuss the use of micromechanical test methods to life estimation and accelerated testing in Sec. 3.5.

Improving the reliability of solder flip chip interconnects can be achieved by several methods. We can reduce the driving force, i.e., stress or the rate of crack growth in the solder, or we can use a solder composition which is more resistant to crack growth damage. In order to reduce the stress, we can (1) reduce the thermal displacements on the solder joint, (2) use a more compliant assembly, or (3) redistribute the stresses over the entire surface of the chip. Thermal displacements can be reduced by a lower distance of the outermost bump from the neutral point of the assembly[1] or by increasing the bump height.[4] Flex substrates or a thinner substrate can reduce the assembly stiffness. Underfills can be used to redistribute the stresses over the surface of the chip and to increase the lifetime of the interconnect system by a factor of 10.[21] Solder joints based on PbIn[3,43] are capable of dramatically increasing the thermal fatigue resistance of flip chip interconnect systems. Their application is, however, limited because of the tendency of such solders to chemically react with the environment.

3.4.2 Other phenomena affecting solder joint reliability

Corrosion-related failures in flip chip interconnect systems occur in two forms: in the solder interconnect or on the chip or substrate metallization. The chemical reactivity of the solder joint and the environment influence the kinetics of corrosion. High-temperature and humidity conditions and reactive halides (e.g., from the underfill or flux residues) can accelerate corrosion-related failures. It has been suggested[22] that in 60/40PbSn solder, moisture may be an important factor affecting thermal fatigue failures. Failures in indium solder flip interconnect systems have been attributed to corrosion in high-humidity environments.[21,44] Ionic contamination such as Cl^- has been identified as one of the factors leading to corrosion of such solders.[43] Modified moisture-resistant underfill or hermetic packages can reduce corrosion-related failures.

Another phenomenon which affects the integrity of PbIn flip chip interconnects is thermomigration.[45] In this mechanical the thermal gradient between the chip and the substrate produces a condensation of voids near the chip bond pad metallization. The consequent failures manifest in increased bump resistance. α-particle emission can sometimes be of concern in memory applications when using Pb-based solders.[21] Lead sources and underfill materials with low α emissions and passivation of the chip with materials which can absorb α particles can reduce these effects.[21]

3.5 Reliability Assessment and Micromechanical Testing

Perhaps one of the most time-consuming tasks in the development of a solder flip chip interconnect system is the assessment of its thermomechanical integrity. In order to address this aspect of accelerated test methodology development, we must be able to explicitly examine all the parameters which affect the structural integrity of the interconnect. The task of assessing the reliability and developing accelerated test methods is then twofold. First, a methodology is required for explicitly incorporating the above parameters. Such a methodology must also outline experimental requirements to independently determine the various parameters used in data analysis and life estimation. Finally, such a methodology must provide a basis for correlating data obtained using a variety of test methods.

In this section, a micromechanical test methodology to measure the deformation behavior of a 90/10PbSn flip chip interconnect is outlined. A damage integral approach is summarized to serve as a basis for further discussion. Deformation and damage behavior of a 90/10PbSn solder flip chip interconnects are then measured. The application of the measured isothermal data to temperature cycling data is then presented for a 90/10PbSn flip chip solder interconnect. Given that the isothermal data can be correlated with thermal cycling, we use isothermal fatigue to establish a relative comparison between several solder flip chip interconnect systems. In particular, we shall compare lead rich PbSn solders produced by two methods: evaporation and plating. These lead based solders are then compared to an InSn flip chip interconnect system.

3.5.1   A damage integral approach

Details of the damage integral approach are summarized in previous publications.[22-25] Some salient aspects of the approach are reviewed below.

Stress calculation.   For any given displacement profile, the stress in the solder joint is dependent on the compliance of the assembly and the inelastic deformation properties of the solder joint. The total displacement rate can be written as the sum of the total elastic displacement rates and the displacement rates corresponding to the inelastic deformation in the solder joint. For a solder joint with a simple geometry (e.g., see Fig. 3.1a), it may be possible to express the load P in the solder joint in the form of the governing equation:[22]

$$\frac{d(P/S)}{dt} = \dot{X} - \dot{X}_i \qquad (3.1)$$

where $\dot{X}$ is the total applied displacement rate. During temperature cycles the applied displacements is determined by the thermal expansion mismatch between the die and the substrate, and the distance of the solder bump from the neutral point. $\dot{X}_i$ is the displacement rate due to inelastic deformation. The effective stiffness S relates the total elastic displacements to the load. It incorporates the stiffness of the solder joint and the connecting members, and is dependent on the geometry, moduli, and loading mode. It may be experimentally determined or calculated.

The calculation of the stresses in the solder joint for an arbitrary loading profile requires the integration of Eq. (3.1). In addition to the assembly stiffness S, the dependence of the stress on the inelastic strain rate is required. The stress–inelastic strain rate behavior is determined in terms of appropriate constitutive relations for inelastic deformation. Experiments such as displacement-controlled cyclic loading, creep, and load relaxation experiments are often used to characterize the inelastic deformation properties of solder joints.[24] Constitutive relations based on a state variable approach which inherently incorporates the temperature and structural dependence of the measured deformation behavior has been used, and can greatly reduce the required experimental effort.

Using such constitutive relations and measured assembly stiffness, it has been shown that it is possible to accurately simulate the stresses during both temperature cycling and isothermal mechanical cycling.[22,24] Since the damage rates (discussed in the next section) are sensitive functions of the stresses, accurate estimation of stresses is of paramount importance in the analysis and data extrapolation.

Damage rate estimation and damage integration. The fatigue crack propagation rate is expressed in terms of a nominal stress intensity factor K such that

$$\frac{dc}{dt} = v^* \times K^r \tag{3.2}$$

where dc/dt is the time rate of crack growth. The nominal stress intensity factor is given by $K = Y(c)\tau_N\sqrt{c}$, where c is the instantaneous crack length and $Y(c)$ is a geometric factor. τ_N is the nominal stress based on the initial loading area. r is the crack growth exponent, which is considered constant over a narrow range of K. The rate constant v^* has an Arrhenius temperature dependence with an activation energy Q_c and a preexponential v_0^*.

Separating variables and integrating Eq. (3.2) over the fatigue life t_f, we obtain

$$\int_{c_i}^{c_f} \frac{dc}{[Y(c)\sqrt{c}]^r} = \int_0^{t_f} dt\, \tau_N^r\, v_0^* \times \exp\left(-\frac{Q_{(c)}}{RT}\right) \tag{3.3}$$

where c_i and c_f are the initial and final crack lengths, respectively. Each side of the integral in Eq. (3.3) is a measure of the total damage at failure; one in terms of the crack length and the other in terms of time. For a given combination of c_i and c_f, the crack length integral will be constant.

Fatigue life estimation.   A complete integration of Eq. (3.2) or a single-cycle approximation [Eq. (3.4)] may be used to estimate the thermal and mechanical fatigue life depending on the nature of the stress cycle. A single-cycle approximation[24] can be used, for example, when the cyclic stress profiles do not change for most of the life. Such a situation is often encountered during temperature cycling, where the assembly stiffness is much smaller that the solder joint stiffness. However, a complete integration of Eq. (3.2) may be necessary when the cyclic stress profile changes significantly throughout the fatigue life.

3.5.2 Micromechanical measurement system

Application of an integral approach such as the damage integral approach requires accurate, well-characterized experimental input. Since the deformation and damage rates can be sensitive to the actual fabrication procedure, the measured properties must be obtained from realistic specimens. Details of a micromechanical testing machine used for testing interconnects are provided in Ref. 24. A schematic of the machine is shown in Fig. 3.6. The displacements to the solder joints are applied through a micrometer-driven translation

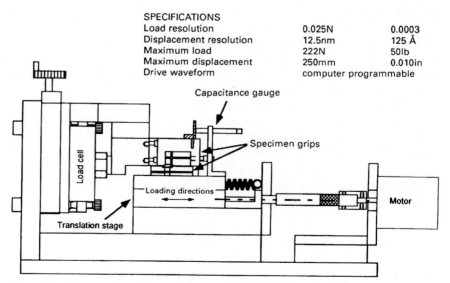

Figure 3.6 Micromechanical measurement system used to determine the mechanical behavior of flip chip interconnect systems.

stage. The loads are measured using a tension compression load cell. The range and sensitivity of the load cell are 444 N (100 lb) and 0.004 N (0.001 lb), respectively. The displacements are measured using capacitive displacement gauges. The range and sensitivity of the gauges are 250 μm and 25 nm, respectively. The effective machine stiffness is 4.5 MN/m; the specimen grip stiffness is 35 MN/m. The load, displacements, temperature, and time data are continuously monitored on a Macintosh platform using a National Instruments A/D board and LabView data acquisition software. These temperatures is controlled by independent temperatures controllers to within ±0.01°C.

Three categories of mechanical behavior may be deduced using the micromechanical test machine described above. Cyclic loading of the specimen provide the load displacement and stress–strain curves. Load relaxation experiments are used to estimate the inelastic strain rate dependence of the flow stress. Displacement-controlled, displacement-limited isothermal cycling experiments are used to estimate the fatigue resistance of the solder interconnect. On the basis of these data,[24] one can deduce the isothermal mechanical behavior of a flip chip interconnect system.

3.5.3 Isothermal mechanical testing and thermal fatigue

A silicon test die with fourteen plated 90/10PbSn solder bumps is bonded to an alumina substrate. The bump geometry and fabrication procedure has been described in Ref. 13. The nominal reflowed bump height and average diameter are 50 and 112.5 μm, respectively.

Two types of experiments are performed. In the first case, the displacements are applied in a triangular waveform. This data provides the hysteresis loops. In the second case, the translation stage is moved to a preset displacement. The ensuing load relaxation data is used to deduce the stress–inelastic strain rate behavior. The shear stress–shear strain behavior (Fig. 3.7a) is used to deduce the cyclic deformation behavior of the solder interconnect as detailed in Ref. 46. The stress–strain rate data measured using load relaxation is shown in Fig. 3.7b. Note that the nonlinear behavior of the measured data on a logarithmic plot indicates that a simple power law may not be able to adequately describe the observed deformation behavior.

For the estimation of crack growth rates, we shall follow the approach given in Refs. 22 and 23. The basis for this approach is as follows. While isothermally cycling the solder joints, a crack propagates through the interconnect. Such cracks reduce the load-bearing capacity of the bump. If the load drop during isothermal cycling is measured, and the relationship between the load drop and the uncracked area is assumed, then it

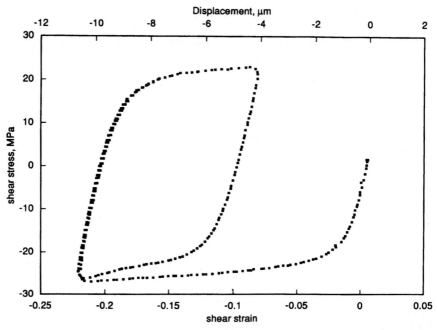

Figure 3.7 Measured isothermal deformation data[24] for a 90/10PbSn solder joint: (*a*) cyclic deformation behavior.

may be possible to estimate the effective crack length as a function of cycle number. If the applied displacements are large, much of the crack growth occurs at the yield stress of the material. Then, the crack growth rate may be estimated by the relation $(dc/dt) = \nu\,(dc/dN)$, where (dc/dt) is the time rate of crack growth, dc/dN is the crack growth per cycle, and $(1/\nu)$ is the time period. The stress intensity factor K is estimated from the nominal stress, $\tau_0 = P/A_0$, where A_0 is the initial area. The damage rates measured by this approach for a 90/10PbSn solder joint is shown in Fig. 3.8.

We now use the deformation and damage data measured using isothermal micromechanical methods to simulate the published data on the deformation and crack growth data during temperature cycles. Pao et al.[47] have used a beam-lead-type specimen in which an aluminum and alumina beam were soldered together on either end using a 90/10PbSn solder. Strain gauges mounted on either side of these beams were used to measure the deflections of these beams during temperature cycling. The measured strains were then converted to the stresses in the solder joint using the geometry and moduli of the beam materials.

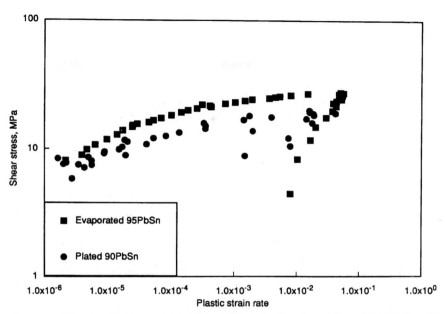

Figure 3.7 (*Continued*) Measured isothermal deformation data[24] for a 90/10PbSn solder joint: (*b*) monotonic stress–strain rate behavior.

Stress simulation. A stress–temperature profile[24,47] for the 90/10PbSn solder interconnect described above is shown in Fig. 3.9 as the solid symbols. We use isothermally measured mechanical data on 90/10PbSn flip chip solder interconnects (Fig. 3.7). The constitutive relations are discussed in detail in Ref. 24. The solder joint is assumed to be loaded in simple shear. The distance from the neutral point and the height of the solder joint are assumed to be 18.5 and 0.381 mm, respectively. The assembly stiffness, calculated according to a bimetallic strip theory,[24] is 0.05 MN/m. The expansion coefficients of aluminum and alumina are 21.1 and 5.8 ppm/°C, respectively. The heating and cooling rates are 10°C/min, and the hold times at peak temperatures are 10 min. The temperature is assumed to be uniform in the entire specimen. The simulated stress–temperature behavior is shown as the solid line. Excellent correlation is observed. Isothermal deformation data appears to be able to describe thermal stresses in a solder joint.

Simulation of crack growth. Pao et al.[47] have also measured the crack length as a function of thermal cycle number for various specimens by visual measurement. These data are represented by the symbols in Fig. 3.10. The crack lengths are calculated using complete integral [Eq. (3.3)]. The stress intensity factor is calculated based on the nomi-

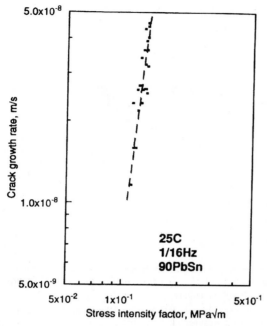

Figure 3.8 Measured crack growth rates in a
90/10PbSn solder joint.[24]

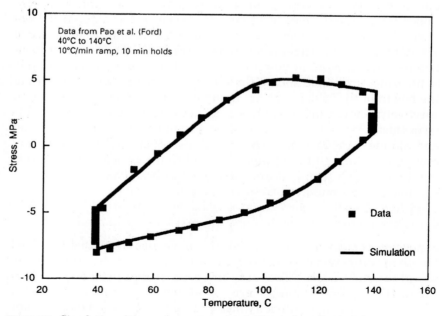

Figure 3.9 Simulation of thermal stresses using isothermal mechanical data.

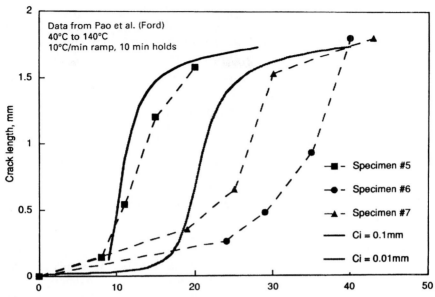

Figure 3.10 Simulation of thermal fatigue crack growth using isothermally measured crack growth behavior.

nal stress. The geometric factor Y(c) is assumed to be unity. The crack growth exponent is assumed to be 4.4, and the crack growth rate constant is 1.00 $(MPa)^{-4.4}$ m/s. The initial crack length was varied between 0.01 mm (the grain size[47]) and 0.1 mm. The temperature dependence of crack growth was assumed to be the same as that measured for a 60/10SnPb solder.[22,25] The simulated crack lengths–cycle numbers for each assumed initial crack length are shown as a solid line in Fig. 3.10. The agreement between the measured crack length and the model appears to be adequate. Here again, crack growth during temperature cycles appears to be consistent with those deduced based on isothermal micromechanical test methods.

3.5.4 Application of micromechanical testing

For the system considered in the preceding example, isothermal mechanical data appear to correlate well with thermal cycling data. Such a correlation may, however, not be universal. Differences in loading mode, effects of temperature gradients, and time-dependent microstructural evolution are some of the factors which cannot be exactly reproduced using the micromechanical measurement system described above. While the measurement of mechanical properties using isothermal methods is more convenient, and such a methodology can potentially reduce the time to assess the reliability of flip chip inter-connect system, several issues must be addressed to deduce correct conclusions. Some considerations in applying micromechanical test methods to reliability assessment are discussed next.

Micromechanical and accelerated testing. If we can correlate isother-mal test data with thermal cycling results, it may be possible to use isothermal test methods alone to establish the reliability of flip chip interconnect systems. One method of application of micromechanical test methods is by calculating acceleration factors for temperature cycle through the damage integral approach outlined in Sec. 3.4.[24] This method uses mechanical properties characteristic of the solder joint to evaluate the life of the solder flip chip interconnect system.

Even if details of the mechanical properties of the solder flip chip interconnect systems are not available, it may be possible to empiri-cally examine their relative performance. The fatigue life of the flip chip interconnect system can be measured as a function of the applied displacements on the solder joint. The data can then be analyzed and presented in the so-called Coffin-Manson plots.[41,42] Although this method may not be able to directly evaluate the durability in service of the interconnect, it can serve as a tool to quickly establish differ-ences in materials and processes used to manufacture a flip chip in-terconnect system.

We shall now examine data measured isothermally using the micro-mechanical measurement system. The objective is twofold. We wish to (1) evaluate the relative performance of flip chip interconnect systems to obtain a figure of merit for the systems of interest and (2) demon-strate the versatility of the isothermal mechanical test method to as-sess the fatigue resistance of the solder flip chip interconnect systems. Although much of the present work emphasizes loading mode in sim-ple shear, tests using four-point bend methods are equally practical.

We have examined three systems of interest. Isothermal displace-ment-controlled micromechanical fatigue experiments were performed on three solder flip chip interconnect systems at room temperature. The first interconnect system is the 90/10PbSn solder flip chip interconnect discussed in Sec. 3.4. These solder joints were produced using a plating process similar to that described by Warrior.[13] The second system uses a 95/5PbSn solder bump evaporated on the wafer. The substrate is FR-4 and consists of a low-temperature 60/40SnPb solder which has been ap-plied using a solder knife. The test chip is 6.5×6.4 mm, and 72 bumps are arranged in staggered pattern on a 300-μm pitch. The third system of interest contains the 50/50InSn solder alloy. These bumps are plated on a test chip as shown in Fig. 3.4. The bumped die is flip chip–bonded to FR-4, alumina, and glass substrates. The nominal heights of each of these interconnect systems is 50 μm. All three interconnect systems had

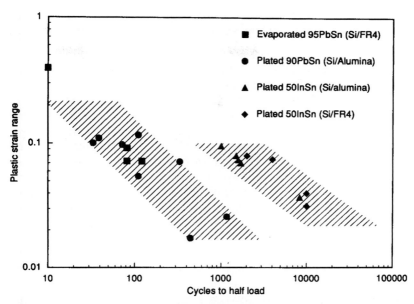

Figure 3.11 A comparison of the room temperature fatigue life behavior for a variety of solder flip chip interconnect systems.

a diameter of 100 μm along the die and 125 μm on the substrate. The interconnect systems had no underfill.

Figure 3.11 shows the measured fatigue life of these interconnect systems at room temperature at a frequency (triangular waveform, no hold periods) of 0.01 Hz. The failures in all the systems were in the bump close to the chip. For this configuration, the fatigue life for the 90/10PbSn solder joint and the 95/5PbSn solder joints do not appear to be significantly different. Furthermore, the fatigue life of the InSn solder is at least a factor of 3 higher than the lead tin solders. Experiments at 70 and 120°C for the Pb-based solders showed a drop of 0.9 and 0.8 in the plot with insignificant change in slope. For the InSn flip chip interconnect systems, the fatigue life drop by a factor 0.8 at 70°C. Since the melting point of InSn solder was 118°C, the 120°C experiments were not performed.

Although the Coffin-Manson-type experiments shown in Fig. 3.11 cannot be directly used to predict thermal fatigue life, they can serve as a valuable tool to assess the effect of process variations on life. Given that these experiments are not very time-consuming, micromechanical test methods can be used to rapidly assess the relative performance flip chip interconnect systems without extensive tooling and fixture designs. Another point to note is that the preceding experiments were performed in simple shear. Such experiments may be able to simulate the loading modes for relative stiff assemblies. For a softer assembly, the fatigue experiments may be performed, for example, in bending.

Failure criterion considerations.   One significant factor to be considered when using mechanical test methods to life estimation and calculation of acceleration factors is the effect of the failure criterion. In mechanical cycling, load drop (i.e., percentage change in load range from the first cycle) is often used as the failure criterion. In thermal cycling, it is common to use electrical resistance of the interconnect. Direct correlation between the fatigue life of the interconnect after isothermal cycling and thermal cycling requires two basic considerations to be addressed: (1) a correlation between the electrical resistance and load drop—this may be determined experimentally or calculated; and (2) the effect of relative stiffness of solder joint with respect to the machine (k_s/k_m) and assembly (k_s/S) (during isothermal mechanical and thermal cycling, respectively). For a given displacement range or temperature range, the relative stiffness affects the stresses in the interconnects, thereby influencing the crack growth rate. In addition, it can impact the sensitivity of load drop to the crack area. Smaller relative stiffness results in smaller sensitivity of the load to the cracked area in the solder joint.[48]

These considerations are incorporated in the damage integral approach. Since, for most temperature cycling applications involving flip chip solder interconnects, the assembly stiffness is much smaller than the stiffness of the solder joint, the nominal stress profile remains relatively constant for most of the life. In such cases, the single-cycle approximation [Eq. (3.4)] may be used. This approach requires the calculation of the crack length integral by using accelerated thermal cycling data. By doing so, the damage integral approach empirically incorporates both the crack length and the relationship between the crack length and the resistance.

3.6   Summary

Solder flip chip interconnect systems account for a significant portion of today's flip chip interconnect systems. A variety of materials, processes, and technologies are currently available today, and new interconnect systems are continuously being developed. The performances of such interconnect systems are comparable to, if not better than, those of most other interconnect systems.

The objective of this chapter was to provide the user with an overview of the performance capabilities of solder flip chip interconnect system. A variety of interconnect materials and process technologies for producing a solder flip chip interconnect system have been discussed. These include both lead-based and lead-free solders, and bumping technologies such as plating, evaporation, wire bumping, dispensing, and printing. Factors which affect electrical thermal performance of solder flip chip interconnect systems have been discussed. The reliability of a solder joint is influenced by its deformation behavior, thermal fatigue resistance, and chemical resistance to the environment. There is no direct correlation between the solder composition-melting point and its reliability. Much of such data must be obtained experimentally. A method is outlined where the mechanical behavior of solder flip chip interconnect system may be deduced. A micromechanical measurement system and its application to solder joint reliability assessment are discussed.

Acknowledgments

The authors would like to acknowledge the constant support of Kent Hansen and Barry Johnson in this effort. We would like to express our sincere appreciation to Bill Marlin, Mohan Warrior, Ken Wasko, Robert Carson, Bill Lytle, Alice Hampsten, and Ben Hileman for their help in producing the requisite modules and micromechanical testing data.

References

1. Goldmann, L. S., and P. A. Totta, "Area Array Solder Interconnections for VLSI," *Solid State Technol.*, 91–97, June 1983.
2. Miller, L. F., "Controlled Collapse Reflow Chip Joining," *IBM J. Research Devel.*, 13:239–250, May 1969.
3. Goldman, L. S., R. J. Herdizk, N. G. Koopman, and V. C. Marcotte, "Lead Indium for Controlled Collapse Chip Joining," *Proceedings of 27th Electronic Components Conference*, p. 25, May 1977.
4. Goldman, L. S., "Geometric Optimization of Controlled Collapse Interconnections," *IBM J. Research Devel.*, 251–269, May 1969.
5. Mori, M., Y. Kizaki, M. Saito, and A. Hongu, "A Fine Pitch COG Technique for a TFT-LCD Panel Using an Indium Alloy," *IEEE Trans. Components Hybrid Manufact. Technol.*, 16(8):852–857, Dec. 1993.
6. Schmid, P., and H. Melchior, "Coplanar Flip-Chip Mounting Techniques for Picosecond Devices," *Rev. Sci. Instrum.*, 55(11):1854, 1984.
7. Estes, R. H., "Fabrication and Assembly Processes for Solderless Flip Chip Assemblies," *Proceedings of ISHM*, pp. 332–335, 1992.
8. Eldring, J., E. Zakel, and H. Reichl, "Flip Chip Attach of Silicon and GaAs Fine Pitch Devices as Well as Inner Lead TAB Attach Using Ball-Bump Technology," *Hybrid Circuits*, 34:20–24, May 1994.
9. Scharr, T. A., R. T. Lee, W. H. Lytle, R. Subrahmanyan, and R. K. Sharma, "Gold Wire Bump Process Development for Flip-Chip Application," *Proceedings of 8th Electronic Materials and Processing Conference*, ASM, Rao, S. T., ed., pp. 97–102, 1994.
10. Chang, D. D., P. A. Crawford, J. A. Fulton, R. McBride, M. B. Schmidt, R. E. Sinitski, and C. P. Wong, "An Overview and Evaluation of Anisotropically Conductive Adhesive Films for Fine Pitch Electronic Assembly," *Proceedings of 43rd IEEE Electronic Components and Technology Conference*, pp. 320–326, 1993.
11. Sugiyama, K., and Y. Atsumi, "Conductive Connecting Structure," U.S. Patent 4,999,460, Casio Computer, March 1991.

12. Chung, K., R. Fleishman, D. Bendorovich, M. Yan, and N. Mescia, "Z-Axis Conductive Adhesives for Fine-Pitch Interconnection," *Proceedings of IEPS,* pp. 678–689, 1992.
13. Warrior, M., "Reliability Improvements in Solder Bump Processing for Flip-Chips," *Proceedings of 40th Electronic Components and Technology Conference,* Las Vegas, Nev., pp. 460–469, 1990.
14. Chmiel, G., J. Wolf, and H. Reichl, "Lead/Tin Bumping for Flip Chip Applications," *Proceedings of ISHM,* pp. 336–341, 1992.
15. Yu, K. K., and F. Tung, "Solder Bump Formation by Electroplating for Flip-Chip Applications," *Proceedings of IEMT,* pp. 277–279, 1993.
16. Lytle, W. H., T. Lee, and B. Hileman, "Application of a CFD Tool in Designing of Fountain Plating Cell for Uniform Plating of Semiconductor Wafers," *Proceedings of Interpack Conference,* 1994.
17. Ogashiwa, T., H. Akimoto, H. Shigyo, Y. Murakami, A. Inoue, and T. Masumoto, "Direct Solder Bump Formation on Al Pad and Its High Reliability," *Jpn. J. Appl. Phys.,* part 1, **31**(3):761–767, March 1992.
18. Scharr, T. A., R. K. Sharma, R. T. Lee, and W. H. Lytle, "Wire Bumping Technology for Gold and Solder Wire Bumping on IC Devices," *Proceedings of 9th European Hybrid Microelectronics Conference,* Nice, pp. 351–357, May 1993.
19. Hayes, D. J., D. B. Wallace, and M. T. Boldman, "Picoliter Solder Droplet Dispensing," *Proceedings of ISHM,* pp. 316–321, 1992.
20. Manchao, X., K. J. Lawless, and N.-C. Lee, "Prospects of Solder Paste in Ultra Fine Pitch Era," *Proceedings of SPIE,* 2105, pp. 69–85, 1993.
21. Subrahmanyan, R., J. R. Wilcox and Che-Yu Li, "A Damage Integral Approach to Thermal Fatigue of Solder Joints," *IEEE Transact. Components Hybrid Manufact. Technol.,* **12**(4):480–491, 1989.
22. Wilcox, J. R., R. Subrahmanyan, and Che-Yu Li, "Thermal Stress Cycles and Inelastic Deformation of Solder Joints," *Proceedings of 2nd ASM International Electronic Materials* and *Proc. Congress (IEMPC),* Philadelphia, pp. 203–211, 1989.
23. Subrahmanyan, R., "Micromechanical and Accelerated Testing of Flip Chip Interconnect Systems," *Electronic Packaging in Materials Science VII, Materials Research Society Symposium Proceeding,* Borgesen, P., K. F. Jensen, and R. A. Pollak, eds., MRS, pp. 395–406, 1994.
24. Subrahmanyan, R., J. R. Wilcox and Che-Yu Li, "A Damage Integral Approach to Solder Joint Fatigue," *Proceedings of 2nd ASM IEMPC,* Philadelphia, pp. 213–221, 1989.
25. Marks, G. T., and J. E. Sergent, "The Electroformed Stencil: A Solution to Printing Solder Paste for Fine Pitch Devices," *Proceedings of ISHM SPIE* 1847, pp. 295–300, 1992.
26. Tummala, R. R., and E. J. Rymaszewski, *Microelectronic Packaging Handbook,* Van Nostrand Reinhold, New York, 1988.
27. Patra, S. K., and Y. C. Lee, "Quasi-static Modeling of the Self Alignment Mechanism in Flip Chip Soldering: Single Solder Joint," *Transact. ASME J. Electron. Packaging,* **113**(4):337–342, 1991.
28. Patra, S. K., and Y. C. Lee, "Modeling the Self Alignment Mechanism in Flip Chip Soldering. II Multichip Solder Joints," *Proceedings of 41st Electronic Components and Technology Conference,* pp. 783–788, 1991.
29. Satoh, R., M. Ohshima, H. Komura, I. Ishi, and K. Serizawa, "Development of a New Micro-Solder Bonding Method for VLSI's," *Proceedings of IEPS,* p. 455, Nov. 1983.
30. Moore, K., S. Machuga, S. Bosserman, and J. Stafford, "Solder Joint Reliability of Fine Pitch Solder Bumped Pad Array Carriers," *Proceedings of NEPCON West,* vol. 1, pp. 264–274, 1990.
31. Kromman, G., D. Gerke, and W. Huang, "A Hi-Density C4/BGA Interconnect Technology for a CMOS Microprocessor," *Proceedings of IEEE Electronic Components and Technology Conference,* pp. 22–28, 1994.
32. Oktay, S., "Parametric Study of Temperature Profiles in Chips Joined by Controlled Collapse Technique," *IBM J. Research Devel.,* 13:272–285, May 1969.
33. Buchanan, R. C., and M. D. Reeber, "Thermal Considerations in the Design of Hybrid Microelectronic Packages," *Solid State Technol.,* **26**(2):39, 1973.

34. Cavanaugh, D. M., "Thermal Comparisons of Flip Chip Relative to Chip-and-Wire Semiconductor Attachment in Hybrid Circuits: An Experimental Approach," *ISHM Proceedings*, p. 214, Sept. 1975.
35. Oktay, S., and H. C. Kammerer, "A Conduction-Cooled Module for High-Performance LSI Devices," *IBM J. Research Devel.*, **26**(1):55–66, Jan. 1982.
36. Simmons, R. E., and R. C. Chu, "Direct Immersion Cooling Techniques for High Density Electronic Packaging and System," *ISHM Proceedings*, p. 314, Sept. 1985.
37. Kurokawa, V., K. Utsumi, H. Takamizawa, T. Kamata, and S. Noguchi, "AlN Substrates with High Thermal Conductivity," *IEEE Transact. Components Hybrid Manufact. Technol.*, **CHMT-8**(2):247, 1985.
38. Okutani, K., K. Otsuka, K. Sahara, and K. Satoh, "Packaging Design of a SiC Ceramic Multi-Chip RAM Module," *Proceedings of IEPS*, p. 299, Nov. 1984.
39. Inoue, H., Y. Kurihara, and H. Hachino, "Pb-Sn Solder for Die Bonding of Silicon Chips," *IEEE Transact. Components Hybrid Manufact. Technol.*, **CHMT-9**(2):190–194, 1986.
40. Doi, H., K. Kawano, R. Minamitani, T. Hatsuda, and T. Hayashida, "Development of Mechanical Fatigue Test Method for Flip Chip Solder Joints," *Proceedings of Electronic Components and Technology Conference*, 167–170, 1992.
41. Coffin , L. F., "Low Cycle Fatigue: A Review," *Appl. Mech. Research* **1**(3):129–141, 1962.
42. Norris, K. C., and A. H. Landzberg, "Reliability of Flip Chip Interconnections," *IBM J. Research Devel.*, 266–271, May 1969.
43. Howard, R. T., "Packaging Reliability and How to Define and Measure it," *Proceedings of 32nd Electronic Components Conference*, pp. 367–384, 1982.
44. Puttlitz, K. J., "Corrosion of Pb-50In Flip-Chip Interconnections Exposed to Harsh Environment," *IEEE Transact. Components Hybrid Manufact. Technol.*, **13**(1):188–193, 1990.
45. Roush, W., and J, Jaspal, "Thermomigration in Lead Indium Solder," *Proceedings of 32nd Electronic Components Conference*, p. 342, 1982.
46. Jackson, M. S., C. W. Cho, P. Alexopoulos, and C.-Y Li, "A Phenomenological Model for Transient Deformation Based on State Variables," *J. Eng. Mat. Technol., ASME*, **103**(4):314–325, 1981.
47. Pao, Y.-H., R. Govila, and S. Badgley, "Thermal Fatigue Fracture of 90Pb/10Sn Solder Joints," *Proceedings of Joint ASME / JSME Conference on Electronic Packaging, Advances in Electronic Packaging*, pp. 291–300, 1992.
48. Wilcox, J. R., R. Subrahmanyan, and C-Y. Li, "Influence of Assembly Stiffness and Failure Criterion on the Apparent Fatigue Life of Solder Joints" *J. Electron. Packag., ASME*, **112**(2):115, 1990.

Large High-I/O Solder-Bumped Flip Chip Technology

Tom Chung, Tom Dolbear, Dick Nelson

4.1 Introduction

Flip Chip (FC) is an integrated-circuit (IC) chip-level interconnect technology that typically uses solder bumps to provide electrical interconnection between a chip and its next level of interconnect. This can be a plastic or ceramic package, or a substrate such as printed-circuit board (PCB) or high-density interconnect (HDI) substrate used in multichip module (MCM)-related applications. (Please refer to Chaps. 6 through 13 of this book for FC using wires or TAB or conductive adhesive.) Flip chip technology includes wafer or chip bumping, wafer probing, assembly and reflow, encapsulation, and thermal design.

Traditionally, FC has been considered as a captive chip-level interconnect technology and has not been widely applied for several reasons: (1) a lack of infrastructure, (2) the expense to license the technology, and (3) the advantages of this technology is not needed for many applications. In addition, almost all of the previous FC-related applications[1–9,18–35] used a relatively small chip (e.g., <10 mm^2). One famous example of FC-related applications is IBM's Controlled Collapse Chip Connection (C4), which was first used in volume production in IBM System/360 in 1964. However, recent demands of product performance and versatility, compounded with increasing chip power and I/O, has increased the die size to a level that poses significant challenges for manufacturers who desire high-performance, cost-effective, reliable packaging solutions for their products.

One example of efforts to develop and demonstrate large, high-I/O FC technology is the Open Systems Project II (OSPII) at the Microelectronics and Computers Technology Corporation (MCC) located in Austin, Texas. This project was initiated in 1992 by more than 10 major electronics companies to address the technical challenges and to find solutions for the next generation of computer workstation technology. The project objectives, approaches, and key technologies were defined and implemented on the basis of target requirements identified and specified by OSPII's participating companies. Flip chip, especially for large, high I/O die, was identified as one of the most important technologies necessary to meet the stringent requirements of the next generation of computer workstations. The advanced workstation packaging requirements, OSPII objectives and approaches, and key technologies under development are discussed in the following sections.

4.2 Typical Technical Requirements of Advanced Computer Workstation

Typical packaging requirements and constraints for advanced computer workstations shown in Table 4.1 were defined by the participants of OSPII and MCC after several iterations. The key challenge was to cost-effectively provide reliable and manufacturable packaging solutions to accommodate large, high-I/O chips with high power density in a high-performance workstation operating at 300 MHz. In addition to the typical packaging requirements shown in Table 4.1, a number of other technical requirements were also specified; for instance, the data bus must be bidirectional and support 300 MHz, there must be more than one large, high-I/O die with high power dissipation, and the data bus must be 128 bits wide. Using these requirements, OSPII objectives, approaches, and key technologies to be developed and demonstrated were identified and implemented. They are described in the following section.

TABLE 4.1 Typical Packaging Requirements and Constraints for Advanced Computer Workstation

Description	Value	Units
Clock frequency	300	MHz
Rise and fall times	≤500	ps
Cache size	4–16	Mbytes
Module power dissipation	100–200	W
Module DC current	30–45	A
Junction temperature (nominal)	<110	°C
Acoustic noise (nominal)	<55	dBA
Signal integrity	<3	dB
(attenuation on any bus)		
Processor VLSI		
Total I/O	<1000	—
Power density	≤20	W/cm^2
Size	20×20	mm
Number of chips	10–25	—
Number of off-module I/O	<1000	—
Air-cooled	Yes	—

4.3   MCC OSPII Objective and Approach

The main objective of the OSPII was to accelerate the development of the domestic, high-performance, high-density packaging infrastructure for commercial applications-especially the microprocessor/cache modules used in advanced computer workstations. On the basis of this objective and the technical requirements described previously, a detailed tradeoff analysis study for a 300-MHz processor module was carried out to identify the problems and issues associated with its design and fabrication. Various packaging alternatives such as the interconnects (PCB, ceramic, thin-film, etc.), materials, bonding techniques (wire bond, TAB, flip chip), and direct-chip-attach (DCA) versus single-chip packaging were analyzed. Examples of the block diagram of the 300-MHz processor module, and the associated packaging alternatives selected for the tradeoff study are shown in Fig. 4.1 and Table 4.2, respectively.

The result of this tradeoff analysis was a complete set of information and solutions for the packaging problems associated with the 300-MHz microprocessor/cache module under investigation. It also delineated three technology-specific test vehicles (TSTVs) that were designed and implemented to validate and optimize the analysis. The three TSTVs are FC thermal management, FC reliability and manufacturability, and electrothermomechanical integration. The objectives, approaches, key technology elements, and some of the test results of the FC thermal management, and reliability and manufac-

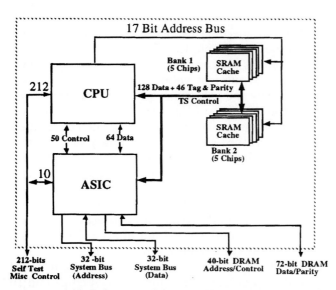

Figure 4.1   A block diagram of the 300-MHz processor module designed for the tradeoff analysis.

TABLE 4.2   Packaging Alternatives Selected for the MCC OSPII Tradeoff Analysis

Case	Large chip assembly technology	SRAM assembly technology	Interconnect technology
1	C4	C4	MCM-D
2	C4	C4	MCM-D/C
3	Adhesive FC	Adhesive FC	MCM-D
4	Adhesive FC	Adhesive FC	MCM-D/C
5	C4	C4	LTCC
6	Microcarrier	OMPAC	LTCC
7	C4	C4	SLC/FRL
8	Microcarrier	OMPAC	SLC/FRL
9	Microcarrier	OMPAC	Fine-line PCB
10	Microcarrier	Wire bond	Fine-line PCB
11	PGA	QFP	PCB

turability TSTVs are described in detail in the following section since they are the key concerns for large high-I/O FC technology.

4.4   Technology-Specific Test Vehicles

4.4.1   Flip chip thermal management TSTV

The objective of the FC thermal management TSTV was to experimentally determine the capabilities of selected air-cooled heat exchangers and MCM technologies for meeting the thermal requirements of the advanced computer workstation specified in Table 4.1. This test vehicle was designed to focus on cooling techniques where the backside of the die is available for direct contact with an air-cooled heat exchanger system (see Fig. 4.2). In addition, the test vehicle builds served to integrate technologies and to verify the performance of a design. Moreover, they were used to screen the impact on FC reliability of selected above-substrate cooling designs that used accelerated testing such as temperature cycling and high-temperature storage. A secondary focus of this test vehicle was to quantify and extend breakpoints of low performance and more established technologies and to investigate higher-risk, high-performance technologies.

The key technology elements involved in this test vehicle are the application and characterization of high-performance thermal interface materials, mechanical tolerance analysis, design and implementation of reliable mechanical reference system, design and characterization of various high-performance heat exchangers, and measurement-based thermal modeling. A multichip module containing 2 VLSI chips and 10

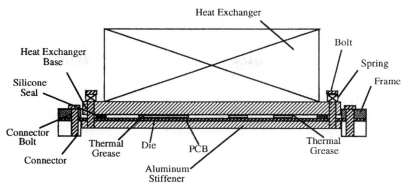

Figure 4.2 Cross section of the flip chip thermal management TSTV.

TABLE 4.3 Summary of Chip Size and Power Dissipation

Chip	Size (mils)	Power, W	Power density, W/cm^2
VLSI 1	800×800	86	21
VLSI 2	670×670	73	25
SRAM	470×300	1.5	1.6

SRAMs that are directly solder ball FC–connected to a fine-line laminate substrate was designed as the thermal TSTV. The size and power dissipation of the chip set are shown in Table 4.3. The results of this TSTV are limited to this chip set specification with each die having uniform power dissipation across its area. A simplified thermal test vehicle, utilizing silicon chips with blanket sputtered metal films for heat sources, was built to investigate the capability of the thermal design to adequately cool an electrically functional module. According to the data listed in Table 4.3, the worst-case power density is ~25 W/cm^2 for VLSI 2.

Because of the high power density, the use of flip chip, and the use of a low-cost, low-conductivity laminate substrate, the module must be cooled using an above-substrate cooling design to meet thermal specifications.[10,11] In this above-substrate design, heat is conducted from the back of the die through a thermal compound or "thermal grease" to a heatsink and then convected from the heatsink by flowing air. The conductivity of the thermal grease as well as the thickness of this connection critically impact its conductance. Two thermal greases were used—one with a thermal conductivity of >2.5 W/m °C and the other with a lower conductivity of ~1.5 W/m °C. The tolerances of the module components must be appropriately controlled or accounted for in the thermal design so that the distribution of chip junction temperatures remains below the temperature specification of 85°C at 30°C ambient or 110°C at a 55°C ambient. In the following sections, the test vehicle design, a tolerance analysis of the test vehicle, and the measured thermal performance of the thermal test vehicle are discussed.

4.4.2 Module design

The thermal test vehicle utilized 0.020-in-thick silicon test chips that had a blanket metal film resistor sputtered on the silicon to provide the power dissipation indicated in Table 4.3. These dies were electrically attached to a PCB (polyimide-glass) using 0.010-in-thick 70/30 InPb solder over a 0.050-in-wide gold stripe on two opposite edges of each die. The dies were then underfilled using a solid-filled liquid epoxy. This structure was used to approximate an underfilled solder FC assembly. The chip layout for the test vehicle is shown in Fig. 4.3. The dimensions indicated are that of the routed area of the board. An additional 0.4 in is required at each edge of the thermal test vehicle for making connections to power the thermal test vehicle and make voltage measurements on the module.

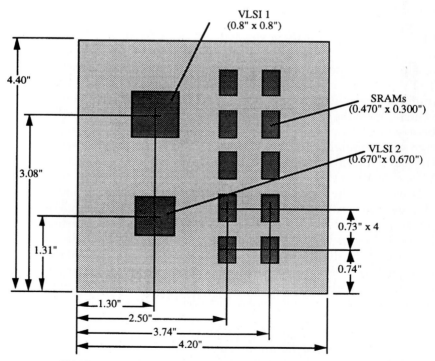

Figure 4.3 Chip layout of thermal test vehicle. Note that dimensions are of routed area only.

Two arrangements were used for the airflow direction: crossflow and impingement flow. For the impingement design, air is directed to flow into the top of the heatsink and exit from its sides. For the crossflow design, air is brought across the heatsink from one face to the other. Because of the fin density of both heatsink designs, the modules will require ducting to force air through the channels when implemented in a system. The height for both modules was limited to be compatible with a 2.5-in centerline spacing between motherboards in a card stack. The overall test vehicle module dimensions are 5.0"×5.2"×1.6" (height) for the impingement design including the connector standoff. The crossflow module has the same footprint but was allowed to be 0.4 in higher than the impingement design since it will not require ducting *above* the module to direct the airflow in a product.

To reduce gap variations during temperature cycling, the module uses springs to hold the heatsink in constant contact with the substrate at its periphery. Furthermore, a stiffener is present under the substrate to provide backing for the PCB and to reduce the effect of substrate bow on thermal grease thickness. Thus, the design sandwiches the die and grease between the stiffener and the heat exchanger. A cross section of the module construction is shown in Fig. 4.2. Figure 4.4 shows the impingement module reported in this chapter. Figure 4.5 shows the crossflow module tested in this chapter.

Figure 4.4 Photograph of impingement test vehicle module.

Figure 4.5 Photograph of crossflow test vehicle module.

4.4.3 Tolerance analysis

Several factors determine the thickness of the thermal grease between the die and the heatsink. For the construction shown in Fig. 4.2, these factors include the chip and chip attach thickness; the flatness of the heat exchanger, die, and substrate; the distance of the heat exchanger standoffs at the edge of the substrate to the die; and the heat exchanger standoff height. To calculate the tolerance of the total thermal grease thickness as a function of each independent dimension, a *root-sum-square* (RSS) equation was used. The theory behind RSS assumes that each of the components in the equation is independent and normally distributed. If this is true, then the final tolerance calculated will represent the same number of standard deviations as that used for each of the inputs in the equation (i.e., $3\sigma_{in} = 3\sigma_{out}$). Figure 4.6 and Eq. (4.1) show how to calculate the nominal thermal gap on the basis of nominal dimensions. Equation (4.2) shows how to calculate the RSS tolerance of the gap from the tolerance of each independent dimension.

$$G = h_{s} - h_{C4} - d_{c} - LF_{hs} - LF_{sub} \tag{4.1}$$

$$T = \sqrt{T_{st}^2 + T_{C4}^2 + T_{c}^2 + (LT_{hsf})^2 + (LT_{subf})^2} \tag{4.2}$$

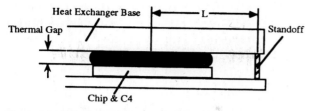

Figure 4.6   Calculating the thermal gap.

where G = gap
$\quad h_s$ = standoff height
$\quad h_{C4}$ = C4 height
$\quad d_c$ = chip thickness
$\quad L$ = length
$\quad F_{hs}$ = heatsink flatness
$\quad F_{sub}$ = substrate flatness
$\quad T$ = tolerance
$\quad T_{st}$ = standoff tolerance
$\quad T_{C4}$ = C4 tolerance
$\quad T_c$ = chip tolerance
$\quad T_{hsf}$ = heatsink flatness tolerance
$\quad T_{subf}$ = substrate flatness tolerance

The module assumptions used in the calculation are listed in Table 4.4. Each dimension and tolerance listed in Table 4.4 is assumed to be independent and normally distributed, and each tolerance represents the 3σ limits of their respective populations. Therefore, the results calculated using these tolerances in Eq. (4.2) also represent the 3σ limit of the distribution for the thermal grease thickness.

Note that since the contribution to the thermal gap tolerance due to the substrate and heat exchanger flatness depends on the distance of the die center to the points where the heat exchanger contacts the substrate, the effect of this flatness depends on the position of the

TABLE 4.4   Component Dimensions and Tolerances (All Measurements in Inches)

Dimension	Nominal Tolerance
VLSI thickness	0.0290 ± 0.0010
SRAM thickness	$0.012 \sim 0.020 \pm 0.0010$
Solder ball FC thickness	0.0032 ± 0.0007
Substrate flatness with stiffener, in/in	0.0000 ± 0.0020
Heat exchanger flatness, in/in	0.0000 ± 0.0010
Standoff height	0.0382 ± 0.0010

contact points on the substrate. If the contact at the substrate edge is positioned within 1.3 in of the die center, the preceding assumptions result in a tolerance on the thermal gap over the VLSI chips of ±0.0033 in about a nominal grease thickness of 0.006 in.

4.4.4 Heatsink design

To account for heat spreading in the fine-finned heatsink base, a three-dimensional ANSYS finite element model was constructed of the heatsink base that models the discrete placement of the die on the substrate (as shown in Fig. 4.3) and models the extended surface as a uniform heat-flux boundary condition on a flat surface. The reciprocal of the product of the heat-transfer coefficient specified on this surface and the area of the surface represents the thermal resistance of the fine-finned heatsink on the physical module. This model was used to perform a number of finite element simulations of the heat equation with different values of the heat-transfer coefficient, heatsink base thickness, and heatsink base thermal conductivity. An example finite element run is shown in Fig. 4.7.

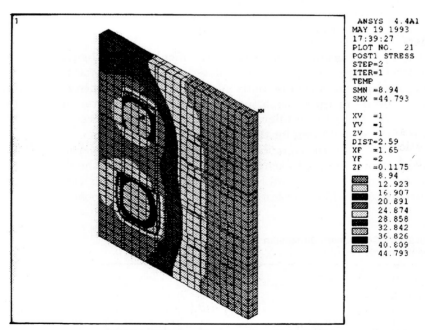

Figure 4.7 Example of temperature distribution in straight channeled heatsink base from finite element analysis. Aluminum base thickness = 0.2 in, heat-transfer coefficient = 0.75 W/in² °C, thermal grease thickness = 0.005 in.

Using a Box-Behnken experimental design matrix,[12] a series of simulations was used to determine a design equation for the maximum chip temperature of the module as a function of grease thickness, heatsink base thickness, heat-transfer coefficient (external thermal resistance), and heatsink base conductivity. A typical design equation for one set of analyses is

$$\Delta T_c = 38.28 - 3.41t^* - 4.91k^* - 22.17h^* + 1.60t^{*2} + 2.10k^{*2}$$
$$+ 14.32h^{*2} + 1.19(t^*)(k^*) + 0.83(t^*)(h^*) + 0.77k^*h^*$$
$$+ (gt - 5)^*2.268 \tag{4.3}$$

where ΔT_c = chip temperature rise above ambient
  t^* = coded base thickness; equal to $(-1,1)$
  k^* = coded base conductivity; equal to $(-1,1)$
  h^* = coded heat transfer coefficient; equal to $(-1,1)$
  gt = ATC2.8 thickness in mils

To transform from the coded values to physical variables, the following relationship is used:

$$x^* = 2\,\frac{(x - \bar{x})}{(x_{-1} - x_{+1})} \tag{4.4}$$

where x^* = coded value of variable; equal to $(-1,1)$
  x = physical value under consideration; equal to (x_{-1}, x_{+1})
  x_{-1} = lower end of physical range under consideration
  x_{+1} = upper end of physical range under consideration
  $\bar{x}$ = average of x_{-1} and x_{+1}

To transform from coded to physical variables, the range for each physical variable must be known. The physical range under consideration for each variable in the above analysis was

  t = 0.100, 0.300 in
  k = 200 W/m °C, 600 W/m °C
  h = 0.25 W/in^2°C, 1.25 W/in^2°C

On the basis of this design equation, the external thermal resistance required to maintain the worst chip junction temperature at less than 50°C above the local ambient temperature could be determined. (A 5°C temperature rise was budgeted for heating the air from the ambient temperature outside the system to the local ambient temperature at the module.) The design was chosen so that modules with a $+3\sigma$ grease gap thickness based on the tolerance analysis would meet the temperature specification.

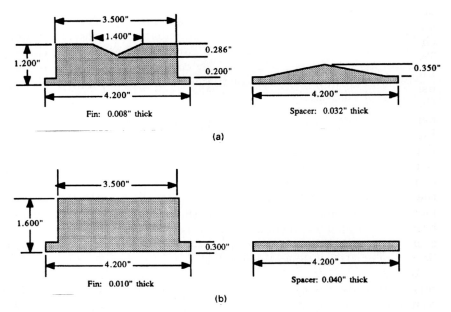

(a)

(b)

Figure 4.8 Summary of dimensions of the two fine-finned heatsink designs. Note that impingement heatsink uses two end plates without the *V* cutout: (*a*) dimensions of impingement heatsink fins and spacers; (*b*) dimensions of crossflow heatsink fins and spacers.

Several different fine-finned heatsink designs in both aluminum and copper for crossflow and impingement were sized using this approach and software previously developed for designing air-cooled heatsinks based on empirical channel flow correlations in Refs. 13 and 14. These designs were fabricated by AAVID Engineering in Laconia, New Hampshire. In this chapter, the results for two aluminum designs, shown in Figs. 4.4 and 4.5, are reported—one impingement design and one crossflow design. The impingement design utilizes an MCC heat sink design reported in Refs. 15 to 17.

Figure 4.8 summarizes the dimensions of the heatsink designs used on the modules reported in this chapter. Multiple fins and spacers are laminated together to form a heatsink. The width of the fin stacks for each heatsink was 4.40 in.

4.4.5 Module testing procedure

The thermal test vehicle used to investigate the design's performance utilized silicon test chips that had a blanket metal film sputtered on the silicon. This film was connected directly to the PCB at two edges using an InPb solder and then underfilled using a solid-filled liquid epoxy. A DC voltage applied across the film supplied the power dissipated by the chip. The circuit board was wired so that each VLSI test chip was powered by an individual power supply and the SRAM test chips were connected in series and powered with a third power supply. Temperature measurement on the die surface was done using a Luxtron Fluoroptic Temperature Measurement System (model 750) using the MEL surface probes supplied by Luxtron (located in Santa Clara, Calif.). Nine 0.062-in-diameter holes through the board and underfill supporting the VLSI test chips allowed the 0.025-in-diameter Luxtron probe to make contact with the heated surface of the die. Contact was assured by use of a spring-loaded probe holder that would clearly deflect upward as contact was made. Each SRAM had one hole at its center for temperature measurement. These holes also allowed the calculation of the thermal gap between the die and the heatsink by measuring the height of the die surface with respect to the heatsink base and subtracting the previously measured die thickness. A Tutor coordinate measurement system by Digital Equipment Automation (located in Livonia, Mich.) was used for these measurements.

Additionally, five thermocouples were inserted into holes (filled with thermal grease) drilled in the heatsink base. One thermocouple was positioned in the heatsink base above the center of each VLSI test chip, and three were positioned above the bank of SRAMs—one at the center and one at each midpoint between the center of the SRAM bank and its edges. The thermocouples were connected to a Fluke datalogger for determining temperature. The datalogger was also used to make voltage measurements on the module and off the module across load resistors to determine the voltage and current in each of the three circuits used to power the module.

Both the datalogger and the Luxtron unit were interfaced to a 386 computer using Quick Basic. Acquired data was averaged over the sampling period, and the mean and sample standard deviation for each measurement were written to a text file. This file could then be imported into an Excel spreadsheet for analysis. After the steady state was reached, randomized temperature measurements were made of the die temperatures and heatsink temperatures by moving the Luxtron probe from one probe location to the next measurement location. The datalogger measurements were made after each Luxtron measurement over the duration of the test. These measurements were averaged and used to judge whether drift was occurring. Typically, the sample standard deviations for voltage and current measurements were less than 0.1 percent of the nominal value. Temperature measurements using thermocouples typically had sample standard deviations of less than 0.1°C, while Luxtron readings typically had sample standard deviations of less than 0.4°C at a given measurement location.

Each module under test was placed on or in a plexiglass fixture to provide uniform airflow to the fins: either an impinging flow or a crossflow through the heatsink. The impingement flow setup utilized a plenum cube 2 ft in length on each edge. The top face of the cube had a cutout that matched the fin top print of the heatsink. The module was placed upside down on this cutout to allow airflow through the heatsink, exiting at two sides. This orientation also conveniently allowed access to the module's bottom for making temperature measurements with the Luxtron system. A similar setup was used for crossflow, except that a duct matching the side of the module's profile was connected to the plenum cube and the module was placed within the duct with the module's bottom accessible for temperature measurements. Each plenum was supplied by two connections from a compressed airline. Flow into the plenums was measured with two Dwyer rotameters. Static pressure loss—from the plenum to ambient in the case of impingement testing, and across the heatsink, in the case of crossflow testing—were made using Dwyer Magnahellic pressure gauges with a 0 to 0.25 in of water range. Reading accuracy for these gauges is ±0.0025 in of water. A similar testing approach is described in Ref. 16.

4.4.6 Experimental results

The two modules indicated in Figs. 4.4 and 4.5 were each tested with the test chips powered according to Table 4.3. Figure 4.9 shows the measured ΔT between the chip and ambient temperature for the impingement module (shown in Fig. 4.4) with an airflow of 30 cubic feet per minute [cfm (ft³/min)]. The associated static pressure loss at this flow is 0.16 in of water. The measurements at different locations on the VLSI 1 and 2 die are indicated. This module used the high-conductivity thermal grease (>2.5 W/m °C). The much lower power SRAMs are well below the 50°C rise above ambient as are the two

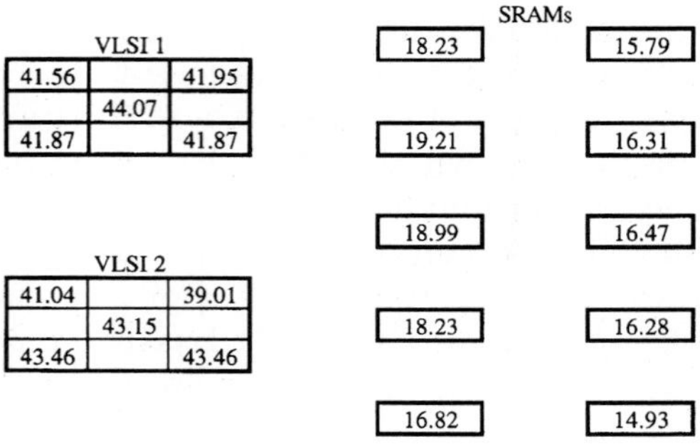

Figure 4.9 Chip temperature rise (°C) above ambient for impingement module with high-conductivity grease at 30-cfm airflow.

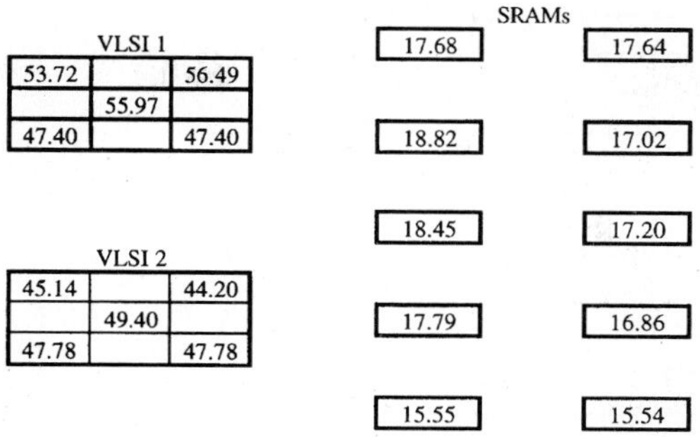

SRAMs

VLSI 1					
53.72		56.49		17.68	17.64
	55.97				
47.40		47.40		18.82	17.02

| | | 18.45 | 17.20 |

VLSI 2			
45.14		44.20	
	49.40		
47.78		47.78	

17.79   16.86

15.55   15.54

Figure 4.10 Chip temperature rise (°C) above ambient for cross-flow module with low-conductivity grease at 30-cfm airflow.

VLSI chips. According to the measurements, the average grease thickness above VLSI 1 was 0.007 in thick and 0.0064 in above VLSI 2. Of the temperature rise shown in Fig. 4.9, approximately 14°C occurs across the grease interface for each VLSI chip; the remainder is associated with spreading in the heatsink base and the heatsink thermal resistance. Thus, at the $+3\sigma$ thickness of 9.3 mils, an additional 4 to 6.5°C rise can be expected. From the data in Fig. 4.9, the module will still operate below the temperature specification of less than a 50°C rise above ambient at 30 cfm.

The same results for the crossflow module (shown in Fig. 4.5) are shown in Fig. 4.10 at an airflow of 30 cfm using the low-conductivity grease (<1.5 W/m °C). The associated pressure loss for this flow rate was 0.44 in of water. The heatsink channels were oriented from left to right; the results shown in Fig. 4.10 are for the flow moving from left to right across the page. From the measurements, the average thickness of grease above VLSI 1 is 0.0053 and 0.0063 in above VLSI 2. Of the temperature rise shown in Fig. 4.10, approximately 22 to 25°C is associated with the grease used. If the grease thickness of the module were at the 3σ thickness of 0.0093 in, an additional 16 to 18°C rise might be expected. Thus, when using the low conductivity grease this module does not meet the 50°C rise above ambient temperature specified. Note that although the power density is greater on VLSI 2 and its greater grease thickness, it is at a lower temperature than VLSI 1. This difference is attributed to variation in the material or the quality of the grease interface between the VLSI 1 test chip and the heatsink.

To evaluate the heatsink designs, two additional tests were performed to determine the thermal resistance of the heatsink alone and each design's pressure–flow characteristic. This test was performed by selecting the chip powers to provide close to an isothermal heatsink base. The heatsink thermal resistance versus flow and the pressure loss versus flow are shown in Fig. 4.11.

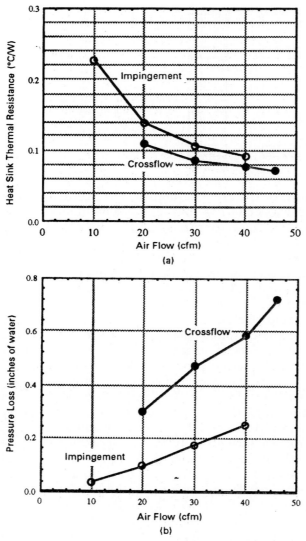

Figure 4.11 Impingement and crossflow heatsink design characteristics: (*a*) thermal resistance versus flow for impingement and crossflow heatsink designs; (*b*) pressure loss versus flow for impingement and crossflow heatsink designs.

On the basis of these curves, the crossflow heatsink requires <20 cfm of airflow to achieve the same thermal resistance as the impingement heatsink at 30 cfm. The pressure loss for the crossflow design at 30 cfm is 0.3 in of water, which is nearly twice the pressure loss associated with the impingement heatsink at 30 cfm. In comparison to a standard axial-flow fan, the operating point of the crossflow design exceeds a single-fan pressure head at 20 cfm. The impingement heatsink is at about 75 percent of the pressure head of the same fan at 30 cfm. The significance of this difference in operating points will depend on the system in which the module is being used.

4.4.7   Flip chip reliability and manufacturability TSTV

The objective of the FC reliability and manufacturability TSTV was to explore selected FC-substrate technology combinations from a performance, routability, reliability, and manufacturability perspective.

Flip chip assembly has been a major focal area within OSPII that affects all the TSTVs. Although both solder- and adhesive-based connection schemes have been identified as possible solutions,[7,18–22] there is significant uncertainty in the application of these technologies and the interactions between FC assemblies and different interconnect substrate technologies in terms of large high-I/O chips.

The escape routing studies were first implemented to understand the limitations and capabilities of escape routing for FC attachment of a large high-I/O chip on a variety of substrate technologies. Its key objective was to allow the use of a minimum bond pad pitch with the most cost-effective substrate technology and to reduce the required number of substrate wiring layers without sacrificing signal integrity. The assumptions of the structure were on the order of 900 I/Os with a signal to reference ratio of roughly 2:1 and a die size of 800×800 mils. The study conducted for escape routing under high-I/O count devices included techniques applicable to surface laminar circuitry (SLC), film redistribution layer (FRL), low temperature cofired ceramic (LTCC), MCM-D, and fine-line PCB. Device pad geometry was considered for both uniform rectangular arrays and for special configurations not confined by a predetermined pad layout. The study also addressed methods of providing for flyby routing, essential for minimizing stub length in high-frequency applications; pad clustering, to reduce signal to ground ratio; and global routing requirements.

As an example, consider escape routing using the SLC technology with a uniform rectangular pad array as shown in Fig. 4.12. SLC has been suggested for escape routing under high-I/O count chips.[24–27] Figure 4.13 shows a typical cross section for SLC. This technology has advantages in that the photolithographic vias are smaller than the plated through-holes, and since they do not extend through the board, are essentially blind and do not interfere with the core routing. The example shown in Fig. 4.12 illustrates the escape routing on the top (FC3) layer; basically the four outermost rows of signal bond pads are routed out on this layer. Note that the signal bond pads in the center

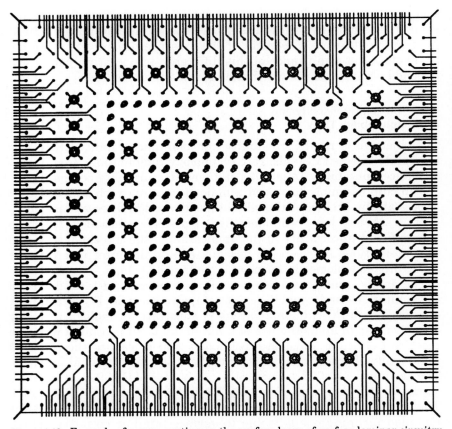

Figure 4.12 Example of escape routing on the surface layer of surface laminar circuitry (containing a regular array of 30×30 bond pads, 5-mil bond pads on 27-mil pitch with three lines between pads with 3-mil line and space).

section are connected to 10-mil SLC vias which terminate on islands in the ground plane (FC2). A second 10-mil SLC via in each island then carries the signal down to the FC1 layer, where the final escape routing is accomplished for the central array of pads. In this example, no core PCB routing is required.

As recently demonstrated by several papers,[2,5,24–28] FC with the C4 stabilized by a resin underfill can significantly increase the fatigue life of the joints. This is especially true for large chips, although the material properties of the underfill and substrate need to be carefully analyzed and designed. Therefore, in addition to the extensive escape routing studies, measurement-based FC modeling was implemented along with several test vehicle designs and builds. These test vehicles also underwent either temperature cycling or power cycling depending on the chip/substrate material cross section. The test results were

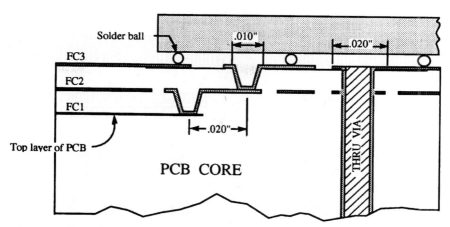

Figure 4.13 Typical cross section of surface laminar circuitry (FC1 and FC3 are used for escape route traces, and FC2 is used as the ground plane).

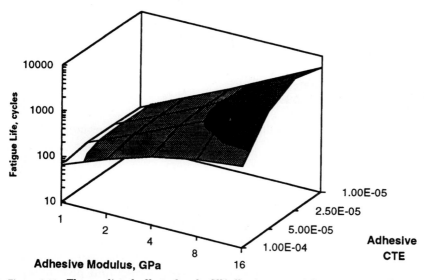

Figure 4.14 The predicted effect of underfill/adhesive material properties on C4 fatigue life (corner site, 10-mil bump pitch, 2-cm chip on FR4, 0 to 125°C, 24 cycles per day). The contour plot gives the lower limit of expected C4 life based on the Coffin-Manson low-cycle fatigue law.

used to calibrate the models and provide a database for fatigue life prediction of both adhesive and C4-based FC assemblies. One example of the modeling work is the 3D plot of FC solder joint fatigue life versus underfill/adhesive modulus and CTE (coefficient of thermal expansion) shown in Fig. 4.14. Both substrate and underfill thermo-mechan-

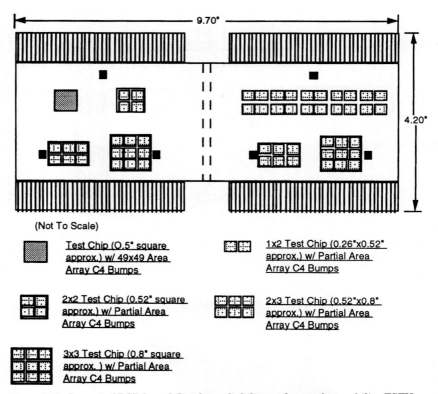

Figure 4.15   Layout of PCB-based flip chip reliability and manufacturability TSTV.

ical properties were found to significantly affect the predicted C4 fatigue life. Also, one of the test board designs is illustrated in Fig. 4.15 with a typical test vehicle cross section shown in Fig. 4.16. Other key technology elements investigated included FC design for manufacturability guidelines, and the feasibility of building a number of test vehicles by more than one source (using standard manufacturing processes) in order to assess manufacturability-related issues and problems for large high-I/O FC builds.

4.4.8   Assembly and test results

All the FC reliability and manufacturability TSTVs were assembled with the process flow shown in Fig. 4.17[7,34] A number of assembly equipment-, process-, and material-related parameters (e.g., reflow profile, underfill material), have been designed into the TSTV builds. The purpose was to understand the effects of those assembly parameters on the reliability of the large high I/O FC assemblies. The assem-

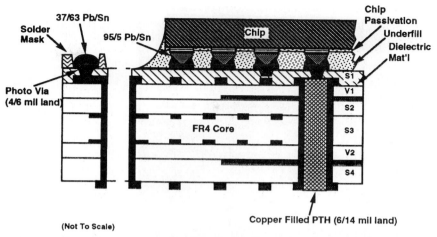

Figure 4.16 Typical cross section of flip-chip reliability and manufacturability TSTV.

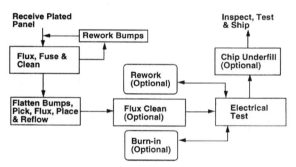

Figure 4.17 A typical flip chip on laminate-based substrate assembly process flow.

bled TSTVs were 100 percent visually and x-ray-inspected, followed by cross sectioning of selective assemblies to verify mechanical integrity of inspected solder FC solder joints. On the basis of inspection results performed on 412 flip chips with more than 250,000 solder joints, it was believed that large (~2 cm square) high-I/O (~800) chips can be successfully FC-assembled onto the laminate-based substrate as demonstrated by the solder joint cross section shown in Fig. 4.18. However, it is important to point out that the shape of a successfully assembled solder joint may vary depending on many factors such as alignment between chip bumps and corresponding substrate pads, coplanarity between chip and substrate, solder pad thickness on substrate, solder mask material, and bump and pad sizes.

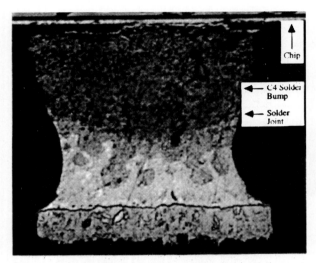

Figure 4.18 This micrograph shows cross section of an acceptable flip chip solder joint made on a laminate-based substrate.

Finally, a number of reliability tests were also performed on successfully assembled TSTVs. The test results indicated that large high-I/O flip chips assembled (with a standard assembly process) on a laminate-based substrate can survive through more than 1000 thermal cycles (0 to 100°C with 20 min of dwell time) without failure if appropriate assembly materials and processes are applied.[34,35]

4.5   Summary

Large high-I/O solder-bumped FC technology is needed to support the growing functionality within devices planned for many next-generation, commercial-grade digital applications. As perimeter I/O attachment technologies exhaust themselves in the face of manufacturing limitations, array flip chip attachment for these devices presents itself as a solution with great potential, together with its unknown performance when supporting the associated higher power, higher I/O count, and dimensionally larger footprint chips. Thermal, mechanical, electrical, and other design questions must be confronted from the perspective of both high-level performance needs, as well as the difficulties associated with the practicalities of high quality and reliability volume production requirements. While many cases of smaller chips have been successfully introduced as flip chip products, larger devices (e.g., >1.0 cm²) have not. A primary limitation to the adoption of large-chip flip chip is that many companies lack comprehensive understanding of its design and performance in statistically relevant samples and from the perspective of design for manufacturability considerations. At one extreme, relevant chips and modeling tools do not exist for the analysis of the associated issues; at the other, multiple knowledgeable suppliers do not exist for volume infrastructure support.

The OSPII charter has been to examine and demonstrate migration paths into solutions and away from fundamentally problematic areas involving large high-I/O solder-bumped flip-chip technology. An OSPII side benefit is to conclude with a sense of multicorporate alignment in terms of what the primary issues are (and are not) and how to confront them, hence improving the efficiency of the emerging related infrastructure. The issues facing this more technically aggressive category of flip chip attached devices can best be addressed through extensive characterization. Primary issues involve the following:

- Design for thermal management in terms of through-substrate or back-of-chip cooling systems
- Design for escape routability of substrate
- CTE mismatch between chip and substrate
- Effects of underfill material in improving reliability and compromising reworkability
- Variation of bump electrical resistance over time
- Breadth of infrastructure support of any given chip/substrate attachment system

These issues, among others, must be carefully examined in the context of the large high-I/O flip chip format—in terms of both modeling and experimental characterization. Once they are understood, the investigation must be extended to explore the ability of multiple suppliers to practically support targeted technologies for volume commercial-grade applications.

Acknowledgments

The authors would like to thank the following people for their support to this work: David Carey (MCC), Bob Gardner (HP), Claude Rathmell (MCC), Scott Sommerfeldt (MCC), Tom Hunter (MCC), Jon Zimmerman (Unisys), Dave deSimone (Tandem), Aurangzeb Khan (Tandem), and John Norris (Tandem). In addition, the authors would like to thank the participants in the MCC OSPII project for their financial support of the project and for allowing this work to be published.

References

1. Patel, C. D., "Backside Cooling Solution for High Power Flip Chip Multi-Chip Modules," *Proceedings of 44th IEEE Electronic Components and Technology Conference* (ECTC), pp. 442–449, May 1994.
2. O'Malley, G., et al., "The Importance of Material Selection for Flip Chip on Board Assemblies," *Proceedings of 44th IEEE ECTC,* pp. 387–394, May 1994.
3. Puttlitz, K., et al., "C4/BGA Comparison with Other MLC Single Chip Package Alternatives," *Proceedings of 44th IEEE ECTC,* pp. 16–21, May 1994.
4. Partridge, J., and P. Viswanadham, "Organic Carrier Requirements for Flip Chip Assemblies," *NEPCON West Proceedings,* pp. 1519–1526, March 1994.
5. White, L., and D. Suryanarayana, "Flip Chip Encapsulation for MCM-L," *NEPCON West Proceedings,* pp. 1493–1502, March 1994.
6. Gregorich T., et al., "Flip Chip on Board Replacement Process," *1994 ITAP & Flip Chip Proceedings,* pp. 13–16, 1994.
7. Hayden, T., and J. Partridge, "Practical Flip Chip Integration into Standard FR-4 Surface Mount Processes: Assembly, Repair and Manufacturing Issues," 1994 ITAP & Flip Chip Proceedings, pp. 1–7, 1994.
8. Chung, T., et al., "Development of Large High I/O Flip-Chip Technology," *NEPCON West Proceedings,* pp. 1527–1536, Feb. 1994.
9. Dolbear, T., et al., "Thermal Management of High-Power Multi-Chip Modules," *NEPCON West Proceedings,* pp. 1570–1579, Feb. 1994.
10. Dolbear, T., "Thermal Management of Multi-Chip Modules," *NEPCON West Proceedings,* pp. 1182–1208, 1990.
11. Dolbear, T., "Thermal Management of Multi-Chip Modules," *Electron. Packag. Product.,* pp. 60–63, June 1992.
12. Schmidt, S., and R. Launsby, *Understanding Industrial Designed Experiments,* CQG Ltd. Printing, Longmont, Colo., 1989.
13. Phillips, R. J., *Forced-Convection Liquid-Cooled Microchannel Heat Sinks,* technical report 787, Lincoln Laboratory, MIT, Lexington Mass., Jan. 1988.
14. Kays, W. M., and M. E. Crawford, *Convective Heat and Mass Transfer,* McGraw-Hill, New York, 1980.
15. Hilbert, C., et al., "High Performance Air Cooled Heat Sinks for Integrated Circuits," *IEEE Transact. CHMT-13,* 1022, 1990.
16. Nelson, R., et al., "Application of Air-Cooled Microchannel Heat Sinks to Card Cage Systems," *Semitherm,* 50–62, 1993.
17. Herrell, D. J., O. Gupta, and C. Hilbert, "Micro-laminar Air-Cooled Heat-Sink," U.S. Patent 4,777,560, Oct. 11, 1988.
18. Estes, R., et al., "Environmental and Reliability Testing of Conductive Polymer Flip Chip Assemblies," *Proceedings of International Electronics Packaging Conference* (IEPC), pp. 328-351, Sept. 1993.
19. Becker, C., et al., "Direct Chip Attach (DCA): The Introduction of a New Packaging Concept for Portable Electronics," *Proceedings of IEPC,* pp. 519–533, Sept. 1993.
20. Tsukada, Y., et al., "Design and Electrical Performance of Surface Laminar Circuit/Flip Chip Attach Packaging Technology," *Proceedings of IEPC,* pp. 1111–1119, Sept. 1993.
21. Clementi, J., et al., "Flip-Chip Encapsulation on Ceramic Substrates," *Proceedings of 43rd IEEE ECTC,* pp. 175–181, June 1993.
22. Powell, D. O., and A. K. Trivedi, "Flip-Chip on FR-4 Integrated Circuit Packaging," *Proceedings of 43rd IEEE ECTC,* pp. 182–186, June 1993.
23. Tsukada, Y., et al., "A Novel Chip Replacement Method for Encapsulated Flip Chip Bonding," *Proceedings of 43rd IEEE ECTC,* pp. 199–204, June 1993.
24. Lau, J. H., "Experimental and Analytical Studies of Encapsulated Flip Chip Solder Bumps on Surface Laminar Circuit Boards," *NEPCON West Proceedings,* pp. 1989–2004, February 1993.
25. Lau, J. H., "Thermal Fatigue Life Prediction of Encapsulated Flip Chip Solder Joints for Surface Laminar Circuit Packaging," ASME paper no. 92W/EEP-34, 1992.
26. Tsukada, Y., et al., "Surface Laminar Circuit Packaging," *Proceedings of 42nd IEEE ECTC,* pp. 22–27, May 1992.

27. Tsukada, Y., et al., "Reliability and Stress Analysis of Encapsulated Flip Chip Joint on Epoxy Base Printed Circuit Board," *Proceedings of 1st ASME/JSME Advances in Electronic Packaging Conference,* pp. 827–835, April 1992.
28. Suryanarayana, D., et al., "Flip-Chip Solder Bump Fatigue Life Enhanced by Polymer Encapsulation," *Proceedings of 40th IEEE ECTC,* pp. 338–344, May 1990.
29. Tummala, R. R., and E. J. Rymaszewski, *Microelectronics Packaging Handbook,* Van Nostrand Reinhold, New York, 1989, pp. 367–391.
30. Pedder, D. J., "Flip Chip Bonding for Microelectronic Applications," *Hybrid Circuits, ISHM-UK,* Jan. 1988.
31. Matsui, N., et al., "VLSI Chip Interconnection Technology Using Stacked Solder Bumps," *Proceedings of 37th IEEE ECC,* pp. 573–578, May 1987.
32. Nagesh, V. K., "Reliability of Flip Chip Solder Bump Joints," *Proceedings of IEEE/IRPS,* 1982.
33. Norris, K. C., and A. H. Landzberg, "Reliability of Controlled Collapse Interconnections," *IBM J. Research Devel.,* **13**(3):266–271, 1969.
34. Jimarez, M., "Manufacturability and Early Reliability Data for Large Size, High I/O Flip Chips," *1995 International Flip Chip, Ball Grid Array, TAB and Advanced Packaging Symposium Proceedings,* pp. 150–163, 1995.
35. Carey, D., et al., "MCM-L Substrates for a 300 MHz Workstation," *Proceedings of the 1995 International Conference on Multichip Modules,* pp. 537–542, 1995.

Microinterconnect Technology

The Large-Volume Fabrication of Cost-Effective Flip Chip MCMs

Y. Degani, T. D. Dudderar, R. C. Frye, K. L. Tai, Maureen Y. Lau, and Byung J. Han

5.1 Introduction

The AT&T microinterconnect technology platform was developed to improve the cost-effectiveness of the flip chip multichip module (MCM) manufacturing process through imaginative materials research and the judicious exploitation of a highly automated, high-capacity manufacturing infrastructure based, as much as possible, on the extensive application of generic tools and processes. In this way the need for developing new, specialized types of equipment is minimized and, through the use of high-speed assembly and batch processing, capital costs can be distributed across many MCM packages. As a result, cost-competitive MCM products manufactured using robust, standardized assembly processes will soon become available for large-volume product applications.

For purposes of discussion, the AT&T microinterconnect MCM assembly process may be divided into two parts: (1) the assembly of the unpackaged chips (bare die) onto the substrates which are diced to form individual MCM "tiles" and (2) the subsequent packaging of these "tiles" to provide interconnection between them and the rest of the world and, at the same time, to both protect them from the environment and facilitate handling. (Here the term "tile" is used to denote the assembled silicon components of an individual MCM. In describing the AT&T microinterconnect technology platform, a silicon MCM tile includes both the bare chips and the silicon interconnection substrate to which they are flip chip–bonded mechanically and electrically.)

Section 5.2 focuses on the former, tile assembly part of the MCM assembly process as developed at AT&T Bell Laboratories for large-volume, low-cost applications. Sections 5.3 through 5.7 describe the components, design, and performance characteristics of these MCMs, and Secs. 5.8 and 5.9 describe the ball grid array (BGA) MCM packaging designed to complement the MCM tile assembly process as well as its performance and qualification.

Since an MCM tile is primarily a high-density interconnection structure in which unpackaged integrated-circuit (IC) chips (or bare die) are connected both mechanically and electrically onto a single substrate of silicon, ceramic, or an epoxy-glass laminate, the choice of interconnection technique plays a major role in defining the assembly process and its influence on MCM cost. Needless to say, a variety of interconnection techniques are available to assemble bare die onto the MCM substrates. These MCM components, which are discussed in detail in Sec. 5.3, must at the very least provide both robust structural support and an efficient fabric of electrical interconnection between the ICs. Today MCMs are being assembled using all the techniques available to interconnect ICs within single-chip packages: wire bonding, tape-automated bonding (TAB), and conventional flip chip soldering. All these are described in detail in the well-known handbook by Tummala and Rymaszewski.[1]

In general, the choice of assembly technique depends on the particular capabilities of the manufacturer, the MCM architecture, the relative cost of materials, and the required I/O configuration and density. In turn, the choice of interconnection technique and associated materials plays a major role in defining the assembly process and its influence on MCM cost.

The most commonly used and nominally the lowest cost interconnection technique is wire bonding. However, wire-bond connections have the disadvantages of having both a larger footprint, which results in a larger substrate and a necessarily less compact MCM, and greater lead inductance, resistance, etc, which result in an unavoidable degradation in electrical performance. Furthermore, the wire bonds must be made one at a time, which even with the high-speed wire bonders available today can make assembly of MCM designs with many I/O relatively time-consuming compared to TAB or flip chip MCM assembly.

On the other hand, TAB bonding has the advantages of both a smaller footprint and of being partially a batch assembly process in which many TAB connections may be bonded at one time. However, TAB assembly generally requires different tooling for each IC design, which is expensive. Furthermore, in common with wire bonding, TAB is limited to the interconnection of ICs with perimeter array I/O configurations, which inevitably requires higher I/O pitches at lower overall I/O densities than can be used with area array I/O configurations. In addition, TAB bond interconnections manifest both higher capacitance and greater parasitic inductance than do flip chip solder interconnections.

As has long been recognized, flip chip bonding provides the best performance at the highest I/O density for either perimeter or area I/O arrays. Furthermore, it is inherently a batch assembly process which facilitates high-speed, high-throughput applications. However, for a variety of reasons flip chip MCM assembly is usually considered to be the most expensive. This is especially true for high-performance MCM designs which often use multilayer cofired ceramics (MCM-C) or deposited thin-film ceramic or silicon substrates (MCM-D) as the interconnection substrate.

The need for increased interconnection density has led to the development of a platform consisting of a composite of substrate, attachment, and packaging that we call *microinterconnect technology*, which exploits the unique capabilities of the silicon MCM-D interconnection architectures. This chapter will describe the design, development, and optimization of these manufacturing processes for the cost-effective large-volume assembly of flip chip silicon-on-silicon (hereafter simply "silicon") MCM-D tiles through the use of a custom solder technology and high-speed automated assembly. As a direct result of this effort, microinterconnect assembly techniques have been successfully demonstrated and factory-qualified for the mass production of mixed-signal silicon MCM-D tiles and packages for applications where cost considerations (i.e., minimum overall system cost in a large-volume production environment with limited capital expense) are paramount.

5.2 Assembly

In classic flip chip assembly as represented by the controlled-collapse chip connection (C4) technology developed originally by IBM, both the IC chips and the interconnection substrate to which they are to be flip chip–soldered are furnished with matching arrays of solder-wettable I/O pads. Usually the pads on the chip are also furnished with solder "bumps," although for various reasons the solder bumps may be placed on both the substrate and the IC chip, or even on the substrate alone.

This solder is typically of a high-lead/low-tin alloy—say, 97/3 PbSn or thereabouts—with a melting point well above that required for a lead/tin eutectic or near-eutectic alloy such as would be used for making printed wiring board (PWB)-level interconnections. Detailed descriptions of how these solder-wettable metallizations and solder bumps are fabricated are available in a number of publications[1,2] and therefore need not be repeated here. It is worth noting, however, that no matter how it is done, solder bumping is inevitably a batch process, with many wafers of many die each being processed at the same time.

At this time it is also instructive to review the conventional flip chip assembly processes used with bumped silicon components, and to compare them with the AT&T microinterconnect printed-paste process developed for the flip chip assembly of unbumped silicon.

5.2.1   Conventional flux, place, and reflow flip chip assembly

In classic C4 assembly a sticky flux is applied to the substrate prior to flip chip placement of the bumped chips. This flux provides the adhesion needed to hold the chips in place while the rest of the substrate is populated with bumped chips and continues through the reflow process to the point when the solder bumps melt to form solder microjoints. (These are termed *microjoints* because at 150 μm in diameter and 50 to 75 μm in height they are an order of magnitude smaller than the solder joints usually encountered in conventional surface-mount technology. However, in this chapter they are hereafter referred to simply as *solder joints*.) The flux is also activated during the reflow process to facilitate solder wetting of the substrate I/O pads and the solder bumps on the chips.

The flux, place, and reflow process can be used with any substrate material: silicon, ceramic, or laminate. Furthermore, like all such reflow-based soldering processes, it supports the surface-tension-driven realignment during reflow of any chips that were partially misaligned during placement. This effectively reduces the required placement accuracy to only a fraction of a wettable pad diameter, often an important advantage when compared to the use of an anisotropic conducting adhesive film as the connecting medium between the chips and the substrate.

5.2.2   Tack, flux, and reflow flip chip tile assembly

In conventional AT&T microinterconnect flip chip assembly, solder-bumped chips are "tacked" to the substrate prior to application of the flux. This tacking step involves the simultaneous application of heat and pressure, and provides a partially bonded joint that will easily withstand subsequent processing. Once the interconnection substrate has been fully populated, the entire substrate is fluxed and reflowed. This tack, flux, and reflow tile assembly process is applicable to assembly onto both silicon and ceramic substrates, but has not been applied successfully to laminate substrates. Additional details of the fabrication of both the solder-wettable pad metallizations and the solder bumps on the chips as developed for this process are described in Ref. 3.

Whenever possible, the flip chip assembly process should be carried out on an entire silicon wafer or ceramic card composed of many interconnection substrates—*not* on a single substrate. This process makes optimal use of a programmed, high-speed automated tacking or placement machine. In order to make maximum use of batch processing and assure maximum tile throughput without compromising yield, this whole-wafer tacking should be followed by whole-wafer fluxing, whole-wafer reflow, and whole-wafer cleaning prior to singulation into individual MCM-D tiles.

In the tack, flux, and reflow tile assembly process the flux can be either a conventional rosin flux in a volatile medium such as isopropanol, or a relatively "nonvolatile" immersion flux. The latter flux works on both silicon and ceramic substrates, and is optimized to assure (1) robust solder wetting, (2) vigorous realignment of even marginally aligned chips during reflow, and (3) ease of cleaning afterwards. The x-ray photographs shown in Fig. 5.1 illustrate the power of the immersion fluxed reflow process with three examples. Here both the chips and the substrates were furnished with solder bumps so as to make them equally visible in the x ray. Consequently, it can easily be seen in the left-hand photographs of Fig. 5.1 that each chip was intentionally severely misaligned during the initial tacking phase of the assembly process. After immersion fluxing and reflow, the assemblies were x-rayed a second time. The results are shown in the right-hand photographs. It is clear from the x-ray photographs in Fig. 5.1 that in every case the previously misaligned chips were fully realigned during reflow.

In fact, the last two photographs of Fig. 5.1 show an example of a misaligned chip and substrate sample with four solder bumps on square pads which was tacked with *only a single pair* of bumps in proper contact, yet the chip fully realigned during reflow. Here, thanks to the enhanced mobility provided by the immersion flux, the energy involved in minimizing the surface area of just this single molten solder joint provided enough force to initiate complete realignment of the entire chip! Extensive experience has clearly demonstrated that with immersion fluxing, so long as a chip is tacked with at least a slight contact between bumps (or bumps and pads), and so long as no chip bump is in contact with the wrong substrate bump (or pad), complete reflow realignment will occur every time. (Needless to say, the immersion flux reflow process is equally effective when only the chip is solder-bumped. It just can't be seen very well in an x ray.)

Self-Alignment of Flip-Chip Solder Joints

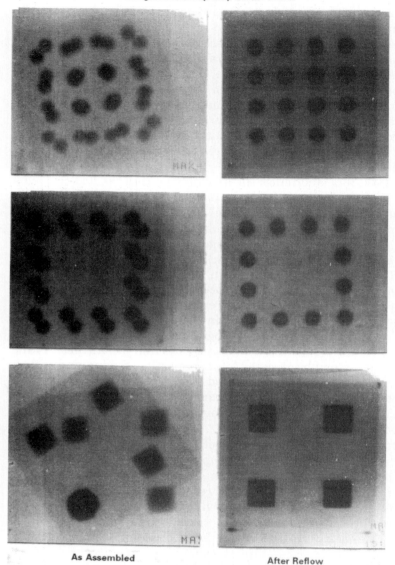

As Assembled   After Reflow

Figure 5.1   x-Ray photos of flip chip solder-bump realignment during reflow.

5.2.3   Flip chip surface-mount printed solder assembly

While it is always necessary to furnish the I/O pads on both the chips
and the substrate with solder-wettable metallizations to fit them for

1. Align Stencil And Print Paste

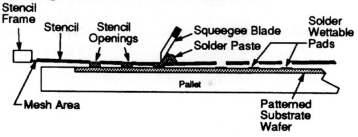

2. Align And Place Chips

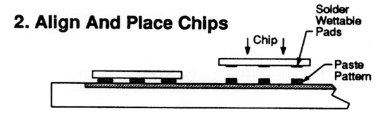

3. Reflow To Form Joints

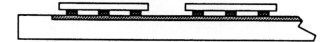

Figure 5.2 The flip chip SMT print, place, and reflow tile assembly process.

solder-based flip chip assembly, it is *not* necessary that either be furnished with solder bumps. As will be explained, the solder can be furnished as part of the flip chip assembly process itself—that is, it can be printed on the substrate just prior to the chip placement process.[4] Figure 5.2 shows a step-by-step representation of this novel AT&T microinterconnect technology printed solder MCM assembly process. (For brevity, we will refer to this as the *printed solder* assembly technique.) The basic idea is to exploit the existing assembly infrastructure for high-speed, low-cost assembly by adapting the well-developed electronic manufacturing capabilities of state-of-the-art *surface-mount technology* (SMT) to the assembly of silicon MCM-D tiles. SMT, which was developed to assemble packaged IC chips, discrete components (resistors, transistors, capacitors, and diodes) and connectors, etc. onto printed-wiring boards (PWBs), is a mature, high-speed, automated technology for increasingly fine-pitch, high-interconnection-density electronics assembly.

However, in order to achieve a high-yield printed solder process for the assembly of silicon MCM-D tiles it was necessary to recognize three factors that limit the interconnection density that can be accommodated by the conventional application of even fine-pitch SMT assembly techniques and components. The first is the relatively large footprint of SMT packaged IC chips with their perimeter arrays of compliant leads, especially when compared to the footprints of the far smaller bare, unleaded die used in the MCMs. The second is the influence of what is referred to as component lead *coplanarity*. Lead coplanarity reflects the geometric variability in the highly compliant metal leads on a high-I/O, fine-pitch SMT package, and the difficulties associated with its resolution drive much of the current industrywide interest in BGA packaging. The third is the limited printing and performance capability of commercially available fine-pitch SMT solder paste when applied to a silicon MCM-D substrate instead of a PWB. As a result, the use of SMT facilities, components, and methodology has not thus far been seriously explored, let alone exploited, for high-density silicon MCM-D assembly applications.

In the printed solder flip chip tile assembly process the first and second considerations—component size and compliant lead coplanarity—are resolved through the use of leadless, bare die with solder-wettable I/O pads, all of which lie well within the outline of the die itself for a minimum-area footprint. Furthermore, since they all lie on and are affixed to a common planar die surface, they certainly manifest optimal coplanarity (albeit with little lead compliance). The third consideration—the limited printability and performance of commercial fine-pitch solder pastes and reflow processes—has been resolved through two steps:

1. Reconfiguring the bare die or IC chip I/O pad perimeter arrays into area arrays by *rerouting* (also called *redistribution*)

2. Redesigning the solder paste for flip chip printed solder assembly onto a patterned silicon substrate wafer

As mentioned earlier, and as described in detail in Sec. 5.3.2, in order to be used for flip chip soldering, the bare die or IC chips must be processed, while in wafer form, to provide solder-wettable I/O pads in the place of each wire-bond I/O pad. As part of this refinishing, and to assure a robust printing process, the I/O bond pads are rerouted from the usual perimeter wire-bond pad array to an area pad array. This increases the pitch from as little as 0.13 mm (5 mils) to 0.4 mm (16 mils) or larger (depending on chip size and the number of I/Os)—well within the intended range of state-of-the-art fine-pitch SMT for the printing process—but without complications of the occasionally bent lead to confound the assembly process, as can occur when placing fine-pitch SMT packaged components.

As is described in detail in Sec. 5.3.3, the second step taken to overcome the limit of printability and performance of commercial fine-pitch SMT solder pastes was to develop a custom flip chip solder paste specifically for the printed solder MCM tile assembly process.

The printed solder assembly sequence.   In microinterconnect printed solder tile assembly a custom solder paste is stencil-printed onto a silicon wafer composed of many MCM interconnection fabric substrates, not all of which need to be of the same design, using a conventional SMT printer. Each substrate incorporates appropriately patterned arrays of solder-wettable I/O pads for all the unpackaged flip chip die required to complete the tile. Silicon chips with matching pad arrays are then automatically placed face down onto the paste-printed substrates in preprogrammed locations using SMT-type placement equipment with appropriate bare-die handling facilities. This process is continued until the silicon substrate wafer is fully populated. Subsequently, the entire wafer is placed in an inert-atmosphere reflow facility and heated to a temperature that is slightly above the melting point of the solder paste to achieve a simultaneous batch reflow of all the solder paste deposits holding the chips. In this step the molten solder wets the I/O pads on the chips and substrates and consolidates into individual joints, thereby bringing the chip pads into precise alignment with the substrate pads and forming uniform, regular-shaped solder joints in the process.

On cooling, the molten solder forms into arrays of electrical, mechanical, and thermal joints connecting the silicon interconnection fabric on each substrate with the IC chips required to make an MCM tile. Printed solder assembly is usually done on an entire wafer of patterned substrates at one time in order to achieve maximum assembly efficiency. (Naturally, smaller sections may be assembled when needed for placement alignment trialing and/or various process development experiments.)

Figure 5.3 shows a photograph of a fully populated and reflowed wafer of 148 tiles, each having six chips of four different sizes. This assembled MCM-D tile wafer is ready to be cleaned, tested, and diced to provide individual MCM-D tiles. These are then packaged and further tested as required. As will be described in Sec. 5.8, finished microinterconnect technology MCMs consist of silicon MCM-D tiles which are die- and wire-bonded to an intermediate-lead frame board and enclosed in bumped, non-hermetic BGA packages which can be SMT reflow–soldered directly onto printed-wiring boards. Naturally, there are alternative packaging options, but they are not described in this chapter.

Figure 5.3 An assembled six-chip MCM wafer.

Substrate wafer and bare-die handling for printed solder assembly. In
many ways the flip chip printed solder MCM-D tile assembly process
shown in Fig. 5.2 follows the steps used in standard SMT assembly, but
with important innovations to accommodate the use of silicon wafers in
the place of multilayer PWBs, and bare silicon die in the place of packaged chips, components, etc. Since the usual surface mount assembly
equipment is designed to handle relatively large, thick, rectangular,
and rugged PWBs, special pallets have been designed to hold (using
vacuum if available) the much smaller, thinner, round-outline, and
more fragile silicon substrate wafers throughout the assembly process.
Each pallet must be provided with a single suitably shaped (round with
appropriate flats for orientation) recess designed to correctly position
and support a silicon substrate wafer for both printing and placement.
In addition to defining the x, y, and θ wafer orientation, these recesses
must be machined to assure that the top surface of a silicon substrate

wafer lies only a mil or so above the top surface of the pallet. Furthermore, whenever the pallet is aligned in the printer or SMT placement machine, the top surface of the wafer should lie in the same plane as would the top surface of a conventional PWB.

Final precise wafer alignment can be accomplished using standard SMT vision system technology as developed for circuit-board assembly. Most of the initial process development for the microinterconnect printed solder flip chip MCM tile assembly technique was carried out using standard, unmodified SMT printing and placement machines that employ vision systems that seek out standard 50-mil alignment fiducials typical of those used on PWBs. Consequently, the MCM-D substrate wafers are always furnished with these fiducials. Such fiducials are, of course, very large features in the silicon IC world. In addition, in order to facilitate electrical test of the assembled MCM tiles in wafer form, additional smaller fiducials of appropriate shapes and sizes must also be provided. Taken together they form an ensemble of fiducials which must be located on the substrate wafer so as to both optimize alignment sensitivity and minimize the loss of usable substrate circuit sites that can be fabricated on a given wafer.

Figure 5.4 shows wettable metal I/O pad layout (on which the solder is printed) of the six-chip electromechanical test vehicle substrate wafer shown in Fig. 5.3. This particular example was developed to match an early prototype of Partner™ Telephone MCM product. However, it used "dummy" silicon (both the substrate and the chips) with daisy-chain circuitry. In this version the individual substrates and alignment fiducials were located so as to maximize the number of usable good substrates that could be patterned by the step-and-repeat process. In this example it was possible to achieve 148 usable substrate sites with four printed solder assembly alignment fiducials located at (approximately) 2, 4, 8, and 10 o'clock positions relative to the major flat.

As described in Sec. 5.4.2, successful assembly trials were run using bare die loaded face down in pocket tape (also know as *cavity-reel* tape). This is an inversion of the usual orientation of the die as they come off the tape and dicing ring and represents a nontrivial change from the die orientation usually associated with handling die for conventional die and wire-bond packaging and assembly. In addition, a specific tape with appropriate pocket dimensions is required for each fairly narrow range of die sizes. Consequently, since bare die are not in any way standardized in size as are, to a large extent, packaged devices, providing the needed variety of tape sizes required for a product with a wide range of die sizes could be expensive. However, at large production volumes none of these issues represent a major challenge or cost disadvantage. In fact, since pocket tape and reel handling is compatible with many types of SMT placement equipment, it readily

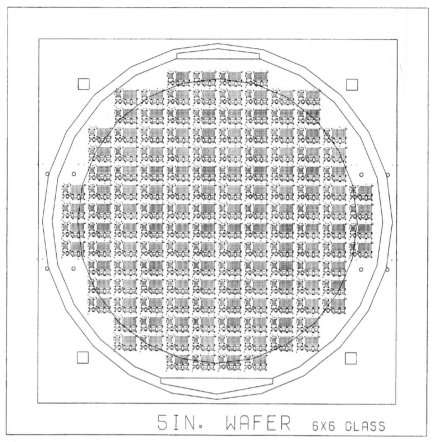

Figure 5.4 The wettable metal I/O pad layout for a six-chip MCM wafer.

supports a highly reliable automated assembly technology for the rapid population of a full wafer—if the flipping is done correctly.

The need to invert the die as they are transferred from the tape and dicing ring to the pocket tape can be deferred to the placement stage, but this requires a nontrivial modification of the machine used to carry out the placements so as to include chip flipping along with the placement. However, if the placement machine can be fitted to pick from waffle trays, as many are, this problem can be eliminated completely through the use of special double-sided "flip chip" waffle trays, which are a variant of the single-sided waffle trays used for bare-die handling throughout the IC industry. This is the approach used in most of the microinterconnect factory assembly trials conducted at AT&T's Shreveport (La.) facility, for which the special flip chip waffle trays were made available by the supplier at no added cost.

Unlike conventional single-sided waffle trays, these flip chip waffle trays were molded with appropriate cavities on both their *top* and *bottom* surfaces. Consequently, they can be loaded from the tape and dicing ring with the die oriented face up using a standard automated chip sorter. After filling, the flip chip waffle tray is removed and a second matching tray stacked on top of it aligned so that when the trays are turned over and tapped robustly a few times, they would affect a batch inversion and transfer of all the die at one time.

This invertible waffle tray approach to chip flipping obviates the need to secure a special chip sorter to pick each die from the tape and dicing ring, individually flip it, and place it into the appropriate tray cavity or reel tape pocket. However, like pocket tape, it still requires the use of a specific waffle tray version for each rather limited range of die sizes to work correctly. So far it appears that the best alternative, especially for short runs appropriate to prototype production, is to use a novel die or chip tray which by its design can significantly reduce the number of tray types required to handle die of widely differing sizes.

This commercially available product, which goes by the trade name of GEL-PAK™, is more expensive than a conventional waffle tray. Fortunately, GEL-PAKs™ are also reusable. Each consists of a pristine, highly compliant polymer membrane stretched over a mesh and mounted in a standard 2×2-in or 4×4-in chip tray. Both the tackiness of the membrane to which the die adhere, and the coarse three-dimensional (3D) texture of the much stiffer underlying mesh are available in several variations to suit the application. Since the dies are "released" by the application of vacuum to the underside of the tray (which allows the membrane to pull it away from the die surface over most of its area so as to release it for pickup), it is also quite practical to transfer dies from one tray to another, conveniently inverting or "flipping" them while simply switching the vacuum. Furthermore, since GEL-PAKs™ have no pockets or cavities, two or three types of them can accommodate a wide range of bare-die sizes and handling conditions. The trick is to be able to pick the flipped die from the tacky membranes at high speed during the assembly process.

Stencil design for printed solder assembly. The solder paste is printed directly onto the silicon substrate wafers through 3-mil-thick stainless-steel stencils with openings somewhat larger than the 6-mil-diameter wettable metallized pads. This assures that even if slightly misaligned, the printed paste will completely cover every pad. At the same time, the oversized print provides sufficient solder volume to assure, after reflow, joint heights of 2.0 to 2.5 mils. Such joint heights are required to facilitate robust cleaning and effective underfill during subsequent encapsulation. Thanks to the optimized wetback capabilities of the flip chip solder paste during reflow, there is little undesirable "nonconsolidated reflow" (NCR) solder ball formation, and only an insignificant trace of each solder deposit consolidates into a single, large-volume joint wholly affixed to both the chip and substrate solder pads.

Die placement programming. Files derived from the CAD (computer-assisted design) data for the individual MCM substrates and describing their arrangement on the substrate wafers are adapted to generate placement data for each chip to be placed by the SMT placement machine. This process involves using the data in the design files to identify each chip or die and its precise x, y, and θ location with respect to the origin on the substrate wafer relative to two of the four alignment fiducials described above. In the initial phases of the microinterconnect technology printed-paste assembly process development program this information was supplied in a flat ASCII file which could be translated to a placement file using a PC (personal computer)-based software package supplied by the placement machine's vendor. The individual die sizes and other relevant information were fed into an individual part description (PD) file for each type of die to be placed. Since there were no peripheral leads to be concerned about, these portions of the PD files were very similar to those used for conventional surface-mountable discrete devices and BGA packages, which are also leadless. The PD files also contained information relative to certain operating aspects of the machine, such as the correct vision algorithm, the type and size of each feeder, and the "carrying data" (nozzle size, prerotation, traverse rates, etc).

The placement sequence can be constructed in any order to achieve the desired assembly process. Programs written in such a way as to build one complete MCM-D tile at a time to allow for interruption of the assembly at almost any step while maintaining the integrity of individual modules. This tile-by-tile assembly sequence greatly facilitates the setup process, but it isn't necessarily the best approach. The alternative approach, which would be more efficient from a machine optimization standpoint, would have been to place all similar devices or chip codes sequentially. This preferred method can be used to minimize cycle time by reducing the time devoted to both (1) pickup and die travel and (2) pickup nozzle changes for different die sizes. Obviously, any program which would allow the operator to readily shift from a tile-by-tile assembly sequence to a chip code by chip code assembly sequence would be the most useful.

Unfortunately, not every MCM site on the silicon substrate wafer can be relied on to function as it should. Consequently, prior to transfer to the MCM-D tile assembly facility it is necessary to automatically test and ink-mark every wafer so as to clearly identify defective substrate sites. Fortunately, most SMT-type placement machines have a vision system–based "block skip" feature which searches the wafer being assembled for ink marks in order to determine which substrate sites should be skipped rather than populated with chips. In this situation, the tile-by-tile chip placement method most readily facilitates assembling only known good sites and was the method used for the first flip chip SMT placement demonstrations, including those conducted at the AT&T Shreveport factory. However, machine control software is now being developed to initially generate, store, and efficiently read a wafer map as often as needed. This arrangement will support the timely identification of bad sites even for a chip code by chip code placement method.

The reflow process, chip realignment, and joint properties. Fully assembled wafers are reflowed in an inert atmosphere IR belt furnace or convection oven using thermal profiles developed for the flip chip solder paste. This reflow process is very important because, as mentioned earlier, it both activates the flux and melts the solder, etc. Furthermore, since an immersion flux cannot be used with a printed solder flip chip assembly process as it can with a bumped chip tack, flux, and reflow assembly process, the flip chip paste itself must assure the accurate realignment of any misaligned chips during reflow.

Finally, the measured electrical resistances of the flip chip solder joints assembled using the printed-paste technique generally fell in a range between 0.1 Ω per joint and 6 Ω per joint, with properly processed and reflowed joints averaging around 1 Ω each. In addition, the measured averaged solder joint shear strengths fell in a fairly narrow range of 41.5 g per joint to 77.6 g per joint, which is satisfactory for this combination of pad metallization, solder alloy, and joint geometry.

5.3 Building Blocks

Silicon MCM-D tiles may be thought of as comprising three components or building blocks:

1. The ICs themselves

2. The silicon substrates

3. The medium used to connect these first two together

In microinterconnect MCM-D designs this medium is solder, and as described in the earlier sections on assembly, this solder may be provided in the form of solder "bumps" already present on the ICs (or even the substrates), or added during the assembly process by printing, as in the present printed solder flip chip MCM tile assembly process. This section describes the unique aspects of each of these three building blocks as developed for low-cost, high-volume microinterconnect technology assembly, beginning with the substrates.

5.3.1 Substrates

One of the basic foundations of microinterconnect technology is the extensive use of the well-developed infrastructure for silicon fabrication.[5–7] Some of its key features are

1. Aluminum metallization
2. Polyimide dielectric
3. Integrated thin-film passive devices
4. Bipolar devices

Figure 5.5 shows representative cross sections of the more important varieties of substrates built from these basic elements. The high-performance digital structure was the first structure to be developed, indicative of the technology's origins in high-speed computing applications. However, as industry trends have moved the focus of MCMs toward higher-volume, lower-cost applications, the microinterconnect silicon substrate structure has evolved to meet these new demands.[8,9]

Several features are common to all these structures. The use of aluminum circuit metallization avoids some of the compatibility problems that copper/polyimide interconnection structures have encountered. Although this results in slightly higher line resistance than copper, it has the considerable advantages of simplicity, well-developed deposition and patterning technologies, and low cost. In the present technology, typical features are 20-μm lines and spaces with a 1- to 2-μm metal layer thickness.

The high-performance digital structure uses a highly doped silicon substrate as a ground plane, separated from the power plane by a layer of thermally grown silicon dioxide. Early structures attempted to use the high capacitance of a thin oxide layer to achieve distributed capacitance between these two planes. These structures, however, suffered from high yield loss from short circuits in the thin dielectric. In the present technology the current minimum oxide layer thickness is 1 μm, which results in a robust structure. Added surface-mount capacitors are used in applications that call for high decoupling capacitance between power and ground. Thin-film resistors are formed directly on the oxide using sputtered TaSi films. These resistors are used for transmission line terminations and for pullup and pulldown resistors. Two signal layers of aluminum metallization complete this structure.

The integrated passive substrate is mainly a stripped-down version of the high-performance digital one. It eliminates the power plane, and has only two layers of aluminum. It is mainly intended for low-cost, mixed-signal applications, but is often used for some digital ap-

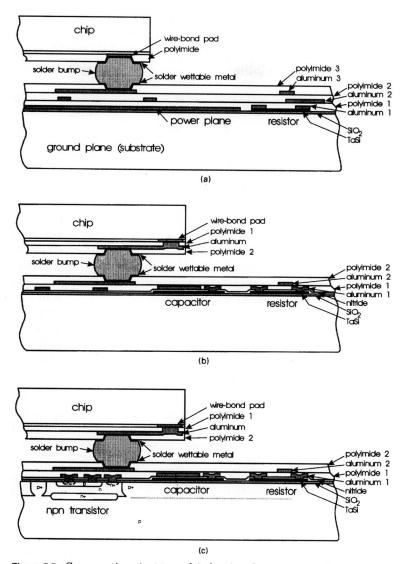

Figure 5.5 Cross sections (not to scale) showing the structures of various microinterconnect silicon substrate types: (*a*) high-performance digital; (*b*) integrated passive; (*c*) active substrate.

plications as well. Miniaturization is one of its principal benefits. Many digital applications that can take advantage of this do not require the very low inductance power supply connections offered by the power and ground planes. One difference is an added silicon nitride dielectric layer that allows capacitors to be formed using the resistor layer as the bottom electrode. This ability to integrate large numbers of passive components in the MCM substrate is a major advantage in terms of both cost and performance.[7]

The advantages of component integration are taken one step farther in the active substrate structure. This basically adds integrated bipolar components to the low-cost, mixed-signal structure. Because the additional substrate processing adds significant cost to the substrate, the technology is cost-effective only in applications that require the extensive use of the bipolar components. In certain niche applications, however, this added functionality can be important.[7]

Cost considerations have driven the fabrication technology in all these structures more and more toward the use of photodefinable polyimide. The effect of this trend is to favor the use of thinner dielectrics. Typical polyimide thicknesses for photodefinable films used in these substrates are 3 to 5 μm. Thinner dielectric layers result in some interconnection speed degradation. The impact that this degradation has on most digital applications is insignificant, especially in view of the cost benefits.[8]

5.3.2   Chip preparation

A distinctive feature of flip chip assembly is the need for chip preparation while the chips are still in wafer form. A disadvantage of this is the weakness of the infrastructure supporting the documentation and procurement of ICs in wafer form. A key advantage, however, is that much of the processing involved with the assembly and interconnection of chips is done in a batch mode. The number of chips on a typical IC wafer ranges from about 100 for large, usually digital chips, to several thousand for small, usually generic building block ICs like operational amplifiers and logic gates. Process steps performed on chips in wafer form are done simultaneously to every IC on the wafer, which results in very high efficiencies.

The process steps used to prepare ICs for printed solder flip chip assembly and for the more conventional tack, flux, and reflow flip chip assembly are illustrated in Figure 5.6. As pointed out above, printed solder assembly requires an additional layer of aluminum to redistribute the solder-bump locations into an area array, increasing the spacing between the bumps. No solder is deposited in this process—only a solder-wettable metal pad or base, usually of Cr/CrCu/Cu/Au or Ti/Ni/Au.

On the other hand, for microinterconnect applications using solder-bumped ICs and the tack, flux, and reflow assembly technique, the flip chip solder bumps are usually located directly over the wire-bond pads. Consequently, there is no need for a redistribution layer. Naturally, a solder-wettable base metallization is still required for each bump, since

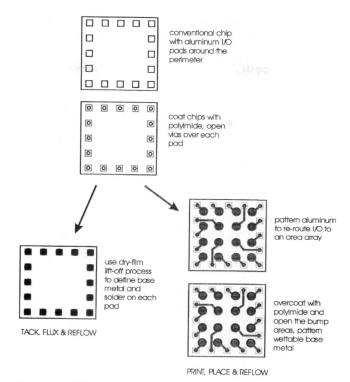

conventional chip with aluminum I/O pads around the perimeter

coat chips with polyimide, open vias over each pad

pattern aluminum to re-route I/O to an area array

use dry-film lift-off process to define base metal and solder on each pad

overcoat with polyimide and open the bump areas, pattern wettable base metal

TACK, FLUX & REFLOW

PRINT, PLACE & REFLOW

Figure 5.6 Chip preparation steps.

an aluminum wire-bond pad is seldom a suitable surface for solder. The solder is deposited directly on this base, typically by evaporation. In the AT&T microinterconnect technology this evaporated solder is patterned using a lift-off process with a thick layer of dry-film resist. The solder thicknesses achieved using this process typically run around 50 to 60 μm. An identical process can be used to deposit and pattern corresponding bumps on the substrate if required.

5.3.3 Solder paste for printed solder flip chip assembly

In addition to the I/O redistribution described above, a second but equally important step in the printed solder assembly process, taken to overcome the limited performance of commercial fine-pitch solder pastes in such an application, is to use a flip chip solder paste and reflow process optimized together for silicon MCM-D tile assembly. In general, all solder pastes are combinations of a viscous liquid flux and a solder alloy in powder form. Any flux appropriate to use in a solder paste for electronic assembly must not only provide the chemical action required to assure wetting of the metal surface by the molten solder but also be able to hold a lot of solder powder, to adhere to a patterned silicon wafer, but not to the metal stencil, and to not contaminate the electronics in any way that might cause long-term reliability problems. The AT&T flip chip solder paste uses an AT&T proprietary YD-series flux formulated to also

1. Assure optimized printing on silicon substrate wafers
2. Hold the die in place throughout the assembly process
3. Exhibit minimum spreading during assembly
4. Facilitate excellent solder wetting
5. Form strong, low-resistance solder joints

Furthermore, it is designed to be reflowed in an inert atmosphere during which it robustly facilitates:

1. Optimum reflow realignment of each and every chip
2. Consistent formation of well consolidated joints
3. Few, if any, nonconsolidated reflow (NCR) solder balls
4. Few, if any, bridges (short circuits)
5. Few, if any, open circuits
6. Minimum reflow voids of any kind

Finally, it is designed to be readily cleanable after reflow to facilitate subsequent die and wire bonding of the tile into the MCM package for next-level interconnection, as mentioned in Sec. 5.2.3, subsection on printed solder assembly sequence, and described in detail in Sec. 5.8.

In choosing the solder alloy, it is important to recognize that the solder used to connect the die to the silicon substrate must be compatible with the soldering technology used to attach the packaged MCM to the appropriate working parentboard, or if it is to be applied before or during the assembly of other components, with the soldering technology used to attach these other components. This may be either through-hole/wave soldering or surface-mount reflow soldering. This means that the flip chip solder joints must not melt or soften during any subsequent next-level interconnect reflow process. Such a situation might cause the tile to fail by coming apart when the MCM is ultimately soldered to a PWB. Consequently a Pb-free solder alloy, 95/5 SnSb, was chosen for the flip chip assembly paste. This alloy melts at a 240°C, well above the 183°C melting point of the eutectic or near-eutectic SnPb solder alloys commonly used for board-level electronic assembly.

Since this higher melting point alloy is already widely used for electronics assembly in the industry, it is readily available in powder form at modest cost for combination with an AT&T YD-series flux into a robust solder paste. In addition, it has respectable if not outstanding fatigue characteristics. Moreover, the use of a silicon-chip on silicon-substrate MCM-D tile architecture essentially eliminates the global thermal expansion mismatch which induces most of the cyclic fatigue damage in solder joints, anyway.

5.4 Printed Solder Flip Chip Assembly Research and Development

Since flip chip surface-mount assembly technology is in its infancy, it is appropriate to review some of the R&D carried out prior to and concurrently with the product design effort involved in the first application: the AT&T Partner™ Telephone mentioned above. In addition to the design of prototype silicon components and the development of the solder paste/stencil printing processes discussed in Sec. 5.3, a variety of flip chip silicon MCM-D tile vehicles were designed, assembled, and evaluated, as described below.

5.4.1 The printed solder flip chip assembly process and solder paste development test vehicles

Most of the initial printed solder flip chip assembly process and paste development experiments were run using electromechanical test vehicles (EMTVs) with simple daisy-chain circuits rather than with working (and relatively expensive) IC chips and active silicon MCM-D substrates. The first two EMTVs had four 3.63×1.9-mm dies per tile each, and were designed to provide a variety of I/O pad array experiments with pad pattern pitches that ranged from 0.34 to 0.4 mm (13.5 to 16 mils). With four chips per tile or module, this required forming 71 quality solder joints per tile, and over 7400 solder joints per wafer. Table 5.1 presents a listing of the substrate dimensions and I/O solder joint counts for all of the EMTVs.

The third EMTV was designed to match an early α_1 prototype of the Partner™ Telephone MCM, and included chips of the same sizes and the same I/O configurations as the working silicon version. These included pitches of slightly more than 0.4 mm (16 mils) and daisy-chain circuits. Each substrate wafer had 112 usable MCM sites with six chips or die of four differing types on each. The die ranged in size from a small dummy differential operational amplifier chip of 1.45×1.22 mm with only eight I/Os to a relatively large dummy microprocessor chip of 5.00×2.85 mm with 39 I/Os giving a total of over 11,530 joints per wafer. Table 5.2 presents a listing of the dimensions of all the die sizes used on these prototypical six-chip EMTVs. The Partner™ α_1-prototype active silicon versions of these chips and sub-

TABLE 5.1   Silicon Fabric Substrate Characteristics

Test vehicle	Length, mm	Width, mm	Thickness, mm	Substrates per wafer	Joints per substrate	Joints per wafer
EMTV-1	8.94	8.41	0.610	104	71	7,384
EMTV-2	9.20	6.88	0.610	140	71	9,940
EMTV-P	10.50	7.52	0.635	112	103	11,536
EMTV-P+	9.40	6.65	0.635	148	103	15,244
α_1 prototype	10.50	7.52	0.635	112	103	11,536
α_3 prototype	9.40	6.65	0.635	148	103	15,244

TABLE 5.2   Electromechanical and α-Prototype Die Characteristics

IC (die)	Length, mm	Width, mm	Thickness, mm	Thickness (0), mm	Joints per die	Die per tile
Micropro-cessor	5.00	2.85	0.635	0.330	39	1
Latch	2.00	1.73	0.635	0.356	20	1
Large op amp*	1.87	1.73	0.635	0.381	14	2
Small op amp	1.45	1.22	0.635	0.381	8	2

*Operational amplifier.

strates had exactly the same I/O arrays and were of very nearly (within ± 20 μm) the same sizes except that they were significantly thinner (on the order of 0.33 to 0.38 mm vs. 0.635 mm for the dummy chips).

A final six-chip EMTV was designed to model the actual Partner™ Telephone α_3- and ß-prototype MCM-D tiles and was configured to simplify routing the active silicon substrates and improve overall electrical performance. Figure 5.7a shows the daisy-chain circuits used in this test vehicle, and Fig. 5.7b shows the active silicon substrate version for comparison. As shown in Table 5.1, the substrates for these were only 9.5×6.75 mm, which permitted the patterning of 148 per silicon substrate wafer. Table 5.1 also gives the joint counts per die and the die counts per module or tile. As shown in Table 5.1, each assembled substrate wafer had 148 six-chip tiles and over 15,200 solder joints.

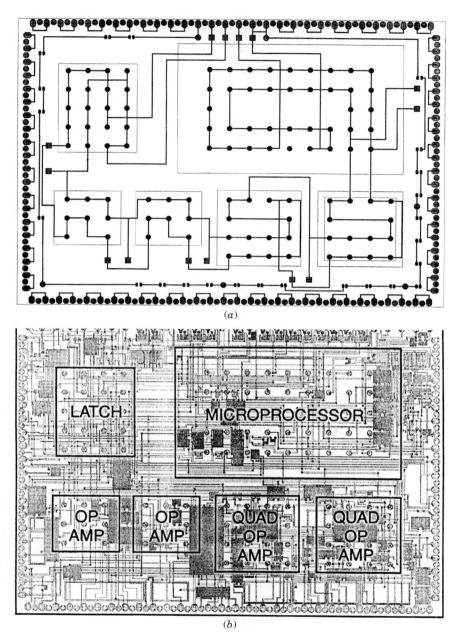

Figure 5.7 An (*a*) electromechanical test vehicle layout and (*b*) matching final prototype working silicon substrate layout.

5.4.2   Printed solder assembly and solder paste development experiments

The first printed solder assembly experiments were run on the four-chip EMTVs using an early YD-series solder paste formulation, and the inverted die in pocket tape technology described in Sec. 5.3 (subsection on substrate wafer and bare-die handling). In most of these initial experiments whole wafers of over 100 tiles each were assembled in a little more than 34 min. This remarkable figure represents the elapsed time as measured from alignment of the palletized silicon substrate wafer in the printer through die placement from cavity or pocket tape into the printed paste by the SMT placement machine to the end of reflow in the inert-atmosphere belt furnace. Such performance is an example of the high-throughput automated process required for low-cost, high-volume MCM assembly.

In this initial, unoptimized experiment 416 dies were placed in slightly over 20 min, which represents less than 3 s per chip! Furthermore, the 34-min input-to-end cycle time represents an overall assembly rate of around 5 s per chip, or 20 s per four-chip tile! X-ray and two-probe continuity testing gave an assembly yield of 96 percent on a per module basis, and well over 99.9 percent on a per-joint basis.

Figure 5.8a shows an x-ray photograph, taken prior to reflow, of six chips placed in solder paste on one of the more prototypical Partner™ Telephone α_1 EMTVs, and Fig. 5.8b shows the same chips after reflow. The spread of the compressed paste under the chips after placement is clearly evident in Figure 5.8a. If everything worked correctly, most of the time the solder was observed to have consolidated onto well-formed solder joints centered on the pads as shown in Fig. 5.8b. Unfortunately, this was not always the case, and early flux and paste limitations required frequent cleaning of the stencil and still resulted in a variety of defects. Consequently, numerous experiments were run to improve solder paste performance and refine the assembly technology in a successful effort to identify, understand and, if necessary, suppress the observed defects.

While many of these defects occurred infrequently or were of a type which would have little effect on the yield as defined by the continuity tests, most could be readily seen on the x-ray, including some which might well adversely affect MCM reliability. These defects included the formation of

1. Undersized or irregular solder joints

2. Large trapped volatile reflow voids

3. Unusually large residual NCR solder balls adjacent to certain joints

Figure 5.8c shows an x-ray photograph (taken prior to reflow) of an early EMTV on which some of the solder paste may be seen to have

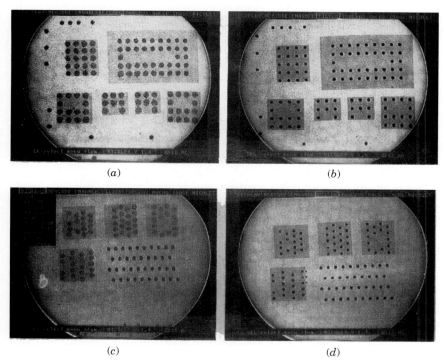

(a)   (b)

(c)   (d)

Figure 5.8 x-Ray micrographs of electromechanical test vehicles taken before and after reflow.

been squeezed out from under one of the die during the placement process. As a consequence, some of molten solder failed to consolidate during reflow and, as shown in Fig. 5.8*d*, left large NCR solder balls on the surface of the substrate next to the chip. At the same time, the adjacent solder joints, deprived of a significant portion of their solder, appear to be much smaller than the rest and were labeled "wimpy." It quickly became obvious that the printed-paste process whose performance is shown in these examples could assure the formation of well-consolidated solder joints under the chips, but not nearly as well when the paste is extruded "out into the open." This problem can be related to (1) locating the wettable pads too near the edge of the die, and (2) overcompressing the solder paste.

Obviously, locating the solder pads well away from the edges of the die will for the most part solve this problem. However, it is not always possible to do so, especially on high I/O die of limited size. In that case it was found that reducing the force used to place such die will significantly reduce the amount of paste compression. Since this will inevitably result in the extrusion of less paste from under the die, it can reduce the occurrence of NCR solder balls and wimpy, undersized solder joints. In many SMT placement machines this can be accomplished by using a more compliant spring in the placement head.

Combined with any and all other adjustments as were possible, these experiments demonstrated that excellent yields (with minimum shorts and wimpy joints due to the excessive formation of NCR solder balls) could be obtained with the final AT&T YD-series solder paste at I/O pad pitches as low as 13.5 mils. Nevertheless, in order to assure robust assembly and to simplify routing on the active silicon substrates used in the working telephone MCMs, all subsequent silicon substrates were designed for chips with solder pads routed to area arrays with pitches of 16 to 17 mils, which are very easy to print.

Subsequent experiments involved the successful flip chip assembly of more than fifteen wafers of the more prototypical six-chip daisy-chain EMTVs and numerous wafers of actual Partner™ Telephone silicon MCM-D tiles using the double-sided invertible waffle or "flip" trays instead of cavity tape. While these experiments weren't always 100 percent successful, as can be seen by careful study of Figure 5.3, many important process refinements were initiated, the Partner™ Telephone MCM product was qualified and the capacity for high-speed, high-yield silicon MCM-D tile assembly using a placement machine configured to pick from trays was confirmed. Final x-ray and electrical test on both electromechanical and working prototype MCM tiles demonstrated assembly yields of over 94 percent on a per-tile basis and 99.94 percent on a per joint basis.

Finally, it should be recognized that unless compromised by some uncontrolled aspect of the subsequent MCM packaging and next-level interconnection structure, solder joints in silicon-on-silicon flip chip tiles have little inherent CTE mismatch related thermal strain to accommodate. Consequently, even if assembled from relatively large chips they will experience much less thermally induced fatigue, especially if deployed in an office rather than outside-plant environment.

5.5   Silicon MCM-D Tile Reliability

If silicon tiles assembled using either the tack, flux, and reflow or the flip chip printed solder approach are evaluated like conventional surface-mount components, their intrinsic reliability (resistance to solder joint fatigue or "wearout" associated with the interaction between global geometry and the CTE mismatch between the chips and the substrate material—also known to as the "global thermal expansion mismatch"), as estimated from the AT&T surface-mount design standards for leadless components, will probably show little to differentiate between them. In general, silicon ICs on silicon substrates will have no global mismatch and should never show such failures.

Consequently, the primary distinctions between them will depend on the occurrence of failures originating in processing irregularities or design defects such as, respectively, the random occurrence of joints with grossly insufficient solder volume, or systematic geometric stress concentrations associated with the particular patterning of the solder wettable metallizations on the lands or pads.

Naturally, with other choices of substrate material such as a ceramic or epoxy-glass laminate for which the global thermal expansion mismatch with silicon will be nonzero and will play a role, the 75-μm joint heights being achieved with the currently more mature AT&T microinterconnect tack, flux, and reflow assembly technique will provide an advantage over the significantly shorter (50-μm) joint heights presently being produced by the microinterconnect printed solder technique. However, it is reasonable to expect that optimization of the latter approach as it matures will be directed at least in part toward increasing joint height and uniformity to the point where for all practical purposes there will be little to differentiate between them in that respect. In this case, as with the case of flip chip ICs on a silicon interconnection substrate, differences in reliability will depend on other details. For example, the evaporated solder deposition approach may offer an advantage in terms of its ability to assure joint-to-joint uniformity, while the somewhat greater joint volumes that result when using the printed solder paste may offer a compensating advantage in terms of increased cycles required to propagate a fatigue crack to failure. Further research will be required to quantitatively evaluate the effects of these and other such processing or design details. However, it is unlikely that significant differences based on the choices of materials or global geometry are likely to be found that will do much to differentiate between their inherent fatigue related reliabilities.

Tables 5.3 and 5.4 present the figures of merit (FM), and associated estimates of cumulative failure probabilities [$F(t)$ (in parts per million

TABLE 5.3 Flip Chip Reliability Estimates Based on March 1990 AT&T Surface-Mount Design Standards for Leadless Components with 75-μm Joint Heights Typical of Tack, Flux, and Reflow Assembly

Material	FM	$F(t)$, ppm	FM	$F(t)$, ppm	FM	$F(t)$, ppm
Si/Si	∞	None	∞	None	∞	None
Si/ceramic or Si/aramid	0.18	5×10^5 (50%)	0.36	5×10^4 (5%)	0.90	5 (0.0005%)
Si/FR-4	0.04	10^6 (100%)	0.08	10^6 (100%)	0.20	5×10^5 (50%)
L_D, mm	10		5		2	

Key: FM = figure of merit; $F(t)$ = cumulative failure probability in 10 years in an office environment; L_D = chip size, characterized by separation, mm.

TABLE 5.4   Flip Chip Reliability Estimates Based on March 1990 AT&T Surface-Mount Design Standards for Leadless Components with 50-μm Joint Heights Typical of First-Generation Printed Solder Flip Chip Assembly

Material	FM	$F(t)$, ppm	FM	$F(t)$, ppm	FM	$F(t)$, ppm
Si/Si	∞	None	∞	None	∞	None
Si/ceramic or Si/aramid	0.12	10^6 (100%)	0.24	5×10^4 (50%)	0.60	50 (0.005%)
Si/FR-4	0.03	10^6 (100%)	0.05	10^6 (100%)	0.13	9×10^5 (90%)
L_D, mm	10		5		2	

Key: FM = figure of merit; $F(t)$ = cumulative failure probability in 10 years in an office environment; L_D = chip size, characterized by separation, mm.

and percent)] for 10-year survival in an office environment for silicon tiles assembled using, respectively, the tack, flux, and reflow process and the print, place, and reflow process. Following the AT&T surface-mount design standards, a typical office environment may be defined as 20°C daily temperature cycles and 60°C monthly temperature cycles.

5.6   Electrical Properties

In general, the parasitic circuit elements associated with flip chip assembly are low in comparison with leaded, wire-bonded components. These parasitics will vary depending on the particular solder joint size, geometry, and assembly method. Typical values of the important parasitic elements for both assembly processes are summarized in Table 5.5.

In microinterconnect tack, flux, and reflow flip chip assembly, the shunt capacitance to the chip (usually ground) is basically the inherent capacitance of the bond pad itself. Shunt capacitance for flip chip solder joints assembled using the microinterconnect print, place, and reflow assembly process is higher, since the additional redistribution routing contributes additional capacitance. This additional capacitance will vary depending on the length of the redistribution path and

TABLE 5.5   Key Electrical and Thermal Elements of Flip Chip Solder Joints (Worst Case) Fabricated Using Tack, Flux, and Reflow Tile Assembly and Print, Place, and Reflow Tile Assembly

	Tack, flux, and reflow assembly	Print, place, and reflow assembly
Shunt capacitance to chip	0.3 pF	1 pF
Shunt capacitance to substrate	0.1 pF	0.2 pF
Series inductance	0.025 nH	1 nH
Series resistance	1 mΩ	1 Ω
Thermal resistance	1000°C/W	1500°C/W

the diameter of the solder joint. The capacitance value in Table 5.5, therefore, represents a worst-case estimate. It should be pointed out that the electrostatic discharge protection circuitry used in most chips will contribute 1 to 2 pF of additional shunt capacitance not accounted for in Table 5.5, and will dominate the total capacitance in most realistic circuits.

The other surface of the solder joint, which anchors to the substrate, contributes an additional shunt capacitance to the substrate (also usually ground). Here, the difference in the two technologies simply reflects differences in the area of the solder joint footprint.

The series inductance and resistance of a flip chip solder joint, which is in reality a short, slightly rounded cylinder of solder, is very low. In the printed solder flip chip structure, the higher inductance and resistance arise mostly from the redistribution metallization, and not from the solder joint itself. For comparison, a typical leaded package will contribute resistance of a few mΩ and inductance of 5 to 10 nH.

5.7 Thermal Properties

Depending on the structure of the final package, the chip's main thermal connection may be through the substrate. Table 5.5 also shows the thermal conductivity of the two joint structures. The flip chip printed solder joint has a larger cross-sectional area and is shorter, but it is located on top of a layer of polyimide. The net result is that its thermal resistance is about 50 percent higher than the more conventional flip chip solder joint structure. To estimate the role of flip chip solder joints in conducting heat from the chip into the substrate, simply divide the thermal resistance of a single joint by the number of joints on the chip. This is a worst-case estimate, since there are invariably other avenues of heat flow that can also contribute. Also, in the final module assembly, the main impedance to heat flow may be at some other interface to the ambient.

In cases where the thermal performance of the system is limited by heat flow from a particular chip into the substrate, it may be advantageous to design the chip to have extra solder joints that are not electrically connected. These purely mechanical joints will provide additional thermal paths to the silicon substrate.

5.8 Packaging

As with any bare silicon device or IC, in order to be used the silicon MCM-D tiles must be packaged to facilitate automated handling for next-level assembly and to protect them from the environment (mechanical, chemical, and thermal). At the same time the packaging must provide the necessary electrical interconnections and assure performance under all anticipated service and specified qualification conditions. To best provide these many interconnections and at the same time meet the objective of minimum overall system cost, two alternative MCM BGA package designs have been developed as described below.

5.8.1   The BGA MCM package designs

In designing the BGA for MCMs, it was decided that the most cost-effective approach would be to explore two printed solder bumped, non-hermetic BGA package designs which could be surface-mounted using reflow soldering. These two designs will be referred to as the "over-molded" MCM package and the "shell and gel" MCM packages. In both of these designs the MCM tile is first adhesively bonded to the top surface of a thin double-sided printed wiring board (the BGA PWB), after which the circuits on the tile are wire bonded to the top surface circuitry on the BGA PWB. This top-surface circuitry is connected by plated-through holes to the bottom surface circuitry and a 10×10-grid array of large I/O pads. Each tile is then enclosed, either by overmolding as pictured in the scaled three-quarter cutaway view of the package alone shown in Fig. 5.9a and the non-scaled cross-sectional view of the package as mounted on a PWB as shown in Fig. 5.9b, or covering it with a premolded plastic shell as shown in Figs. 5.9c and 5.9d. In both designs each I/O pad on the BGA PWB bottom surface is furnished with a large (on the order of 0.8 mm in diameter) solder ball.

The concept behind developing two BGA designs was to explore the capabilities of a "monolithic" structure, represented by the overmolded package, versus a "loosely coupled" structure, represented by the shell-and-gel package. As in overmolded BGA packages used for single-chip packaging, the overmolded MCM BGA package depends on its mechanical strength, inherent stiffness, and interfacial adhesion for its robustness and long-term reliability. On the other hand, because such a design emphasizes the effects of the CTE mismatches between MCM components, low-cycle thermal fatigue may cause wearout of the solder joints, especially those connecting the BGA to its parent board. This problem becomes more severe as the MCM becomes larger, so that an overmolded package design which can provide the needed long-term service life for a small tile may not be able to do so for a much larger tile. However, a more loosely coupled design (like the shell and gel) should not have these limitations, provided it meets the other requirements for handling, resistance to moisture, and cost.

5.8.2   Design for BGA reliability

As discussed above, in order to optimize BGA MCM reliability, a shell-and-gel BGA structure was used to achieve a packaging design in

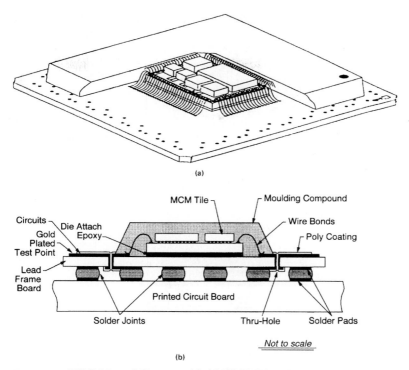

Figure 5.9 AT&T (a) and (b) overmolded MCM BGA packages.

which the thermal expansion mismatches between the tile, the shell, the BGA PWB, and the parentboard are almost completely accommodated by minimizing the mechanical coupling between them. This is accomplished by using a highly compliant epoxy adhesive to attach the shell to the BGA PWB, by filling it with an even more (indeed far more) compliant, commercial-grade silicone gel, and by using the same highly compliant epoxy adhesive to couple the silicon tile to the BGA PWB. This leaves the balance of the bottom surface of the silicon tile and the BGA PWB relatively free to expand and contract independently during changes in temperature. Considering that the CTE of the BGA PWB is in the range of 12 to 15 ppm/°C and the CTE of the silicon tile is much lower—being on the order of 2 to 4 ppm/°C—this behavior is highly desirable.

As noted in Sec. 5.5, by itself the tile is a very low-risk structure (for thermal fatigue) because both the die and the substrate are of the same material, silicon. Consequently, unless the packaging itself interferes with the mechanical relationship between them, the solder joints which interconnect the die with the silicon substrate will experience

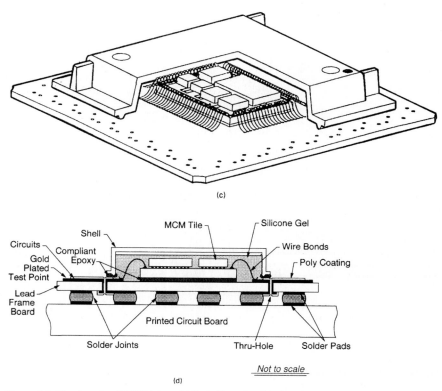

(c)

(d)

Figure 5.9 (*Continued*) AT&T (*c*) and (*d*) shell-and-gel MCM BGA packages.

little stress or strain during thermal cycling and should be highly reliable. However, it is possible that next-level interconnection to the parentboard could be at some risk. Intrinsically, the BGA PWB and the parentboard to which it will be soldered are of laminate materials which have nearly the same CTE, and if there were no other components to consider the interconnections between them would be nearly free of stress or strain so long as there were no major thermal gradients to contend with. Both the shell and the tile have rather different CTE values and unless the effort to uncouple them from the BGA PWB is successful, thermal stresses will develop in the BGA to parentboard solder joints, with an ultimately negative impact on reliability. This would certainly be the case with the monolithic overmolded BGA structure. However, so long as the design of the shell-and-gel structure is successful in uncoupling these components so that the effects of the CTE mismatches among the silicon tile, the BGA PWB and the shell and gel are minimized, the BGA-to-parentboard solder joints should also be relatively free of thermally induced stress or strain. This is discussed further in Sec. 5.9.

5.8.3 The BGA MCM packaging assembly sequence

The BGA assembly process begins with a supply of cleaned and ready-to-process BGA PWBs in panel form and cleaned, tested MCM tiles ready for packaging, as well as the appropriate adhesives, cleaners, and solder pastes as will be described. First, an appropriate pattern of die attachment adhesive is printed on the BGA PWB panel. A strong, stiff adhesive with a high glass transition temperature T_g is used for the overmolded version, and a compliant adhesive with a low T_g is used for the shell-and-gel version. Tiles are then placed on each of the MCM sites until the panel is fully populated. Once a batch of panels has been populated, they are loaded into an oven for curing at 150°C for 10 min, followed by environmentally compatible solvent and plasma cleaning steps. The electrical connections are formed between the tile circuitry and the BGA PWB circuitry by wire bonding, and the panels are ready for nonhermetic encapsulation.

If they are to be overmolded, they are placed in a heated mold and the molding compound is forced around the tiles to form the package shown in Figs. 5.9a and 5.9b. In this process it is important to assure that there is sufficient standoff height under the chips to permit the molding compound to completely underfill even the largest chips; otherwise cavities will form under the chips, which can seriously degrade reliability by trapping moisture. After overmolding, the panels are removed from the mold, cleaned, and prepared for bumping.

If the panels are to be packaged using the shell and gel approach, either of two assembly processes may be used. In one the shells are dipped into the compliant shell-attach adhesive so as to coat the shell flange and are then placed on the appropriate site. In the other approach, the shell-attach adhesive is dispensed directly onto each BGA PWB site in a square pattern around the wire-bond array, and the shell is placed into this adhesive frame. The "shelled" panels are then cured at 150°C for 10 min to fix the shells in place, after which each is filled with sufficient gel to cover the tile and the wire bonds. Once all the shells are filled, the entire batch of BGA MCM panels is returned to the oven for 15 min at 150°C to cure the gel. (Both of these cures can be carried out at lower temperatures, but the cure times will be significantly longer and the throughput reduced correspondingly.)

Final processing is the same for both the overmolded and shell-and-gel versions. Both are furnished with solder balls by a printed solder/reflow process which is applied to the I/O pads on the reverse side of the inverted panels as described below. This "solder bumping" process is extremely efficient and postreflow cleaning is not necessary. MCMs are separated from the panel, tested on a bed-of-nails tester, marked, sorted, and boxed for shipment to the surface mount line for assembly.

5.9   BGA Reliability Evaluations

To demonstrate the comparison between the overmolded and the shell-and-gel MCM packages, and to quantify their responses to thermal loadings under various conditions, a variety of studies were conducted as described below.

5.9.1   Thermal strain measurements

In Sec. 5.8.2 it was noted that the reliability of the shell-and-gel design depends on the design's success in mechanically uncoupling the various MCM components so that they can expand and contract independently. Since the flip chip joints join silicon to silicon and the wire-bond connections can easily flex in the compliant gel, this is a concern primarily for the BGA solder joints which connect the MCM to the PWB parentboard. To assess the shell and gel design in this regard, a study of the induced thermal strains was made by B. T. Han and Y. Guo of IBM Endicott using a highly sensitive optical technique called *Moiré interferometry*.[10] This study also included a variety of PWB mounted 225 I/O and 400 I/O single-chip BGA packages by other manufacturers, as well as one each of the two AT&T MCM BGAs, also mounted on PWBs.

Moiré interferometry is a whole-field optical technique, which is described in detail in the recently published text by Post et al.[11] The specimen to be studied is first sectioned through the region of interest and a high-frequency (1200 lines/mm) diffraction grating is replicated on the exposed planar surface. In most studies a 90° crossed grating is used so that both the u- and v-displacement fields can be measured.

When the specimen is deformed, in the present instance by cooling from a steady-state, uniform reference temperature of 82°C to a steady-state, uniform room temperature of 22°C, the specimen grating deforms with it. In order to reveal the entire field of deformation, dual-beam laser illumination is used to generate a virtual grating of twice the frequency of the specimen grating on the test surface. The resulting high-contrast Moiré interference fringes can readily be seen and recorded. If the virtual grating is aligned with the x direction, the u-displacement fringe field is recorded, and if it is aligned with the y direction, the v-displacement fringe field is recorded.

In either case, Moiré interferometry is a highly sensitive technique, with each fringe representing 0.417 μm of in-plane displacement or deformation. Obviously, if the specimen is of uniform CTE and tem-

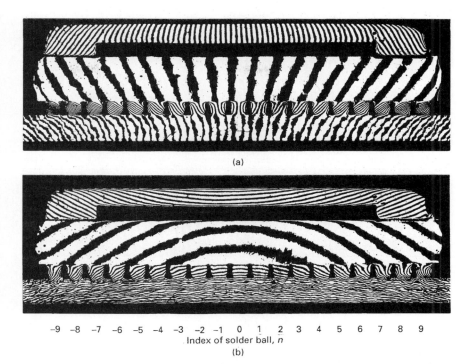

(a)

-9   -8   -7   -6   -5   -4   -3   -2   -1   0   1   2   3   4   5   6   7   8   9
Index of solder ball, n

(b)

Figure 5.10 Moiré patterns for both u- and v-displacement fields on an IBM SBC single-chip package from Guo et al.[12] (*Copyright 1993 by International Business Machines Corporation, reprinted with permission.*)

perature, the Moiré fringe field would appear as uniformly spaced, parallel straight lines—vertical lines for the u-displacement field and horizontal lines for the v-displacement field. For purposes of comparison, Fig. 5.10 shows the resulting u- and v-displacement fields for an SBC (for solder ball connect) single-chip package by IBM which was sectioned through the center of a row of solder balls by Guo et al.[12] This monolithic package, which is exactly the same size as the AT&T BGA MCM packages and has the same size solder balls (albeit at a closer spacing), was subjected to the 60°C thermal cycle described above. Because of differences in the CTE values of the various components of the SBC package, which is ceramic and plastic surrounding a silicon chip, the fringes are hardly straight, parallel, or uniform except in the middle of the package. In fact, the fringes on the solder joints themselves become increasingly inclined and distorted row by row as the distance from the center of the package increases.

It can be readily seen that the solder joints become especially strained as they approach the outer edges of the package, with the v-displacement field which forms a vertical fringe pattern in the center

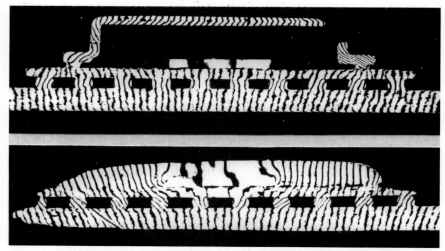

Figure 5.11 Moiré patterns for the u-displacement fields on both AT&T BGA MCM packages.

joint becoming a horizontal pattern in the outer joints, and the u-displacement field which forms a horizontal fringe pattern in the center joint becoming almost vertical in the outer joints. This indicates extreme shear strain as a result of the thermal deformation mismatch between the package and the PWB to which it is attached. It can also be seen that the ceramic substrate and the plastic covering experience considerable bending. Unfortunately, no grating was applied to the chip, so its response is not observable.

As for the AT&T BGA MCM packages, Fig. 5.11 shows the u-displacement field for the conditions described above for both the shell-and-gel package (upper photograph) and the overmolded package (lower photograph), and Fig. 5.12 shows the corresponding v-displacement fields. Naturally, no specimen grating could be replicated on the gel, so the corresponding region of that moiré fringe field is dark, as are the chips on the tile which lay out of the plane of the section, which was taken through the fourth row of BGA solder joints. Also there is a small corner portion of the shell, as well as the upper corners of the overmolding, to which the grating failed to adhere. The short range irregularities seen in the v-displacement field fringes formed on the BGA PWB and the parentboard in Fig. 5.12 are due to the fiber laminate structure of the composite itself, and have no influence on reliability. With this in mind, the relatively uniform and parallel fringe patterns clearly demonstrate that, as intended, the critical components of the shell-and-gel package, the silicon substrate and the BGA PWB, move independently and with little unwanted in-

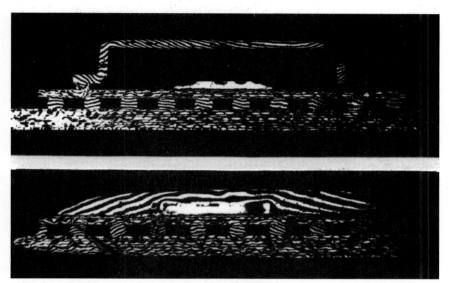

Figure 5.12 Moiré patterns for the v-displacement fields on both AT&T BGA MCM packages.

teraction—indeed, the silicon tile substrate, the BGA PWB, and the parentboard are all essentially free of thermally induced strain. At the same time, there is some inevitable flexing or bending of the shell, but thanks to the compliance of its coupling to the PWB, very little of this is transmitted to the PWB. As a result, except for minor distortions due to the local mismatch between the CTE of the solder and the laminate, the BGA solder joints are also virtually strain-free, not just at the center of the package but all the way to the outermost row.

As for the overmolded MCM package, clearly there is a good coupling between both the MCM tile substrate and the epoxy molding compound and the BGA PWB, as would be expected. As a result, there is a CTE mismatch strain between the BGA MCM and the parentboard which manifests itself in some flexing of the BGA PWB and corresponding rotations of the both fringe fields in solder joints away from the center of the overmolded package. This indicates some shear strain in those solder joints, but the worst case, which appears in the joint nearest the edge of the tile,[10] is only 0.28 percent, which is quite acceptable. Not surprisingly, local fringe distortions due to the solder-to-laminate CTE mismatch are also evident, much as they were for the shell-and-gel BGA package. In addition, there was virtually no shear strain in the solder ball joints nearest the outer edge of either AT&T package, and the joints nearest the edge of the tile in the shell-and-gel design showed a shear strain of only 0.075 percent, which was by far the smallest of any package ever tested. This edge represents a discontinuity beneath which the joints on all BGA structures inevitably exhibit the greatest shear strains.

5.9.2 Reliability testing

A range of reliability and qualification tests have been run on both of the AT&T MCM BGA designs, including thermal cycle tests of early and intermediate BGA electromechanical BGA development samples. Results of these tests indicated that, aside from a few infant mortality failures resulting from the immaturity of the assembly process, the attachment reliability for both the shell-and-gel design and the over-molded MCM packaging is very good, and after over 4500 (0 to 100°C) thermal cycles have experienced no wearout failures of the shell-and-gel samples and only one possible wearout failure of the overmolded (at 1967 cycles). These results demonstrate that both of the packages are very robust, particularly the shell-and-gel one. On the basis of the appropriate acceleration factors, the survival of 1200 thermal cycles indicates a 5-year service life—which is an important attachment reliability milestone.

Finally, the shell-and-gel package itself has been qualified by meeting or surpassing the required performance specifications for a broad variety of tests. These include a −55- to 125°C temperature cycle test, the MIL-STD 883 (methods 2002 and 2007) mechanical shock and vibration test, the MIL-STD method 2019 die shear test, the IEC 695 flammability test, the 1000-h THB (85°C, 85%-RH, 5-V) test, and the HTOB [high-temperature (125°C) operating bias] life test.

5.9.3 "Popcorn effect" failure-mode evaluation

Because of its loosely interconnected mechanical structure and its relatively open design, which results from having to provide fill/bleeder ports in the shell for the silicon gel injection process, the shell-and-gel package is far from hermetic and can readily take up water. Not surprisingly, concern has been expressed regarding the "popcorn effect," which is the term used to describe a catastrophic mechanical failure mode related to the abrupt vaporization of entrapped water in a non-hermetic package or component during a reflow soldering process. Thanks to the design and materials choices exercised in developing the shell-and-gel package described above, repeated experiments in which shell-and-gel BGA MCMs have been immersed in water for hours prior to being removed and put through a standard (to well over 200°C) reflow process have consistently demonstrated no "popcorn" failures. In addition, samples of both the shell-and-gel and the overmolded packages were subjected to over 100 h of 85/85 THB followed by five reflow cycles. They showed no failures even after four repetitions. These results clearly demonstrate that both the AT&T overmolded BGA MCM packages and the AT&T BGA shell-and-gel MCM packages can be stored and handled without having to provide special protection from either atmospheric or processing-derived moisture.

As a result of these studies, it is apparent that the shell-and-gel design was successful in meeting its design objectives, and offers an effective approach to the design of packaging for extremely large chips and MCM tiles where the problem of CTE mismatch is compounded by size. In addition, it appears that the choices of materials and design of the AT&T overmolded MCM package also contribute to good long-term reliability, both as measured in corresponding reliability and thermal cycle tests and as will be experienced in service.

5.10   Summary and Conclusions

This chapter presents the results of research and development on a flip chip MCM manufacturing platform which point the way to the establishment of a cost-effective, industrywide silicon MCM-D manufacturing capability in the very near future. As described, the application of a printed solder paste technology, batch processing and generic high-speed automated SMT assembly tools and equipment can assure the throughput required to distribute capital expense over many units and thereby eliminate the silicon MCM-D tile assembly as a significant cost factor. Indeed, most of the required processes and material resources are already known and/or exist, and as demonstrated in the factory, standard wafer and chip handling, dicing, sorting, fluxing, and reflowing equipment can be readily adapted to the high-volume manufacturing and packaging of flip chip MCMs. Consequently, assembly costs need not present a barrier to the development of MCMs for use in less expensive, high-volume consumer or commercial electronics products.

In addition, a novel packaging technology has been developed for initial applications. These finished MCMs take the form of solder-bumped, nonhermetic ball grid array (BGA) packages which can be reflow-soldered directly onto printed-circuit boards using standard SMT assembly facilities and processes. Two designs are presented. One of these exploits standard processes for overmolding as used for conventional monolithic single-chip BGA packages and offers a minimum-cost solution for applications where size restrictions and reliability requirements are of limited concern. In the second, more radical version, the reliability problems resulting from the thermally induced stresses and strains that are inherent in conventional monolithic electronic packages are avoided by means of a unique design which allows for relatively independent thermal expansion and contraction of the packaged MCM components without compromising electrical integrity.

Finally, this chapter has described how, by combining innovative designs and solder ball application technologies which are compatible with the existing surface-mount assembly infrastructure, it is possible to reduce the cost of assembly significantly and to enhance the reliability of even large, difficult-to-package MCMs. Both full-field thermal stress and strain measurements and thermal cycle fatigue experiments demonstrate the success of this approach to high-reliability MCM packaging. This technology also has potential applications to the packaging of individual large devices for applications where long-term reliability under challenging thermal conditions is important.

Acknowledgments

The authors wish to express their gratitude to T. I. Ejim, Y. Guo, C. K. Lim, S.-C. Chen, A. W. Lin, B. T. Han, B. J. Han, P.-Y. Lu, W. L. Woods, F. P. Hrycenko, M. McCormack, C. D. Hruska, R.-L. Day, D. D. Bacon, J. C. Conway, J. R. Morris, J.-P. Clech, M. G. Johnson, A. R. Storm, D. I. Kossives, E. K. Sorensen, J. E. Barreto, P. Johnson, T. A. Hajec, V. R. Raju, F. X. Ventrice, D. R. Cokely, P. A. Sullivan, C. A. Strittmatter, R. A. Gottscho, and the many, many others who contributed to this effort.

References

1. Tummala, R. R., and E. J. Rymaszewski, eds., *Microelectronic Packaging Handbook,* Van Nostrand Reinhold, New York, 1989.
2. Doane, D. A., and P. A. Franzon, eds., *Multichip Module Technologies and Alternatives,* Van Nostrand Reinhold, New York, 1993.
3. Blonder, G. E., R. A. Gottscho, and K. L. Tai, *Interconnection Processes and Materials, AT&T Technol. J.,* **69**(6): 46–59, Nov./Dec. 1990.
4. Dudderar, T. D., Y. Degani, J. Spadafora, K. L. Tai, and R. C. Frye, "AT&T μSMT Assembly: A New Technology for the Large Volume Fabrication of Cost Effective Flip Chip MCMs," *Proceedings of the 1994 International Conference on Multichip Modules,* Denver, pp. 266–269, April 13–15, 1994.
5. Lau, M. Y., K. L. Tai, R. C. Frye, M. Saito, and D. D. Bacon, "A Versatile, IC Process Compatible MCM-D for High Performance and Low Cost Applications," *Proceedings of the 1993 International Conference on Multichip Modules,* Denver, pp. 107–112, April 14–16, 1993.
6. Frye, R. C., K. L. Tai, M. Y. Lau, and A. W. C. Lin, "Silicon-on-Silicon MCMs with Integrated Passive Components," *Proceedings of the 1992 IEEE Multi-Chip Module Conference MCMC-92,* Santa Cruz, Calif., pp. 155–158, March 18–20, 1992.
7. Day, R. L., C. D. Hruska, K. L. Tai, R. C. Frye, M. Y. Lau, and P. A. Sullivan, "A Silicon-on-Silicon Multichip Module Technology with Integrated Bipolar Components in the Substrates," *Proceedings of the 1994 IEEE Multi-Chip Module Conference MCMC-94,* Santa Cruz, Calif., pp. 64–67, March 15–17, 1994.
8. Frye, R. C., "Balancing Performance and Cost in CMOS-Based Thin Film Multichip Modules," *Proceedings of the 1993 IEEE Multi-Chip Module Conference MCMC-93,* Santa Cruz, Calif., pp. 6–11, March 15–18, 1993.
9. Frye, R. C., and A. V. Shah, "Targeting Low-Cost, High-Volume MCM Applications," *Internatl. J. Microcircuits Electron. Packag.,* **16**: 285, 1993.

10. Elim, T. I., J.-P. Clech, T. D. Dudderar, Y. Guo, and B. T. Han, "Measurement of Thermomechanical Deformations in Plastic Ball Grid Array Electronic Assemblies Using Moire Interferometry," *Proceedings of the 1995 SEM Spring Conference,* Grand Rapids, Mich., June 12–15, pp. 121–125, 1995.
11. Post, D., B. T. Han, and P. Ifju, *High Sensitivity Moire: Experimental Analysis for Mechanics and Materials,* Springer-Verlag, New York, 1994.
12. Guo, Y., C. K. Liem, W. T. Chen, and C. G. Woychik, "Solder Ball Connect (SBC) Assemblies under Thermal Loading: I. Deformation Measurement via Moire Interferometry and Its Interpretation," *IBM J. Research Devel.,* **37**(5):635–647, Sept. 1993.

10. Chen, [illegible], [illegible], and [illegible] "[illegible] of [illegible] Defects [illegible]," [illegible] 1981 [illegible], pp. [illegible]–[illegible], 1981.
11. [illegible] and [illegible], *[illegible]*, Springer-Verlag, New York, 1986.
12. [illegible] "[illegible] [illegible] of [illegible] and its Interpretation," [illegible], Sept. 1985.

Conductive Adhesive Polymer Materials in Flip Chip Applications

Richard H. Estes and Frank W. Kulesza

6.1 Introduction

The evolution of electronic packaging has resulted in an increasing number of I/O connections on integrated circuits as semiconductor devices become more sophisticated. As the number of interconnections increases, there are equally difficult demands placed on packaging engineers to develop small, high-density packaging for today's sophisticated microelectronics technologies. This entails significant changes in circuit design, bond pad layout and metallization, and the optimum method of electrical interconnection. As microelectronics packaging continues its move toward miniaturization it is apparent that conductive polymer flip chip interconnection can provide reliable, high-density packaging at a lower cost than alternative technologies such as tape-automated bonding (TAB). Although flip chip technology was developed more than 25 years ago, this form of electrical interconnect is only now beginning to be recognized for high-volume, commercial applications in microelectronics.

The electronics industry adopted the use of isotropic, electrically conductive adhesives for IC chip attach in the early 1970s for attaching semiconductor chips and other passive devices onto ceramic substrates in the fabrication of hybrid microcircuits. Epoxy Technology pioneered the use of isotropic conductive adhesives for die attach applications in 1966. These early adhesives were filled with gold and silver powders and flakes which provided electrical and/or thermal pathways for the bonded devices. Gold adhesives were used almost exclusively for high-reliability, military, and aerospace applications, while silver adhesives, formulated for commercial applications, were slower to be accepted. As these conductive adhesives evolved, and reliability data demonstrated that conductive adhesives had reliability equivalent, or superior, to that of solder and eutectic bonding alloys, the use of isotropic, Ag-filled epoxies expanded into high-volume semiconductor applications. Silver-filled epoxies proved to be a highly reliable materials for attaching semiconductor chips to Ag-spotted lead frames for plastic package applications. Electrically conductive, Ag-filled epoxies have proved to be excellent performing die attach materials in the harsh environment of steam autoclave testing demanded by nonhermetic packaging.

Virtually all conductive adhesives have been used for die attach, grounding and shielding, conductive coatings for ESD, and for grounding purposes in low-voltage transistors. The applications have historically been to use Ag-filled epoxies for passive die attach, but almost never as the active electrical interconnection for IC devices in microelectronics. This potential use of electrically conductive epoxies has remained dormant for more than two decades while wire bonding, TAB, and solder flip chip have been pursued, the latter almost exclusively by IBM. However the new challenges of reducing the size, weight, and cost of microelectronics packages has paved the way for a new generation of advanced polymeric conductors, substrate materials, and manufacturing processes to be developed. Many of these are now being evaluated and strongly considered for use in new applications, both for the fabrication of high-reliability military and aerospace MCM packaging and a myriad of applications for automotive and consumer electronics. The use of conductive polymers continues to grow, with unprecedented development of applications for these materials in the areas of solder replacement for surface-mount technology (SMT) and chips on board (COB) and flip chip.

6.2 Polymer Flip Chip: PFC

6.2.1 Solderless flip chip technology

The *polymer flip chip* (PFC) process involves the use of dielectric, screen-printable organic polymer coatings, isotropic, silver-filled conductive polymers, and epoxy underfill encapsulants. These three types of materials are coupled with advanced, high-resolution and high-definition screen and stencil printers which feature automation and pattern recognition, vision alignment systems. In its simplest context the advanced wafer passivation (AWP) and PFC processes are used to apply a dielectric coating to the wafers, leaving the metallized bond pads open, after which the conductive polymer is applied to the bond pads, which become the electrical and mechanical connections of the IC to its corresponding substrate. The use of isotropic, conductive adhesives for flip chip applications is today recognized as a viable alternative to solder-bump processes. In addition to lower-temperature processing and the lack of environmental concerns, the conductive adhesive, or polymer bumps, offer electrical and thermal performance comparable to that of solder bumps, and flip chip devices can be produced at lower cost.

The information in Table 6.1 gives a direct comparison of the the relative advantages of solder bump versus polymer bump for flip chip applications, and Table 6.2 provides an overview of the various chip interconnect features.

TABLE 6.1 Solder-Bump versus Polymer-Bump Process

PFC process	Solder-bump process
Low capital equipment cost	High capital equipment cost
Clean process; no solvents or flux	Dirty process; solvents and flux
Two-step; simple processing	Multistep; complex and slow process
Versatile; all types of substrates	Restricted to high-temperature substrates
High electrical and thermal conductivity	High electrical and thermal conductivity
Processed parts need no cleaning	Processed parts need careful cleaning
Low process temperatures	High process temperatures
No environmental problems	Serious environmental problems
No waste treatment and disposal	Waste treatment and disposal necessary
Versatile process	Restricted process
Conductive polymer bumps are compliant	Solder bumps are brittle
Array and peripheral bonding	Array and peripheral bonding

TABLE 6.2 Chip Interconnect Technology Features

Feature	Wire bond	TAB	Flip TAB	Flip chip solder	Flip chip adhesive
Maturity	Very good	Good	Limited	Very good	Improving
Die avail-ability	Very good	Fair	Fair	Poor	Very good
Edge bond pitch	4–7 mils	3–4 mils	3–4 mils	8–10 mils	8–10 mils
Maximum I/O	400–500	800–1000	800–1000	>1000	>1000
Footprint[1]	20–100 mils	80–800 mils	80–100 mils	<20 mils	<30 mils
Assembly rate	Slow	Good (gang)	Good (gang)	Good (gang)	Good (gang) and cure
Repairability	Poor	Poor	Poor	Moderate	Good–poor
Chip Bi/Test	Difficult	Good	Good	Improving	Improved
Cost	$0.001	$0.003–$0.01	$0.003+	$0.002	$0.0001

The conductive adhesive bumping process, coined PFC, is a patented process technology for the fabrication of conductive adhesive bumped devices which are to be used in flip chip applications. This process was initially developed in 1989 at Epoxy Technology, and feasibility studies of both the process and the integrity of using conductive adhesive polymer bumps for flip chip were conducted for 2 years. Having determined that conductive adhesive polymeric bumps meet the electrical performance criteria for electrical interconnect, this flip chip bumping process was introduced to the industry in 1991. Since that time, all studies conducted on conductive adhesive bumps, for flip chip applications, has focused on obtaining reliability data on adhesive bumps as opposed to alternative interconnect methods. At the same time both the process and materials have been optimized for better performance and processing characteristics in flip chip applications. The remainder of this discussion focuses on the work that has been accomplished to date.

6.2.2   Metallization

Before entering into a discussion of the the process for making conductive polymer bumps, the issue of bond pad metallization must be addressed. Typically IC devices utilize aluminum metal for epitaxial layer connections and top-layer, bond pad metal. Aluminum metal has been, and continues to be, the metal of choice because of electrical conductivity, low cost, ability to wire-bond to Al, as well as other performance and reliability factors. In the case of flip chip, however, whether solder flip chip bumping or polymer flip chip bumping, aluminum is clearly not the metal of choice. Aluminum metal oxidizes readily; these native oxides being present in 40 to 200 Å thickness. The aluminum oxide coating on the bond pad is a dielectric, and quite difficult to conduct through unless the oxide can be penetrated and the underlying Al metal exposed. Wire bonding eliminates this problem as the wire is ultrasonically bonded, or thermal-compression-bonded to the bond pad. The AlO barrier is penetrated, and a stable low-resistance connection is effected.

In one set of experiments conducted with Al bond pads and isotropically conducting polymer bumps, this high-resistance interface was clearly demonstrated. An IC wafer was bumped using the Epo-Tek H20E-PFC conductive epoxy. The bumps were applied directly to the Al bond pads, cured at 150°C, and contact resistance measurements were then taken. The wafer surface and bumps were probed using a Ruck and Kolls prober using micropositioners and 0.001-in-diameter tungsten probe tips. The HP4145 semiconductor parametric analyzer was used as a power source and measuring instrument.

The HP4145 was connected as a constant source to sweep from 0 to 10 mA with a limit of 10 V. Probes connected to this source were placed across a bump to the substrate, or between the bumps to make the connection. A probe connected to the voltage monitor input was used to monitor the voltage from the bump to the substrate or from bump to bump. Some measurements were taken using a 500-Ω resistor paralleled with the test load to expand the plot. Figure 6.1 shows a cross section of the test vehicle used to make the four-point kelvin probe.

The results of the test (Fig. 6.2) confirm that aluminum metal cannot be used as the bond pad metallization for flip chip. The epoxy bump/Al pad contact has an initial high resistance in excess of 100 Ω; increasing the voltage up to 3.172 V causes the current [intermediate frequency (IF)] to increase. At the 3.172-V threshold the aluminum oxide barrier is penetrated, and the resistance goes down to near zero. Figure 6.2 shows the corresponding drop in voltage. However, the interface is not stable, and the oxide will form again, causing con-

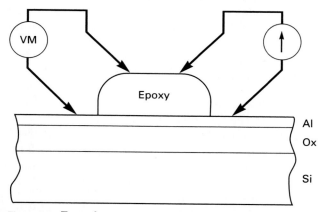

Figure 6.1   Epoxy bump contact resistance measurement.

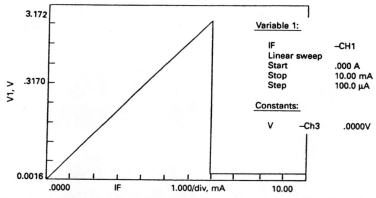

Figure 6.2   Plot of input voltage versus amperage for conductive adhesive bumps.

tact resistance to increase to unacceptable levels. Clearly the Al bond pads must be cleaned of oxide, and noble metals, preferably gold, must be deposited over the aluminum in order for the polymer bumps to be effective as the electrically conducting interconnect from the IC to the circuit on the substrate to which it is bonded.

Several methods can be used to overcome the AlO problem, two of which are discussed here. The first method, which is used extensively, is to sputter-etch the AlO off the bond pads in a vacuumtight sputtering chamber. After removal of AlO, a barrier metal is sputtered onto the aluminum; common barrier metals include TiW, TiN, and NiCr. The final step is to sputter pure gold onto the barrier metal to a thickness of 1000 Å or more. This metallization scheme, coupled with polymer bumps, yields contact resistance values of 10 to 25 mΩ.

The second method, which is relatively new and is commonly referred to as the *zincate process,* is an electroless plating process in which the wafers (Al bond pads) are subjected to an activation of the Al using a *zincate solution,* which consists of zinc oxide and sodium hydroxide. The alkaline solution dissolves the AlO, and a thin layer of zinc is deposited over the aluminum. The second step is to deposit a nickel phosphorus barrier layer of approximately 5 μm. Finally immersion plating is used to coat the nickel with about 0.1 μm of gold. This structure (Fig. 6.3) has excellent adhesion to the Al bond pad, and contact resistances of less than 10 mΩ are achieved.

The electroless deposition method may prove to be the preferred plating process for flip chip applications as it is less expensive and requires no masking or etching. Reliability data is still being developed for devices Au-plated with this method, but it appears to be an attractive alternative for bond pad metallization.

6.2.3 Screen and stencil fabrication

As a prelude to discussing the actual process by which polymer bumps are deposited, one needs to understand the process by which

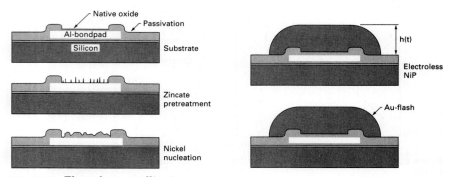

Figure 6.3 Electroless metallization process.

the screens and/or stencils are produced. The final configuration of the dielectric layer and conductive bumps will depend, to a large degree, on the quality of the screen and stencil which will produce the desired dielectric and/or bump pattern.

Screen fabrication is as follows. A screen is fabricated using a stainless-steel mesh cloth which is tightly stretched onto a metal frame. The mesh size varies, but if small vias are to be left open, the wire size should not exceed 0.9 mil, the mesh count should be between 200 and 325, and the percent open area should exceed 65 percent in order to achieve the best definition of the print. The thickness of the dielectric to be printed is determined by the thickness of the polymer emulsion which is applied to the back of the screen mesh.

The pattern to be printed is formed by aligning the pattern mask over the emulsion, or a mylar image produced from the mask, using light to polymerize the photosensitive emulsion, and then washing out the unpolymerized emulsion to produce the desired pattern. An example of a screen used for printed dielectric is shown in Fig. 6.4.

A stencil is preferable to a screen for formation of the conductive adhesive bumps, and the fabrication of the stencil may be different making a screen. The primary difference in construction between a screen and a stencil is that the stencil typically is a metal foil which has openings etched into it to produce the desired bump pattern. A stencil is preferable to the screen for forming bumps because the conductive paste undergoes less shear in the process, thereby producing better-formed bumps which do not spread nearly as much as in screen printing. This is of particular importance for forming 3-mil-diameter bumps on a 5- or 6-mil center-to-center pitch.

There are three basic methods for producing stencils.

Chemical etching.   A metal foil is coated with a photosensitive emulsion on both sides. Molar patterns generated from the mask or computer data, having the bond pad layout, are aligned over both sides of the metal foil, and light is used to expose the mask pattern and develop the emulsion. The unreacted emulsion, and underlying metal, is then chemically etched away to produce the bond pad pattern. This type of etching is typically used when large bumps (6 to 10 mils in diameter) and a large pitch (center-to-center pitch >250 μm) are to be formed on the semiconductor wafers. The height of the bump to be produced is dictated by the thickness of the metal foil which is used. A 0.001-in-thick stencil generally is not durable enough for multiple printing, while 0.002-in stainless-steel foils are quite adequate for producing more than 1000 prints with excellent reproducibility. The optimum stencil thickness will be determined by the diameter of the bumps to be printed as well as the center-to-center spacing (pitch) required by the specific application.

Figure 6.4 Screen for printing dielectric paste.

Actual bump heights are also controlled by the aspect ratio that can be achieved. For most stencils the aspect ratio is 3:1. This means that the highest bump to be produced from a stencil with a 3-mil, 75-μm opening is 1 mil, or a 2 mil maximum height for a bump which does not exceed 6 mils in diameter.

Laser etching. Laser etching of the metal foil is used when smaller bumps, more closely spaced together, are to be formed in the stencil. A computer file, representing the actual bond pad pattern, and stepped and repeated to the appropriate size, is loaded into a computer which controls the laser. A yttrium aluminum garnet (YAG) laser is then used to drill the hole pattern in the metal foil and produce the desired bond pad pattern. A laser stencil takes longer to produce than does a chemically etched stencil, but it yields a stencil with better-defined holes, less undercutting, and a truer 1:1 reproduction of the mask pattern desired. This, in turn, allows for more accurate placement of the conductive adhesive on the bond pads of the wafer. There is also less overetch than with chemical etching, and this means that the adhesive bump will more accurately reflect the actual size of the bond pad.

Electroform stencil. The electroforming process for stencil formation involves applying photoresist onto a flat carrier film, and then aligning the bond pad mask over the photoresist and exposing with light to develop the resist and form the bond pad pattern. The undeveloped resist is then washed away, and the bond pad pattern is produced. An electrodeposition technique is then used to deposit a nickel alloy to the carrier film. The electrodeposition process is continued until the desired height of the stencil is achieved. The remaining resist is washed away to yield the bond pad pattern, and the stencil foil is then mounted to a frame and is ready for use. In theory the electroforming technique will give a true 1:1 reproduction of the bond pad pattern on the wafer. Walls are straight, and there is no underetch or overetch to contend with. This relatively new technique of stencil formation is being aggressively pursued for the stencil printing of "very fine-pitch" SMT devices, solder bumps, and conductive adhesive bumps for flip chip. An example of the quality of the hole formed by these stencils is shown in Fig. 6.5.

6.3   Why Polymer Bumps for Flip Chip?

As the microelectronics industry strives to meet the requirements of tomorrow's packaging technologies, there is a high likelihood that flip chip will play a major role as a preferred method of IC attach to a substrate. To meet the requirements of improved reliability, lower

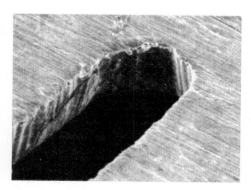

Figure 6.5 Stencil for polymer bumps.

costs, and high-density, high-speed circuitry in a smaller package, flip chip technology provides the radical change required for connecting an increasing number of I/Os in a smaller space. Conventional wire-bonding techniques cannot meet this challenge.

Specifically the rationale for the use of conductive adhesive flip chip bumps are as follows:

- Small size and weight of the bumps

- Reduction of processing costs associated with solder flip chip bump formation

- Electrical and thermal performance comparable to those of solder bumps

- No flux, no cleaning after bumping; reduced environmental impact

- Ability to deposit conductive polymer bumps in peripheral, staggered, or array configuration.

- Ability to cure or process conductive adhesives at relatively low temperatures (≤150°C)

- Ability to easily rework flip chip devices bumped with conductive adhesives

The actual process of forming conductive polymer bumps on the bond pads off a wafer is accomplished by using a one- or two-step process. The process can be used with single chips or wafers, but the lowest cost is realized by producing "polymer-bumped" semiconductor chips in wafer form. This allows for high-volume throughput using automated manufacturing processes. The solderless bumping method involves the deposition of screen printable dielectric polymer onto the surface of the silicon chips in wafer form. leaving the metallized bond pads open.

6.3.1 Dielectric paste application and function

The solderless bumping process, coined PFC, involves two process steps to coat the wafer with dielectric and subsequently form the conductive adhesives bumps on the remetallized bond pads. The first step of the process, called *advanced wafer passivation* (AWP), and utilizes high-precision, automatic screen printers that are equipped with a high-resolution pattern recognition alignment system. The work described here has been conducted with the MPM-SPM printer and/or the MPM-AP21 printer. Both of these printers (shown in Fig. 6.6a, 6.6b have reregistration accuracies of less than ±10 μm, and both are available with automatic alignment, automatic height measurement feature, automatic load/unload feature, and backside stencil wiping to

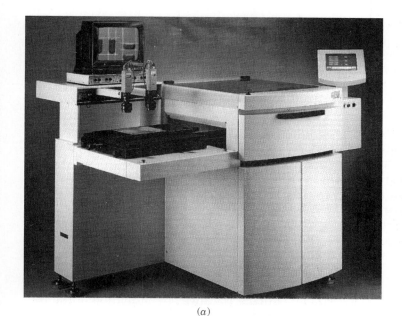

(a)

(b)

Figure 6.6   Semiautomatic (a) and (b) fully automatic stencil printer.

ensure that the bottom side of the stencil remains free of residual conductive paste.

In the first step a screen is mounted into the printer, test prints are made to optimize print parameters, and the semiconductor wafers are then coated with a dielectric polymer such as the Epo-Tek 600 or Epo-Tek 688PFC paste. These highly thixotropic pastes are screened through the openings in the screen to form a 25-μm film on the wafer, leaving open the bond pad areas where the bumps are to be deposited. After printing the wafers are removed to nitrogen-purged belt furnaces, IR furnaces, or convection ovens where curing of the dielectric polymer takes place, typically at 250 to 275°C. The wafers are inspected to make sure that there is no "bleeding" into the bond pad areas, where this can become a reliability problem.

The function of the dielectric is to provide a secondary passivation layer for mechanical protection of the underlying passivation, and also to provide additional environmental protection. Also, special formulations with low-α-particle-emitting fillers are very effective as α particle barriers when printed over memory devices. It should be noted that the dielectric barrier is an optional coating. The formation of the conductive adhesive bumps can be accomplished with or without the use of the dielectric coating. In fact, there are applications where the IC cannot be coated, such as photoemitting or photodetector devices. Figure 6.7 shows a wafer which has been coated with the dielectric polymer. The pattern is that of a transistor chip needing a low-CTE coating to minimize wafer stresses, and to reduce current leakage on

Figure 6.7 A 6-in wafer coated with polymer dielectric 8-mil vias.

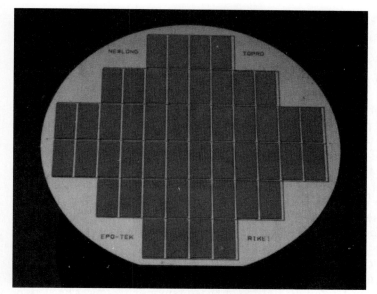

Figure 6.8   Polyimide film on 4-Mbyte DRAM.

the surface of the device. The coating thickness is 25 μm, and the bond pad openings are 0.008 in in diameter. An 8-mil via opening is the minimum opening possible with current screens and materials.

When smaller openings are desired, the entire wafer or device can be coated with the dielectric polymer and an excimer laser can be used to reopen bond pad lands as small as 2 mils square. Figure 6.8 shows an example of cured dielectric polymer which has been screen-printed onto a 4-Mbyte DRAM (dynamic random-access memory) wafer to protect the memory cells from α particles. The dielectric coating can be printed to yield a thickness of 100 μm for this type of application where bond pads are on the edges of the chip and only the middle is coated with dielectric.

6.3.2   Conductive polymer bump deposition

Following application of the dielectric coating to the surface of the wafer, the conductive adhesive paste is now deposited onto the bond pad lands using stencil printing techniques. The metal stencil is mounted into the screen printer, the conductive paste is applied to the stencil, test prints are made, and the bump quality is optimized. The parameters which are established and stored into the computer's memory in the printer are as follows:

- Height of stencil above wafer
- Down stop pressure applied to squeegee

- Squeegee type (metal or polymer)
- Speed of squeegee
- Angle of squeegee with respect to stencil surface
- Accuracy of alignment of stencil with bond pads

When all process parameters are set, the vision alignment system of the screen printer verifies the precise alignment of the openings in the stencil to the bond pads of the IC devices. The conductive paste is then stencil-printed through the openings in the stencil and deposited onto the bond pads of the IC devices. Bump diameter and height are determined by the size of the openings in the stencil and the thickness of the stencil foil, and also by the thixotropic nature of the conductive paste and the shear forces applied during the printing process. The ideal conductive paste for screen printing has a high thixotropic index, in excess of 5.0, in order to achieve fine-pitch bumping for flip chip. A paste with a low thixotropic index will flow readily when pushed through the stencil onto the bond pad surface. This is due to the nonnewtonian nature of these polymer pastes. The flow or spread can exceed 0.001 in when the thixotropic index is less than 3.0. A conductive adhesive paste, having a thixotropic index greater than 5.0, will reduce in thixotropy when the shear forces caused by printing force the paste through the small vias in the stencil. However, as soon as the shear force is removed, the thixotropy of the paste is restored, and the bump spreads less than 5 to 10 μm; this depends on the actual bump diameter and height of the bump.

The stencil print process currently is capable of producing bumps which are as small as 75 μm in diameter on a center-to-center pitch of 130 μm. The printer and stencil print process will accommodate 2- to 8-in wafers having lapped thicknesses of 0.10 to 0.30 in. Typical bump configurations for many applications are 4.0-mil-diameter bumps, 40 to 50 μm high, on a center-to-center pitch of 200 μm or 8 mils. The stencil print process is capable of producing finer-pitch devices than the solder-bump processes currently provide. Solder, when heated to the >200°C reflow temperature, will flow, and this can cause bridging of adjacent bumps if the pitch is too small. The isotropic conductive adhesive pastes, because of their thixotropic nature, tend not to flow, but to stay positioned where they are deposited. Figure 6.9a and 6.9b show examples of a CATV alignment system and the squeegee, stencil setup on the stencil printer. As you can see from the photo, the top-layer IC bond pad layer mask pattern has been created on the metal stencil. This precise formation of the stencil mask, and the thixotropic nature of the paste, allows the engineer to design and produce flip

(a)

(b)

Figure 6.9 (a) Stencil, squeegee, conductive adhesive paste; (b) cable TV alignment.

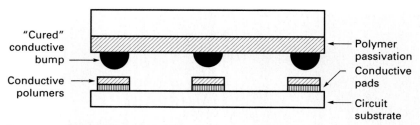

Figure 6.10   Polymer flip chip profile.

chip devices which will have a fine pitch. Currently there is ongoing development work to optimize the stencil print process for 2×2-mil bond pads and a 4-mil pitch. The ultimate goal is to produce a 1-mil-diameter bump on a 2-mil pitch. This will require new screen and stencil technology to produce such small features.

One difference between conductive adhesive bumps and solder bumps is that the conductive adhesive bumps do not have the self-aligning tendency of solder alloys. While self-aligning is sometimes seen as an advantage in the manufacturing process, it must be noted that this occurs only because the solder reflows on heating into a molten, low-viscosity liquid. Engineers must be aware of the pitch between bond pads when forming solder bumps, as the molten solder can bridge, and short-circuit, adjacent contacts. On the other hand, the conductive adhesives have no self-aligning tendency, and therefore high-precision alignment equipment must be used to align the bumps precisely onto the bond pads of a substrate. The adhesive bumps will not flow, but only soften, and this can be a significant advantage when very fine-pitch flip chip devices are desirable. Using polymer bumps, it is possible to form 2-mil-square bumps on a 4-mil pitch, although this is not commercially ready yet. Presently the best alignment equipment available can offer registration accuracy of ± 5 μm, which is adequate for most flip chip applications. The cross section in Fig. 6.10 shows the typical dielectric and conductive adhesive bump profile after both have been printed onto the IC. Figure 6.11 shows an example of an array of conductive adhesive bumps which have been printed onto the IC bond pads. Bump diameter is 80 μm and edge pitch is 200 μm. Note the repeatability of the bump sizes, lack of bleed, and precise alignment of each bump.

6.4   Classes of Conductive Adhesives

6.4.1   Thermoset

Three types of conductive adhesive pastes can be used to form polymer bumps; these are depicted in Fig. 6.12 as thermoset conductive adhe-

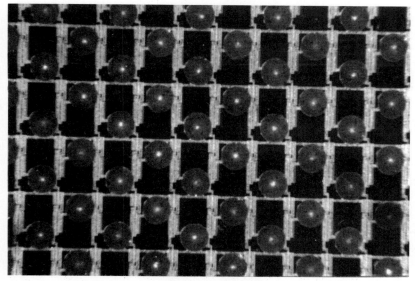

Figure 6.11 Array of 3-mil polymer bumps, 8-mil pitch.

sives, B-stage conductive adhesives, and thermoplastic conductive adhesives. The thermoset conductive adhesives contain no solvents and must be polymerized after being applied to the wafer. The adhesive needs to be in the cured state before the wafer can be sawed into individual IC chips. Silver-filled epoxies are the choice of polymers to be used for forming thermoset bumps. These adhesives have excellent adhesion to gold, excellent thermal stability, and very low outgassing for use in hermetic packages. The glass transition temperature, or softening point of these adhesives varies according to formulation and cure schedule, ranging from 50 to 180°C. The contact resistance of bumps formed using the thermoset adhesives is less than 20 mΩ for 4-mil bumps, and the resistance is stable over wide temperature ranges.

Since the thermoset adhesives cannot be reflowed after cure, an additional process step is required when the bumped IC is to be connected to the contact pads of the circuit. A layer of the conductive adhesive paste must be applied to the substrate, and the cured bumps are then pushed into this paste, after which the paste is cured. An alternative method is to push the cured bumps into a film of wet conductive adhesive paste, wetting the cured bumps, and then attach the flip chip to the substrate and cure the epoxy to effect the mechanical and electrical connection of the flip chip to the substrate. This "two-bump" system is more labor-intensive than use of the B-stage or thermoplastic materials, but the thermoset materials offer high levels of thermal and structural integrity.

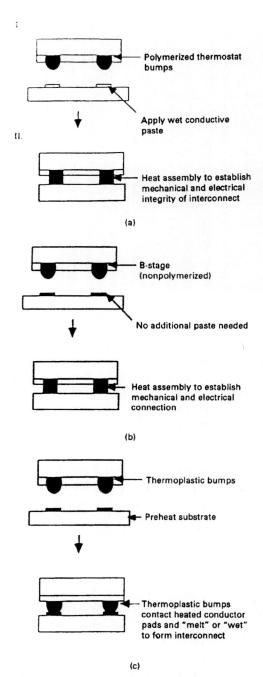

Polymerized thermostat bumps

Apply wet conductive paste

Heat assembly to establish mechanical and electrical integrity of interconnect

(a)

B-stage (nonpolymerized)

No additional paste needed

Heat assembly to establish mechanical and electrical connection

(b)

Thermoplastic bumps

Preheat substrate

Thermoplastic bumps contact heated conductor pads and "melt" or "wet" to form interconnect

(c)

Figure 6.12 Types of conductive polymers: (*a*) thermoset; (*b*) B-stage; (*c*) thermoplastic.

6.4.2 B-stage conductive adhesives

The B-stage conductive paste is a single-component, solvent-containing conductive material which is stencil-printed onto the wafer to form bumps. The advantage of this adhesive is that there is no need to print conductive adhesive on the substrate. The paste is stencil-printed to form bumps, and the adhesive bumps are dried in an oven at low temperature, typically 70 to 90°C, to evaporate the solvent. The wafer can now be handled as the bumps are adhered to the bond pads, but not yet polymerized. When individual IC chips are ready to be flipped and mounted to the substrate, the substrate is preheated, on the flip chip aligner-bonder stage to 150 to 170°C, and the bumps are contacted to the preheated surface. Light pressure is applied and the B-stage adhesive wets the contact pads, and begins to cure. It is not necessary to fully cure the B-stage adhesive at this point. Once some mechanical strength is developed, the flip chip assembly can be removed to another oven for completion of cure.

6.4.3 Thermoplastic

The thermoplastic conductive paste is processed in the same manner as the B-stage adhesive paste. The bumps are stenciled onto the IC chip, and solvent is evaporated to dry the bump. The IC wafer or chips can be handled at this point, and the shelf life of the bumps is several months long. When the flip chip is to be mounted, the chip and/or substrate are heated to approximately 20°C above the melt temperature of the thermoplastic polymer. Using the aligner-bonder, the chip is aligned and placed onto the contact pads of the substrate, using just enough pressure to facilitate wetting of the bumps. This pressure will vary from 30 to 100 g, depending on the mass being heated and the temperature being used to wet the conductive thermoplastic to the circuit. The thermoplastic may be the most user-friendly conductive paste to use for high-volume flip chip applications for the following reasons:

- Single-bump process is ideal for MCM cavities and also for B-stage conductive paste.

- Thermoplastic is more easily reworked than is thermoset or B-stage epoxy.

- Thermoplastic bump connection can be made in seconds, depending on temperature.

- Chip can be temporarily connected to circuit for testing, and then removed.

- IC chip, with thermoplastic bumps, can be stored for long periods with stability.

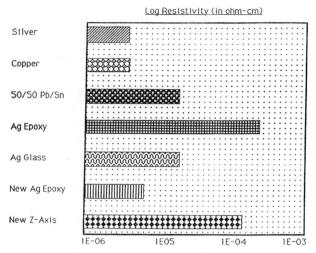

Figure 6.13   Volume electrical resistivity of selected materials.

All three types of conductive paste are filled with silver particles as the conductive filler to effect the electrical connection. Silver offers excellent electrical conductivity, as do the oxides of silver (Fig. 6.13), and silver is easily processed to provide excellent rheology for conductive pastes. Additionally, silver-filled conductive pastes are well accepted in microelectronics industry and volumes of reliability data is available which attests to their excellent electrical, thermal, and mechanical performance. Gold and silver/platinum fillers have been advanced as potential fillers for conductive pastes. While both may address any potential concerns one might have about electromigration, neither is as conductive as silver and both are difficult to process into smooth thixotropic pastes suitable for stencil printing of bumps.

Finally all three conductive polymer types can be formulated to meet the specific requirements of an application. Properties such as levels of mobile ions, glass transition, elastic modulus, creep, and rheology can be tailored to ensure that reliable results are obtained.

6.5   Conductive Adhesives and Known Good Die

Once the selection of the ideal conductive paste is complete and IC devices have been bumped, the IC devices are ready to be electrically tested and burned in where the application dictates that known good die (KGD) be used. High-reliability MCM applications, and "high end" consumer applications like microprocessor ICs are examples where fully tested die are needed before committing the chip to the package. All three conductive adhesive bump types—thermosets, B-stage, and thermoplastic—can be tested once they have been cured or dried. The B-stage and thermoplastic conductive adhesives are electrically conductive when dried, and of course the cured thermoset is electrically conductive after cure. In one study 1-Mbyte DRAM devices were bumped and subjected to a KGD test matrix. The memory die were then burned in at 7.0 V and 125°C for 20 h. Post-burn-in electrical testing was performed to select DRAM devices with a <70-ns access time. The KGD burn-in and electrical testing of DRAM devices was performed by EPI Technologies of Richardson, Texas, as this company appears to have a viable KGD solution which is automated and compatible with many types of IC semiconductor devices, and which provides repeatable results. There are many KGD approaches in the written literature, and packaging engineers need to utilize KGD techniques which are repeatable, flexible enough to be applied to all kinds of ICs, and can be accomplished economically.

The ability to test polymer-bumped IC devices is critical to their use in flip chip applications. Performing KGD pretesting with unbumped devices will not ensure that the bumped devices will meet performance criteria because the bond pad–adhesive bump interface and IC integrity has not been verified. The result may be that "once good die" may no longer be viable. This leads to potential reliability problems with flip chip devices and costly rework processes to replace the die. One advantage of polymeric conductive bump materials for flip chip is the fact that they are reworked readily by heating the bumps to a temperature which is higher than their glass transition temperature T_g or softening point. For example, a conductive adhesive having a $T_g = 100°C$ can be heated to 140°C, where its flexibility and adhesive strength are considerably weaker. In this state the die are easily removed from the substrate with minimum force. In the case of conductive thermoplastic formulations, the bumps are heated to a temperature above the melting point of the thermoplastic and the IC is readily removed from the circuit. The ability to formulate a bump adhesive to ensure a specific T_g range makes the adhesives attractive for applications where it is desirable to match the rework or operating temperature of the flip chip assemblies to a specific range of rework, or even cure temperatures of the conductive adhesives.

6.6 Assembly of Flip Chip Devices with Polymer Bumps

The process by which IC devices, having either solder bumps or conductive adhesive bumps, are attached to the circuit is quite uniform.

Figure 6.14 Flip chip aligner-bonder.

A flip chip aligner-bonder, such as the MRSI model 503TCS or 505 (Fig. 6.14), is used to pick up bumped die from waffles packs, orient and precisely align the chip to the substrate, and then accurately place the bumps onto the corresponding contact pads of the circuit. These aligner-bonders for flip chip applications utilize precision optics, manual or automatic alignment, and utilize both heat and pressure to effect the attachment of the bumps to the substrate conductor pads. The amounts of heat and pressure are dictated by the materials used and the size of the thermal mass which is being heated. Conductive adhesive bumps require considerably less pressure than do solder bumps to ensure good wetting of the bumps onto the substrate. ICs which are 120×120 mils, and have two hundred bumps, require bonding pressures which do not exceed 100 g. The adhesive bumps are more compliant than solder bumps and easily wet onto a surface when heated. In fact, one must use caution not to apply excess pressure, or the adhesive bumps will be squashed flat. This will adversely affect the mechanical and electrical integrity of the bumps.

The final step in the assembly of flip chip devices with conductive adhesive bumps is the underfill process. The use of an underfill is mandatory when using conductive adhesive bumps. While individual solder bumps have pushoff strengths in excess of 200 g, polymer bumps rarely exceed 80 g of strength for 4-mil-diameter bumps. Therefore the underfill is needed to enhance the mechanical strength

LOW STRESS EPOXY SEALANT

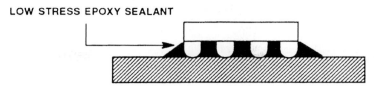

Figure 6.15 Flip chip with underfill.

of the flip chip assembly. The other major reasons for using underfill are to prevent electromigration from occurring, and to provide dimensional stability to the flip chip assembly during thermal excursion. The underfill provides environmental protection, keeping moisture from the bumps, thus minimizing the possibility of corrosion or electromigration. Figure 6.15 is a cross-sectional diagram of a flip chip assembly with underfill. The role of underfill encapsulants will be discussed further in the reliability sections.

As a final note regarding the process by which flip chip assemblies are made using conductive adhesives, Fig. 6.16 shows a process flow diagram which can be used as a model for the process sequence required to manufacture conductive adhesive bump flip chip assemblies.

6.7 Properties of Conductive Adhesive Bumps

Prior to a discussion of the reliability of conductive adhesive bumps in flip chip applications, it is important to understand the properties, shape, and structure of the conductive bumps, and the capabilities of the processes used to form the bumps. Table 6.3 summarizes the results of a study on process capability and bump properties. A 2-mil-thick stencil with 110-μm-diameter openings was used to form bumps on a silicon wafer. A random sampling of 100 bumps was made in order to study the properties of conductive adhesive bumps as well as the repeatability of the process.

Given a theoretical 50-μm bump height, an average bump height of 50.4 μm was realized, with a tolerance of $\pm$ 3 μm. This would certainly seem to be adequate coplanarity for a majority of flip chip applications. The data taken on bump diameter did not indicate tolerances which were as tight as for bump height. The average bump diameter of 113 μm is acceptable, given the 110-μm theoretical diameter, but the spread from 105 to 120 μm may be too large for fine-pitch devices. However, it should be noted that the stencil used was chemically etched stainless steel, and the chemical etching process does not give the same degree of reproducibility as a laser etch or electroform stencil.

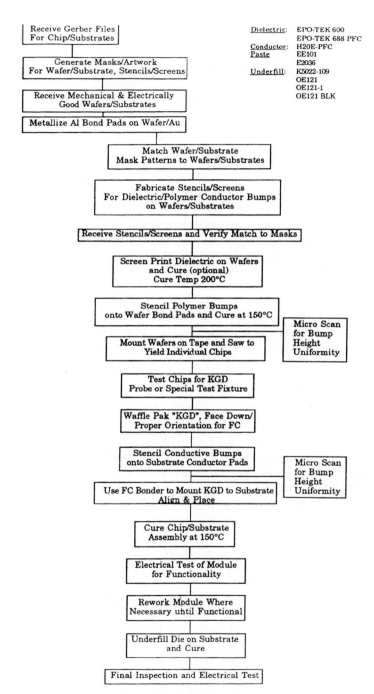

Figure 6.16 Process flowchart for polymer flip chip.

TABLE 6.3   Properties of Conductive Adhesive Bumps [Polymer Bumps (PFC)]

Bump height, μm (100 bumps)
Theoretical, 50
Average, 50.4
Range, 47–53
Bump diameter, μm (theoretical from mask, 110 μm)
Average 113
Smallest 105
Largest 120
Ball shear, g
Aluminum
High, 37.5
Low, 19.5
Average, 27.6
Mostly cohesive failures
Gold
High, 80
Low, 36.5
Average, 59.4
All cohesive failures
Bump resistance, mΩ
Aluminum: 40–80 average
Gold: 10–20 average

Ball shear data was generated using a Royce MSB200 system to push the bumps off of aluminum and gold surfaces. As expected, the adhesion of the bumps to the aluminum was inferior to that of gold. High values (80 g) were achieved with gold, and a low value (36.5 g) occurred on only one bump. This may have been due to inadequate wetting of the polymer bump to the gold or perhaps to experimental error. In all cases the adhesion to gold is superior to that achieved with aluminum. Solder bumps, tested with the same Royce system, will consistently yield ball shear values in excess of 200 g. This further reinforces the need for underfill encapsulants with polymer flip chip assemblies.

The final test made was to test bump resistance; this was the electrical resistance from the top of a bump to the edge of a bond pad on which the bump rests. As expected, low stable resistance values in the 10- to 20-mΩ range were obtained on gold. However, on aluminum metallization, the values for resistance vary from 40 to 80 mΩ to as high as several megohms of resistance. This was expected and confirms that the oxide coating on aluminum makes it impossible to achieve a low-resistance, stable electrical contact when used in conjunction with electrically conductive adhesives. Inductance measurements have also been made for polymer bumps which show performance comparable to that of solder bumps, with inductance readings typically <0.1 nH (Table 6.4).

TABLE 6.4   Interconnect Resistance Wire Bond versus Polymer Bump

Interconnect	Length, in	Resistance, mΩ
Wire bond (Au)	0.022	55–80
Conductive polymer bump (Ag)	0.003	25–50

Figure 6.17 Side profile of polymer bump.

6.7.1 Bump morphology before and during assembly

The adhesive bumps themselves are quite spherical in nature when stenciled, but a side profile of a bump, made by SEM techniques, shows that the bumps tend to be shaped more like a mountain peak than a true sphere. Top and side views of adhesive bumps, taken at 860× and 480× magnifications, depicted in Figs. 6.17 and 6.18, show the typical morphology of the bumps. The pictures also show the fine-grain structure of the silver particles used to fill the conductive adhesive. The SEM photo shown in Fig. 6.19 shows a 1000× photo of the conductive particles used; these are a mixture of both small and large silver flakes. The tight packing of the conductive silver flake and the shrinkage of the polymer during curing create a high-density silver matrix where electrical conductivity is thought to occur via electron tunneling between adjacent conductive particles.

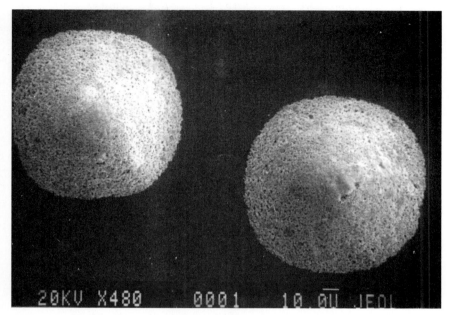

Figure 6.18 Conductive polymer bump, top view.

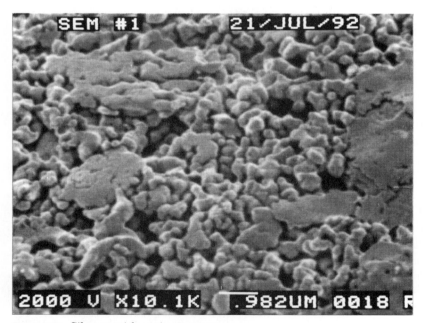

Figure 6.19 Silver particle grain structure.

The percentage of conductive flake used in the adhesive formulation to achieve optimum electrical and thermal conductivity is highly dependent on the chemistry used and the desired rheological properties for processing.

6.7.2 Bump morphology after assembly

When a bumped IC is attached to the substrate, the shape of the bump changes according to the amount of pressure and heat used in the bonding process. The photograph in Fig. 6.20a and 6.20b depicts a bonded cross section showing the bumps attached to both the chip and the substrate. The bump no longer looks like a mountain peak, as it has been compressed during the bonding process to take on more a collimated shape. This type of collimated interconnect is always formed when a B-stage or thermoplastic polymer bump is used in the flip chip process. When thermoset bumps are used, and the cured bumps are interconnected to the substrate, the mountain shape of the bump is maintained. This leads to an interconnect which has the shape of an inverted funnel as opposed to a straight column.

6.8   Conductive Adhesive Bump Reliability

The following discussion details the results of reliability studies that are currently available for flip chip assemblies produced with conductive adhesive bumps. These studies have been conducted using a variety of different test vehicles, in order to compare conductive adhesive bump technology to comparable wire-bond and solder-bump interconnect techniques. The use of polymeric conductive materials for electrical interconnect is relatively new to the industry, and therefore a great deal of experimentation is still being conducted to establish design rules, process parameters, and performance thresholds for the use of conductive adhesives as the active electrical interconnect in flip chip devices. The following discussion focuses on the results of studies already completed.

6.8.1   Resistor networks

The first multipurpose test vehicle designed for determination of conductive polymer bump reliability was a thin-film resistor network. The power ratings of the resistors varied from 35 mW to 1 W. The resistor metallization in the network was Au deposited on alumina, and comparisons of polymer-bumped resistors to wire-bonded resistors, both in the same package, were made. A schematic of the resistor network test vehicle is shown in Fig. 6.21, and the polymer-bumped and wire-bonded resistor networks were assembled onto the same substrate and her-

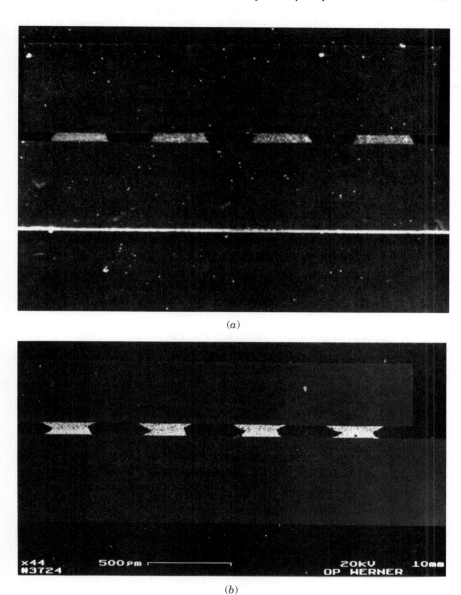

(a)

(b)

Figure 6.20 Conductive adhesive bump assembly: (a) column; (b) hourglass shape.

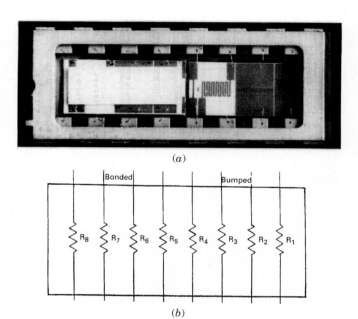

Figure 6.21 R values from 200 mΩ to 80 Ω, thin-film resistor network.

metically sealed in the same package cavity. All resistor devices were underfilled with a liquid encapsulant to improve strength of the bonded networks. Testing both the polymer bump and wire-bonded resistors in the same package cavity was done to ensure reduction of the number of variables in the test matrix, and also that the conductive adhesive–bumped resistors and wire-bonded resistors would be equally stressed in the sequential environmental test matrix.

Table 6.5 shows the environmental tests to which the resistor network packages were subjected. Where feasible, the testing was done in accordance with current military specifications. As depicted in Fig. 6.21, the resistor values R_1 through R_4 represent polymer-bumped resistors, while R_5 through R_8 represent wire-bonded resistors. The sequential environmental test data is shown in Tables 6.6 to 6.9 and is summarized below. The change in resistance values are in milliohms, and represent the change from initial readings of the resistors before the environmental testing.

With the exception of the thermal shock testing, the performance of the conductive adhesive bump and wire-bonded resistor networks appears to be comparable after environmental stressing. The anomaly in the test data is the thermal shock data. Posttest failure analysis reveals that some of the silver adhesive bumps were delaminated from the gold on the substrate, as though they were pulled or pushed away

TABLE 6.5   Environmental Tests*

Test description	Conditions	Duration	Measurement points
High-temperature operation life	125°C with bias	1000 h	0, 40, 250, 500, 1000
High-temperature storage	150°C	1000 h	0, 40, 250, 500, 1000
Temperature shock	−55 to 150°C air–air 10-min dwell time, 10-s transition	100 cycles	0, 20, 50, 100 cycles
Power cycling	25°C–ambient bias on/off —5 min	500 h	0, 40, 250, 500
Particle impact noise detection	MIL-STD-883, method 2020, condition B	10 passes	2, 4, 6, 8, 10

Results: The average (ΔR) change in resistance is based on readings recorded prior to environmental testing. R_1 through R_4 are the bumped devices; R_5 through R_8 are the corresponding bonded devices.

TABLE 6.6   High-Temperature Operation Life,* ΔR, mΩ

Time, h	40	250	500	1000
R_1	6	115*	7	12
R_5	8	7	8	9
R_2	5	6	5	14
R_6	7	6	6	8
R_3	12	10	9	10
R_7	11	9	8	4
R_4	25	18	15	22
R_8	20	19	16	25

*Comparable changes in resistance values were obtained for adhesive and wire bonds.

from the gold. These failures have been traced to the high CTE (>30 ppm/°C) of the underfill encapsulant used in the experiments. The use of an underfill with a lower CTE value, preferably less than 20 ppm/°C, will help alleviate these types of failures in flip chip assemblies.

As a final test, a comparison of the conductive adhesive "bump" resistance versus "wire bond" resistance has been made. The inductance in a 0.020 in length of 1.0-mil-thick gold wire has an average resistance of 55 to 80 mΩ, while polymer bumps, having a 100 μm diameter and 75 μm height, have a resistance of 25 to 50 mΩ. The conductive adhesive bumps, having a shorter length, exhibit a lower resistance pathway as shown in Table 6.10.

TABLE 6.7   High-Temperature Storage,* ΔR, mΩ

Time, h	40	250	500	1000
R_1	7	6	7*	10
R_5	5	7	5	8
R_2	15	11	10	14
R_6	11	7	8	15
R_3	18	14	11	13
R_7	14	13	12	11
R_4	31	33	30	25
R_8	20	35	31	22

*Comparable change in resistance readings for conductive adhesive bumps and wire-bonded resistors.

TABLE 6.8   Power Cycling,* ΔR, mΩ

Time, h	40	250	500
R_1	5	6	10
R_5	4	6	12
R_2	7	10	14
R_6	5	9	15
R_3	15	11	20
R_7	11	12	18
R_4	22	20	25
R_8	24	19	19

*Comparable change in resistance for polymer bumped and wire-bonded devices.

TABLE 6.9   Temperature Shock,* ΔR, mΩ

Cycles	20	50	100
R_1	4	5	>-(2) open
R_5	3	6	22
R_2	7	5	>-(2) open
R_6	5	8	20
R_3	11	10	>-(1) open
R_7	13	9	27
R_4	22	19	>-(2) open
R_8	19	18	32

*All resistors show significant increase in resistance. At least one conductive adhesive bump connection opens up for each resistor type. This type of failure indicates that the adhesive bumps have cracked or delaminated from the substrate.

TABLE 6.10   Comparison of Chip Electrical Specifications

Method	Cross section, mils	Length, mils	Inductance, nH
Wire bond	1 (diameter)	100	1.8
TAB	1×3	100	1.4
Flip chip (solder)	4×4	4	< 0.1
Flip chip (conductive polymer)	5×5	0.5	< 0.1

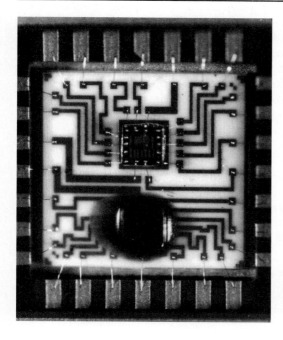

Figure 6.22 Two-chip MCM wire bond and polymer bump.

6.8.2 Two-chip MCM module

A second test vehicle was designed, fabricated, and tested, again utilizing the comparison between conductive polymer bumps and wire-bonded devices. The design in this study was a dual, isolated IC interconnect circuit which was plated into a standard A&B, 28-lead "dual in-line" ceramic package as shown in Fig. 6.22. The thin-film metallization was deposited onto alumina, and efforts were made to equalize the NiCrAu signal etch lines for both devices. The IC chips used were triple-input NAND gates, 4023AH semiconductors with the bond pads having 4-mil-square geometries.

The Al bond pads were metallized with a TiN as a barrier metal, and Au was sputtered over the TiN to form the top-layer, noble metal conductor. This was done to eliminate increases in contact resistance due to aluminum oxide. The objective of this study was to again evaluate the performance and reliability of wire-bonded versus conductive adhesive bump interconnects. A Sentry 21 test program was created to meet the characteristics of a standard CD4023AH device. The test program included an electrical functionality test, high- and low-level input leakage (I_{iH}, I_{iL}), high- and low-level output voltage, (V_{oH}, V_{oL}), and propagation delay (T_{PHL}, T_{PLH}). The initial electrical test measurements, shown in Table 6.11, depict equivalent performance for polymer-bumped and wire-bonded devices, with the exception of V_{oH} measurements where the wire-bonded devices have lower values. However, all measurements for both interconnect types are within the expected operational parameters for this type of semiconductor IC. All measurements were taken at ambient temperature and at a V_{dd} of 5.0 V.

Having established baseline electrical parameters for the IC device performance in the two-chip, MCM test vehicle, more extensive testing was conducted on the packages. A total of 25 packages each were subjected to the following environmental tests:

High-temperature storage	150°C for 1000 h
Thermal shock	−25 to +125°C for 100 cycles
Thermal cycling	−55 to +150°C for 1000 cycles
Temperature and humidity bias	85°C, 85% RH for 1000 h

The results of this testing (see Tables 6.12 to 6.15) show fairly com-

TABLE 6.11   Electrical Test Results, Preenvironment Testing of Two-Chip MCM

Parameter	Polymer-bumped	Wire-bonded
T_{PHL}, ns	61.0–68.7	60.4–69.2
T_{PLH}, ns	82.0–84.8	81.1–84.2
I_{iH}, nA	1–3	1–3
I_{iL}, nA	0–(−10)	0–(−10)
*V_{oH}, V	3.20–3.31	2.63–2.89
V_{oL} (MV), mV	386–404	384–400

TABLE 6.12   Two-Chip MCM High-Temperature Storage*

Parameter	Polymer bump (PB)	Wire bond (WB)	Change, % PB	Change, % WB
T_{PHL}, ns	61.0–69.0	60.4–69.4	0.44	0.29
T_{PLH}, ns	82.0–85.3	81.1–84.4	0.59	0.24
I_{iH}, nA	1.1–3.2	1.0–3.1	6.6	3.3
I_{iL}, nA	0–(−10)	0–(−10)	—	—
V_{oH}, V	3.20–3.34	2.63–2.94	0.91	1.7
V_{oL}, mV	386–407	384–403	0.95	0.75

*$T = 150°C$; 1000-h storage.

TABLE 6.13 Two-Chip MCM Thermal Shock per 15-s Dwell Time*

Parameter	Polymer bump (PB)	Wire bond (WB)	Change, % PB	WB
T_{PHL}, ns	61.0–70.2	60.5–70.7	2.2	2.1
T_{PLH}, ns	82.2–85.6	81.3–85.1	0.94	1.06
I_{iH}, nA	1.0–3.3	1.0–3.2	10.0	6.67
I_{iL}, nA	0–(−10)	0–(−10)	—	—
V_{oH}, V	3.20–3.39	2.64–2.95	2.41	2.07
V_{oL}, mV	386–411	384–406	1.73	1.50

*$T = -25$ to $+125°C$; 100 cycles.

TABLE 6.14 Two-Chip MCM Thermal Cycling per 15-min Dwell Time*

Parameter	Polymer bump (PB)	Wire bond (WB)	Change, % PB	WB
T_{PHL}, ns	61.3–69.2	60.6–69.7	0.72	0.71
T_{PLH}, ns	82.0–85.3	81.2–84.9	0.59	0.83
I_{iH}, nA	1.1–3.2	1.0–3.1	6.7	3.3
I_{iL}, nA	0–(−10)	0–(−10)	—	—
V_{oH}, V	3.20–3.34	2.63–3.02	0.91	4.5
V_{oL}, mV	387–409	384–403	1.2	0.75

*$T = -55$ to $+150°C$; 1000 cycles.

TABLE 6.15 Two-Chip MCM Temperature-Humidity Bias*

Parameter	Polymer bump (PB)	Wire bond (WB)	Change, % PB	WB
T_{PHL}, ns	61.2–70.1	60.3–70.2	2.03	1.45
T_{PLH}, ns	82.0–85.9	81.1–85.3	1.29	1.30
I_{iH}, nA	1.1–3.3	1.0–3.2	10.0	6.6
I_{iL}, nA	0–(−10)	0–(−10)	—	—
V_{oH}, V	3.20–3.41	2.63–2.97	3.02	2.77
V_{oL}, mV	386–412	384–407	1.98	1.75

*$T = 85°C$; RH = 85%; 1000 h.

parable results between the wire-bonded ICs and the polymer-bumped chips. The most notable difference is seen in the I_H (nanoampere) readings where the leakage current for the polymer bumped devices is consistently higher than that of the wire-bonded devices. At the present time there is no clear explanation for this difference. The >6 percent shift in the leakage current is probably a function of an interface problem between the gold pad–epoxy bump, which has not yet been determined. However, overall results show good performance of the polymer bumped devices in this flip chip application.

6.8.3 Four-Chip MCM Module

As an extension of this test program, a four-chip MCM has been designed and fabricated using the same 4023AH, triple-input NAND gate semiconductor devices. Thin-film metallization was deposited onto alumina, and again efforts were made to equalize the NiCrAu etch lines of all devices. Four circuit test sites were deposited onto the alumina substrate, and chips were mounted. One IC was wire-bonded, one device was mounted as a flip chip using polymer bumps, and the last device was mounted as a flip chip using solder bumps as the interconnect. The alumina substrate was then mounted into a ceramic PGA package, and circuit connections were wire-bonded from the alumina substrate to the package (Fig. 6.23).

Initial readings have been taken for propagation delay, input leakage, and output voltage as with the two-chip MCM. These electrical test readings, shown in Table 6.16, show that the three different interconnect types exhibit little difference in electrical performance. The wire-bonded

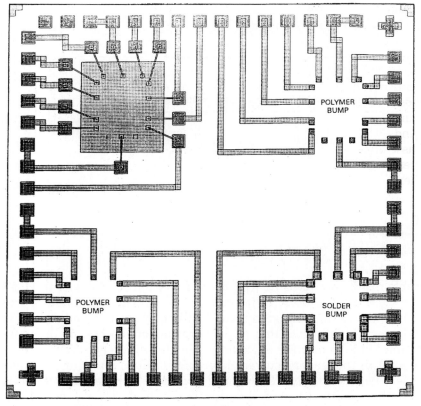

Figure 6.23 Four-chip MCM circuit layout: polymer bump, solder wire bond.

TABLE 6.16   Four-Chip MCM: Initial Electrical Data*—Solder Bump, Polymer Bump, Wire Bond

Parameter	Polymer bump	Wire bond	Solder bump
T_{PHL}, ns	62.6–69.4	61.7–68.8	62.6–69.8
T_{PLH}, ns	82.6–85.8	81.9–85.1	82.3–85.4
I_{iH}, nA	1.0–3.0	1.0–3.0	1.0–3.0
I_{iL}, nA	0–(−10)	0–(−10)	0–(−10)
V_{oH}, V	3.25–3.39	2.84–3.01	3.19–3.34
V_{oL}, mV	381–403	378–399	383–401

*$T = 20°C$; $V_{dd} = 5.0$ V.

devices show slightly lower propagation delay values than do the polymer-bumped or solder-bumped ICs, but all devices function within the operational devices expected for this NAND gate semiconductor device. Currently these PGA packages are undergoing the same sequence of environmental tests as the two-chip MCM module were exposed to. The results obtained thus far indicate that the conductive adhesive bumps provide mechanical and electrical performance similar to that of wire-bonded and solder-bumped devices. The test results obtained in the studies described above reflect flip chip assemblies which have gold metallization on bond pad and also on the circuit pad. Also, the CTE mismatch is minimized as silicon IC chips (CTE = 6 ppm/°C) are bonded to ceramic substrates (CTE = 4 ppm/°C). As the CTE mismatch increases, one must become more conscious of the strain exerted on the adhesive bumps and select conductive adhesives with lower modulus of elasticity, and underfill encapsulants which have low CTE, high modulus of elasticity, and T_g values which exceed the maximum operating temperature of the assembled circuit. Flip chip bonding of silicon ICs to high CTE substrates such as FR-4, or to flexible substrates, continues to be the most challenging applications for conductive adhesive flip chip applications.

6.8.4   Chemical sensor IC packaging

In a study recently completed at the University of Michigan a polymer flip chip process using conductive adhesive bumps has been successfully applied to silicon-based chemical sensor arrays. Passive sensors were fabricated on silicon, and sensor-specific polymer membranes were screen-printed directly onto the electrode sites. The sensor chips were mounted onto the substrate using the flip chip interconnect approach in which fluid channels were sealed with a polymer gasket. The electrical connections between chip and substrate were made using conductive adhesive bumps. Figure 6.24 shows the structure of the silicon sensor and bumps which were fabricated in this study. The silver epoxy bumps were applied to 15×15-mm chips; the bond pads were 4 mils square on an 8-mil pitch. Bump height deposited onto the

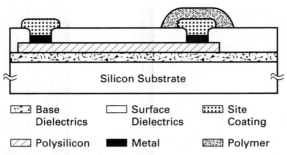

Base
Dielectrics   Surface
Dielectrics   Site
Coating

Polysilicon   Metal   Polymer

Figure 6.24   Silicon sensor with adhesive bumps.

silicon sensors was 25 µm, Fully functional chemical sensor devices
were fabricated using the conductive adhesive bump, flip chip technology to provide the electrical contact between the flow sensor chips and
the circuit.

In a similar test matrix adhesive bumps were used to flip chip fluid
logic-ISFET sensors in work conducted by the Technical University of
Dresden in Germany. A multiple-ion analyzer was developed for the
sensing of liquid media using a microsensor array, microactuator, mi-
crofluidics, and electronic analyzing equipment. The low processing
temperature of the conductive adhesives bumping adhesive (<150°C)
proved to be an optimum solution for attaching the ISFET signal lines
to the spacer chip of the FIM (fluidic-ISFET-micro system). The ISFET
has high thermal sensitivity, and therefore PbSn solder bump connections could not be chosen for this applications. Figures 6.25 and 6.26
show the FIM and flip chip cross section of the assembly. The isotropically conductive adhesive was stencil-printed to form the bumps which
provide the electrical connection. The bumps were 50 µm in height
and 5 mils in diameter, and center-to-center pitch was 10 mils.
Transition resistance data obtained through the interconnect were less
than 25 mΩ, and there were no failures of sensing devices which were
attributed to ionic impurities in the conductive adhesive. Figure 6.27

ISFET and Spacer chip

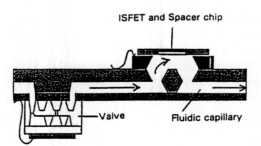

Valve   Fluidic capillary

Figure 6.25   Cross section of
FIM.

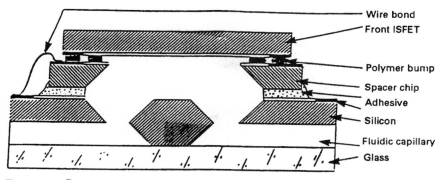

Figure 6.26 Cross section of ISFET, spacer chips, silicon glass with polymer bumps.

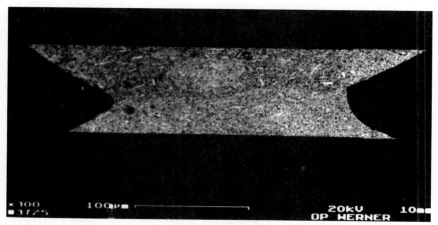

Figure 6.27 Two-bump conductive adhesive flip chip assembly.

shows a cross section of the conductive adhesive bump interconnection. Note the hourglass shape of the connection. As previously discussed, this is the expected shape of the conductive adhesive bump connection when a cured adhesive bump is pushed into wet conductive adhesive dispensed or printed onto the contact pad of the substrate. The column-shaped bump provided by B-stage or thermoplastic bumps may be more ideal for conductive bumps as they provide for more conductive pathways within the columnar structure. The results of the study clearly show that conductive adhesives can be used in flip chip applications to provide stable electrical and mechanical connections in complex microsystems.

TABLE 6.17   Substrate Materials Metallization and CTE

Substrate material	Metallization	Thermal expansion coefficient, ppm/K
FR-4	Laminated Cu, electroless plated NiAu	16
Ceramics Al_2O_3	Thick-film ink	7
Glass	Thin-film CrPdAu	5
Si	Thin film Al, electroless plated NiAu	3

6.9   Conductive Adhesive Bumps on Various Substrates

Another flip chip study made of conductive adhesive bumps has been carried out in a collaborative effort between the Technical University of Budapest and Deutsche Thomson in Germany. This study dealt specifically with the potential use of conductive adhesives for flip chip applications in consumer electronics. Daisy-chain silicon IC devices were fabricated, gold-metallized, bumped using electrically conductive polymers, and finally flip chip–attached to four different substrate types. The substrate materials used in this study were FR-4, ceramic, glass, and silicon as shown in Table 6.17. The daisy-chain IC measured 2.4×2.4 mm in size, and 4×4-mil bond pads were arranged in an array having a pitch of 300 μm.

After attaching the chips to the four substrate types, the Kelvin method was employed to measure the transition resistance (TRR) of the devices. The initial results showed that the 10-μm-high bumps yielded a TRR of ≤25 mΩ or less. The flip chip samples were then subjected to a series of environmental stresses. The transition resistance was measured as a function of

- Hours of storage at 150°C
- Humidity testing: 85°C, 85% RH
- Thermal cycling: −25 to +125°C or 0 to +100°C

The results obtained in these experiments are summarized below.

6.9.1   Silicon substrates

150°C storage.   There is virtually no change in TRR average value of 25 mΩ after 250 h of storage.

Thermal cycling (−25 to +125°C).   After 100 cycles the TRR values show no significant shift from the 25-mΩ initial value, even on 6×6-mm chips.

6.9.2 Glass substrates

150°C storage. The transition resistance increases with storage time. Up to 65 h the TRR is still at acceptable levels, but increases to unacceptable levels at longer storage times. This phenomenon is attributed to poor adhesion of the adhesive bump to the shiny metallization, and also to the fact that the polymer bump must have a thickness of >10 μm in order to minimize thermal stresses.

Thermal cycling (−25 to +125°C). The TRR does not increase from the initial low reading after 130 thermal cycles, including larger 6×6-mm silicon devices.

Humidity test (85°C, 85% RH). For chip sizes of ≤4×4 mm there is no increase in the TRR after 300 h of storage. Larger chips exhibit positive shifts in the TRR, presumably due to the larger number of chips involved in the TRR measurements.

6.9.3 Ceramic substrates

150°C storage. There is no change in TRR after 260 h at 150°C. The adhesion of the polymer bumps to the thick-film metallization is excellent.

Thermal cycling (−25 to +125°C). For 6×6-mm chips the TRR increases significantly after 10 thermal cycles. The 10-μm-thick bump is too thin, the stresses create too great a strain, and adhesion of the bumps is lost. The 4×4-mm chips survive 60 thermal cycles before the TRR increases, and both chips sizes survive 100 thermal cycles between 0 and 100°C. The differences in CTE of chip and substrate, and the thin polymer bump (10 microns) are responsible for the thermal cycle failures.

Humidity storage. There is no deviation from the initial 25 milli-ohm TRR after 250 hours for chip sizes up to 5 mm×5 mm.

6.9.4 FR-4 substrates

150°C storage. There is no shift in the TRR after 250 h for chip sizes of ≤ 6×6 mm.

Thermal cycle (−25 to +125°C). After five thermal cycles all polymer-bump connections are broken for chip sizes > 4×4 mm. Thermal cycling between 0 and 100°C causes a gradual increase in the transition resistance. As an example, 3×3-mm chips show a tenfold increase in TRR after 100 cycles.

Humidity test (85°C, 85% RH). There is little change in TRR for chip sizes of <3×3 mm. For larger chip sizes the TRR gradually increases

TABLE 6.18   Electrical Failures versus Chip Size Thermal
Cycling

Chip size, mm^2	Silicon, %	Glass, %	Ceramic, %
2.4×2.4	0	0	0
4.8×4.8		0	25
7.2×7.2	0		
9.6×9.6	7	7	43
12.0×12.0	16		
14.4×14.4	14	71	100

as a result of the thermal fatigue of the bumps, which is caused by the CTE mismatch between the FR-4 and silicon.

In all the data taken above it should be noted that no underfill was used to encapsulate the flip chips. The use of an underfill which has a low CTE, T_g value of >140°C, and a high modulus, will impart better strength to the unit and absorb much of thermal stress exerted on the bumps. This would lead to lower numbers of failures occurring especially during thermal cycling.

Additionally the data in Table 6.18 shows the results of thermal cycling tests, 100 cycles from −25 to +125°C for silicon chips with increasingly large chip sizes mounted onto the different substrate types. It is clear from these results that the mortality rate of the flip chip connections increases as the size of the chip increases and also as the CTE mismatch increases. The failure rates for silicon chips to FR-4 was not included, but comparable results are to be expected.

The results of this experimentation show that conductive adhesive flip chip connections are viable and should be studied further for use in consumer electronic applications. The results are the best for silicon-to-silicon, where no CTE mismatch exists, but the study also shows that the adhesive bumps can be used for smaller chips attached to higher-CTE, FR-4 substrates. In addition, optimizing bump heights, material properties, process parameters, and underfill materials can help alleviate potential problems incurred when flip chip–bonding silicon chips to high CTE substrates.

In finite elemental analysis (FEA) modeling it has been determined that as much as a 10 percent strain is exerted on polymer bumps, 50 μm high and 4 mils in diameter, when 90×90-mil chips are flip chip–attached to FR4 substrates. Given the average tensile strength of 10,000 psi (lb/in^2) for electrically conductive adhesives, the 10 percent strain is too great and the conductive adhesive bonds will break. The solution to this problem lies in the selection of the proper underfill which can absorb the majority of the thermal stress, leaving the polymer bump connections intact so that electrical performance is not compromised.

6.10 Electrically Conductive Polymer Flip Chip Applications

Electrically conductive adhesives have been employed for several applications, and this continues to increase as more data is collected on the process capability and reliability of polymer-bump connections. Some of these applications are briefly described and summarized below.

6.10.1 Optocoupler devices

Infrared detector/emitter pairs have been fabricated using electrically conductive bumps for flip chip interconnect. Special IR-transmitting underfill materials have been formulated to ensure efficient data transfer when this system is used to couple computers and transfer information between them.

6.10.2 High-frequency applications

Gallium arsenide IC devices have been polymer-bumped and attached to ceramic substrates. These devices have been operated successfully at frequencies up to 10 GHz with no significant noise detected in the circuit.

6.10.3 Memory module applications

One-megabyte DRAM devices have been polymer-bumped with electrically conductive adhesives and flip chip–attached directly to copper pads on FR-4 substrates. Standard 1×9 SIMM devices have been fabricated which operate at 60 ns at 25°C and 65 ns at 70°C. KGD techniques have been applied to test and burn-in the DRAM devices prior to assembly.

6.10.4 LCD driver chip for COG applications

Electrically conductive adhesives have been used to bump LCD driver chips and then to attach these to glass substrates. The silver-filled epoxy bumps exhibit contact resistance, on ITO (indium tin oxide), of 2-4 Ω per bump, and the connections survive thermal shock and humidity aging tests with little or no degradation of the electrical and mechanical connection of the chip to the ITO on the glass substrate.

6.10.5 RFID applications

Electrically conductive adhesive bumps have been used to flip chip–mount RFID devices directly onto conductive films on Mylar. Functional devices have been obtained using this technique to mount the chip.

6.10.6 High-density display technology

Conductive polymers have been successfully employed to form 80,000 connections having 3-mil-diameter bumps on an 8-mil pitch. Electrical and mechanical integrity of the flip chip connections have been established and operational units are routinely fabricated.

6.11 Conclusion

As the technological infrastructure continues to develop for the commercialization of flip chip interconnect in microelectronics applications, the role of electrically conductive adhesives is increasing. Polymer-bumped, flip chip assemblies are being fabricated with low-cost processes, and reliability data generated indicates that electrically conductive adhesives offer excellent reliability as the active electrical interconnect medium.

The role of electrically conductive adhesives will increase as the flip chip is more widely adopted for packaging applications. The low contact resistance achieved with conductive adhesives and the ability to process at low temperatures affords the opportunity to apply flip chip technology to a variety of low-temperature substrate types. In addition the ability to rework the conductive adhesives without sacrificing the integrity of package reliability makes the use of conductive adhesives more attractive.

Reliability data continues to be developed, and the number of applications for polymer flip chip continues to expand as the industry seeks new methods to successfully interconnect IC chips in smaller and smaller packages without compromising performance of the circuits. Design rules and process parameters are being established for the use of conductive adhesives in flip chip applications where solder bumps cannot be applied because of the high process temperature or environmental issues involved with cleaning of flux after flip chip attach.

Presently it appears that silver-filled adhesives will dominate the market because of their excellent electrical conductivity and their excellent reliability in microelectronics applications for the last two decades.

References

1. Epoxy Technology, "Design Predictions for the 70's," *EDN,* 38, Jan. 1970.
2. Kulesza, F. W., "Stuck by Conductive Epoxies," *EDN,* 33, March 1971.
3. Epoxy Technology, *Testing Program Proves Superiority of Electrically Conductive Adhesives over Eutectic Die Bonding,* in-house Test Program, July 1971.
4. Kulesza, F. W., "Epoxies vs. Eutectic Chip Bonding," *Electron. Packag. Product.* **2:** 99, 1971.
5. David, R. F. S., "Advances in Epoxy Die Attach," *Solid State Technol.,* 40, 1975.
6. Kulesza, F. W., R. Estes, and K. Spanjer, "A Screen Printable Polyimide Coating for Silicon Wafers," *Solid State Technol.,* 1988.

7. Estes, R., and R. Pernice, "Die Attach Adhesives Evaluation of VceSat and Thermal Resistance Performance in Power Devices," *ISHM Proceedings,* 1989.
8. Mitchell, C., and H. Berg, *Use of Conductive Epoxies for Die Attach,* Motorola, Inc.
9. Ginsberg, G., and D. Schnorr, *Multichip Modules and Related Technologies,* McGraw-Hill, New York, 1994.
10. Estes, R., and F. W. Kulesza, "Solderless Flip Chip Technology," *Hybrid Circuit Technol.,* 1992.
11. Estes, R., and F. W. Kulesza, D. Baczek, and G. Riley, "Environmental and Reliability Testing of Conductive Polymer Flip Chip Assemblies," *IEPS Proceedings,* Sept. 1993.
12. Estes, R., and F. W. Kulesza, "Fabrication and Assembly Processes for Solderless Flip Chip Assemblies," *IEPS Proceedings,* 1992.
13. Estes, R., "Polymer Flip Chip—PFC, a Technology Assessment of Solderless Bump Processes and Reliability," *IEPS Proceedings,* 1994.
14. Poplawski, M. E., R. H. Hower, and R. B. Brown, *A Simple Packaging Process for Chemical Sensors,* Dept. Electrical Engineering and Computer Science, University of Michigan.
15. Kriebel, F., T. Seidowski, T. Univ. Dresden, Germany, "Flip Chip with Conductive Adhesives in Multichip Modules," *VDE/VDI Symposium on Conductive Adhesives in Microelectronics,* Berlin, 1994.
16. Vierti, B., B. Rosner, and D. Thomson-Brandt, "Flip Chips with Polymer Bumps on Various Substrates," *Symposium on Conductive Adhesives in Microelectronics,* Berlin, 1994.
17. Schafer, H. E. A., and H. M. van Noort, "Conductive Adhesives Processing for Chip-On Glass," *Symposium on Conductive Adhesives in Microelectronics,* Berlin, 1994.
18. "Flip Chip Using Electrically Conductive Polymers and Dielectrics," U.S. Patent 5,074,947, Dec. 1991.
19. "Flip Chip Bonding Using Electrically Conductive Polymer Bumps," U.S. Patent 5,196,371, March 1993.
20. "Flip Chip Technology Using Electrically Conductive Polymers and Dielectrics," U.S. Patent 5,237,130, Aug. 1993.
21. PFC-R, Epoxy Technology Inc., Registered Trademark of Epoxy Technology.

Compliant Bumps for Adhesive Flip Chip Assembly

**Mark Breen, Diana Duane, Randy German,
Kathryn Keswick, and Rick Nolan**

7.1 Introduction

Compliant bumps are polymer-core bumps with a metal coating. Compliant bump technology was originally developed for flip chip on glass, but has been demonstrated with flip chip connection to printed-circuit boards (PCBs) and laminates. The advantages of compliant bump technology are not yet fully explored, and many other applications of this "flexible" technology are under investigation.

Flip chip on glass (FCOG) is susceptible to electrical open circuits for a variety of reasons including, but not limited to, movement in the z axis caused by flip chip adhesive CTE and water absorption of the adhesive. Flip chip assembly to cofired ceramic and laminate substrates suffers from these problems as well as others, such as bow or twist in the substrate and bond pad height irregularities. Success with adhesive flip chip connections to these substrates has, to date, been limited. Commercially available adhesives have either failed to produce reliable bonds, or have suffered from long cure time or a lack of reworkability.

A solution to these problems has been demonstrated by forming compliant bumps on the chip or substrate bond pads using a photoimagible polymer coated with a thin layer of gold. Bumps 17 μm high with diameters between 17 and 93 μm have been fabricated and bonded. The resulting compliant bump structure provides 30 percent of the bump height (5 μm) within the elastic compression range. This compliance eliminates many of the demands placed on the assembly adhesives by other electrical contacting methods (such as solid metal bumps or particles). Compliant bumps allow the use of commercially available, fast-curing, easily reworkable adhesives for reliable flip chip assembly.

MCC's compliant bumps have been mechanically cycled from minimum compression (needed for electrical contact) to maximum compression (based on diminishing compression distance versus applied force) 1000 times, with minimal degradation of the polymer core or metal overcoat. Assemblies have been subjected to temperature cycling and steam pot aging with substantial improvement in reliability when compared to assemblies using solid metal bumps. Using compliant bump technology, low-temperature rework has been demonstrated with compliant bumped chips on glass, laminate, and MCM-C substrates. Chips or substrates with compliant bumps are reusable, which is a significant advantage over conventional gold bump processes where the bump structure is permanently deformed by the bonding process.

7.2   Problem to Be Solved

For the past 3 years, MCC and its member companies have led several efforts to commercialize adhesive flip chip technology. As part of these efforts, we have identified several reliability and manufacturability barriers to widespread use of adhesive flip chip.

Failures occur in adhesive flip chip assemblies on glass or ceramic for a variety of reasons. Electrical connections in most adhesive flip chip technologies are achieved through pressure contact between chip and substrate pads. This contact pressure is maintained throughout the product's life by tensile stress in the adhesive. Loss of electrical contact can occur when the adhesive expands or swells in the z axis direction, due to thermal expansion or moisture absorption. Gross delamination and tensile stress cracking can occur in the adhesive when the bond stresses are excessive.[1]

Adhesive flip chip systems commonly use solid metal bumps (see Fig. 7.1), or solid metal or metal-coated polymer particles [anisotropically conducting adhesives (ACA); see Fig. 7.2] to achieve electrical connections from the chip to the substrate.[2] This is because the chip pad is typically recessed 1 μm or more in the chip passivation coating. Other chip or substrate topography may also preclude electrical contact without some sort of bump or particle between the chip and the substrate.

In the case of ACA flip chip technology, the particles are relatively small (<10 μm diameter).[3] This is so that they can be spaced as closely as possible without touching to ensure reliable z axis conductivity without lateral conductivity. The resulting bond line thickness is also less than 10 μm, resulting in high tensile stresses—particularly at the corners of the chip (see Figs. 7.3 to 7.5). Technologies using solid metal bumps have an advantage in that they can optimize bond line thickness for optimum stress levels by adjusting bump height. Unfortunately, taller bumps and thicker bond lines aggravate another

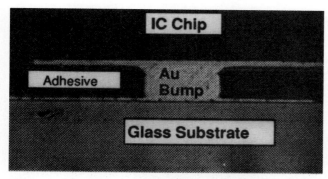

Figure 7.1 Microbump adhesive flip chip technology is fine-pitch-capable, but the gold bumps' lack of compliance limits reliability.

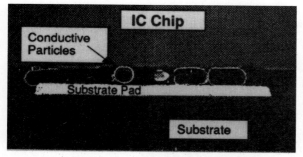

Figure 7.2 z-Axis adhesive flip chip is limited to relatively large I/O pitch, and the small particle size produces a thin adhesive bond line that produces excessive stresses in the bonded assembly.

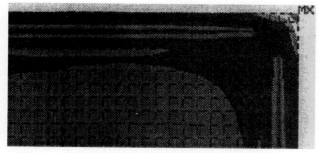

Figure 7.3 3D finite element modeling predicts excessive stresses at the corners of this z-axis adhesive flip chip assembly.

Figure 7.4 Test vehicle results confirm modeling predictions. This sample exhibits delamination and tensile stress cracking in the corners and around the bond pads after exposure to MIL-STD 883, schedule B conditions.

Figure 7.5 z-Axis adhesive flip chip on glass assembly. The arrow points to a crack in the silicon chip that propagates along an isostress plane predicted by the finite element models.

problem. The CTE of the metal bump is typically much lower than that of the adhesive. As the assembly increases in temperature, the adhesive expands faster than the bump, resulting in an open circuit.

Because of these issues, adhesive flip chip technology requires an adhesive with extraordinary mechanical and electrical properties. CTE and moisture absorption must be very low, while glass transition temperature, storage modulus, and tensile and adhesion strengths must be high. Few adhesives have been found to meet these requirements. When one is found, it typically has manufacturability problems such as a short pot life, lack of reworkability, residual ionics, or a long cure time. To date, successful adhesive flip chip bonding has been achieved mainly with heat-curing epoxies which are far from ideal. When bonding on opaque ceramic or laminate substrates, heat-curing adhesives are required. For display driver applications, UV-curing adhesives are more desirable because they can be cured at low temperatures.[4] UV curing also eliminates the bonder "dwell" time needed to raise the bonder thermode to curing temperature and then cool it down again.

7.3 Solution: Compliant Bump

A solution to these problems has been demonstrated by forming compliant bumps on the chip or substrate bond pads (see Figs. 7.6 and 7.7). Standard tape-automated bumping (TAB) processes and equipment are used to form polymer bumps on IC wire-bond pads. The pad and bump are then overcoated with gold metallization. The bumps are elastically compressible up to 30 percent of the total bump height, which dramatically alters the mechanics of adhesive flip chip bonding (see Figs. 7.8 and 7.9). The result is reliable, cost-effective flip chip assembly and rework using commercially available adhesives.

The compliant bump process can be used for bumping either substrates or die. Chip wafer bumping is generally the most cost-effective, since many chips are prepared for each piece processed. Substrate bumping results in fewer chip sites bumped per piece processed, but is attractive if the assembly uses chips from many different suppliers (or chips not available in wafer form). Demonstrations thus far have been

Figure 7.6 Oblique view of an MCC compliant bump. The concave top surface is due to polymer shrinkage during cure.

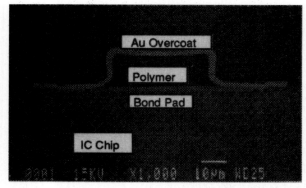

Figure 7.7   Cross-sectional view of a compliant bump.

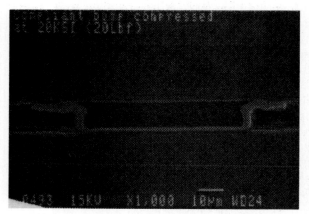

Figure 7.8   Compliant bump assembled to borosilicate glass, using an assembly force of 20 lb. The bump is compressed to about 85 percent of its original height.

on silicon wafers with aluminum pads and silicon nitride passivation (see Fig. 7.10). Finished bump height is currently 17 μm for an assembled height of around 14 μm (determined to be optimum through finite element analysis and test vehicle results). Bump diameters range from 18 to 93 μm.

The compliance of the polymer provides a mechanism by which the bumps can adjust for warp or bow or pad height nonuniformities in the substrate. Better bump height uniformity can be achieved with polymer bump processing compared to conventional electroplating. Polymer bump uniformity was measured at 3 percent, whereas gold bump uniformities average 10 percent. Polymers respond elastically

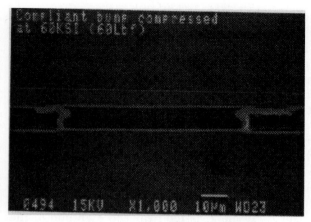

Figure 7.9 Another compliant bump flip chip assembly. Assembly force was 60 lb, resulting in an approximately 30 percent compression of the bump.

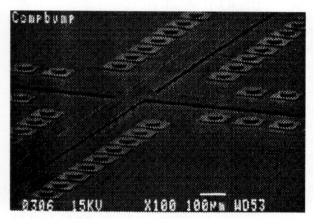

Figure 7.10 MCC compliant bumps fabricated on a wafer of 216 I/O, 150-μm-pitch silicon test chips.

over a larger range of strain loads than do metals. These factors allow bonding forces to be lower than with solid metal bumps, because there is no need to plastically deform metal bumps to correct for bump nonuniformity. Therefore, a wider range of bonding parameters can be used, while ensuring electrical continuity.

Process parameters can also be optimized to reduce as-bonded stress. In conventional adhesive flip chip systems, as-bonded stress is maximized to provide protection against loss of electrical contact due to moisture-induced swelling, creep relaxation, and thermal expansion. With compliant bumps, the bump tends to spring back to its original height. Even if the bump's composite CTE is lower than that of the bonding adhesive, or if the adhesive absorbs moisture and swells, the bump will maintain a contact force commensurate with its effective compression modulus until it has returned to its original height. The effective compression modulus of the bump can be tailored by adjusting the polymer composition and/or the metal plating thickness.

Desired polymer properties are high glass transition temperature T_g, high yield stress, linear elastic response over a large stress–strain region, and a high compressive strength. Many polymers available today satisfy these requirements. A high T_g is optimum for assemblies that operate at elevated temperatures because material creep occurs at temperatures near the T_g. If the T_g is too near the high end of the temperature operating region, then material creep could cause an electrical open circuit. Young's or elastic modulus varies slightly with temperature, but it is not markedly temperature-dependent—therefore a high modulus is desirable, although not required. However, too high of a modulus could result in a very brittle material that would fracture under large bonding forces.

Compliant bump technology provides for a wide range of bump sizes and shapes. Figures 7.6 and 7.7 show compliant bumps that have a concave bump shape. The shape of the bump can play an important role in the bonding process, contributing to both mechanical contact and electrical connection. For example, the polymer core of the bump may be concave such that the top of the bump edge "scrubs out" under the bonding pressure to enhance mechanical contact . The scrubbing action, in concert with the application of a hard-metal coating on the bump bond surface, may allow direct connection to oxidized aluminum pads. As a further example, a donut-shaped bump (with a center connection to the pad rather than the perimeter-pad connection shown in Figs. 7.6 and 7.7) may enhance electrical connection by allowing a larger contact area.

7.4 Process Description: Polymer Bump Formation

Three approaches to bump formation are proposed: dry-etch, wet-etch, and photoimaging processes. Each process uses a polymer material as the core of the bump, but the bumps are formed by different processing sequences. First, the polymer material is applied to the wafer/substrate, then the material is imaged or masked and removed from the field areas so that a polymer bump is formed on top of the bond pads. Alternative proposed process sequences for metallization will be described later.

7.4.1   Dry-etch process

The dry-etch approach to compliant bump formation uses plasma etch processing. The polymer material is coated onto the substrate and then cured. An etch mask is used for pattern definition—the mask material can be an erodable mask made of resist or a hard-metal mask. The mask is patterned either by photolithography or wet etching. After the mask is imaged, the polymer material is etched to form the bumps.

There are advantages and disadvantages to the dry-etch process approach. One advantage is that the sidewall profile can be modified by adjusting the etch process parameters. One problem is that very large aspect ratios are difficult to achieve. This is because the undercut is large with hard masked, dry-etch processes when etching very thick films. Etch rates for polymer dry-etch processes are 0.5 to 2.0 μm/min. Subsequently, for very thick films the etch process is lengthy and throughput is low.

7.4.2   Wet-etch process

Wet-etch processing of the polymer is similar to the dry-etch approach, but pattern definition is achieved by wet chemical etch instead of plasma. The polymer material is coated onto the substrate and partially cured. Photoresist is then coated over the polymer, and the photoresist is exposed and developed to form the etch mask. Finally, the polymer is etched to form the bumps. The advantages to this approach over dry etch are that no costly dry-etching equipment is necessary and there are fewer steps. This approach may be limited by geometry, and very large aspect ratios are difficult to fabricate. Additionally, sidewall profiles are not easily modified when using wet-etch materials. Bump height uniformity is very good when using the wet-etch approach.

7.4.3   Photoimagible process

Bumps are formed using a photoimagible polymer, resulting in significant process simplification. As with the other two approaches the substrate is coated with the polymer material, but no etch mask is required. The bumps are formed by directly exposing and developing the polymer material. The advantages to this approach are that bump size and shape can be modified by varying the exposure parameters, large aspect ratios are obtainable, and processing is quick and simple and can be performed using conventional photolithography equipment. Bump height uniformities are very good with this approach. For these reasons, MCC's development work has focused on photoimagible polymers. Most photoimagible materials shrink by ≤ 50 percent during cure, resulting in a slight dish-shaped bump after cure. During the bonding process, the bump's top surface conforms to the mating pad's topography.

7.5   Process Description: Metallization

Several processing options are available for coating metal onto compliant polymer bumps, including direct wet metallization processing using conventional PCB techniques, subtractive metallization using sputter or evaporation deposition, or additive processing using a sputtered or evaporated plating bus with a plating mask. There are advantages and disadvantages with each process.

7.5.1   Direct metallization

The wet metallization process makes use of standard techniques used in the PCB industry. The process consists of applying a plating catalyst to the polymer bump and electrolessly plating a ductile metal, such as gold, onto the bump and the surrounding bond pad. Wet metallization processes permit high throughput of parts and do not require a vacuum metallization step. However, care must be taken to ensure that the aluminum electrodes of the IC device are not damaged by the wet metallization process. A barrier metal may be needed to protect the aluminum from the wet metallization process. In addition, a resist mask may be needed to prevent plating on the IC passivation.

7.5.2   Pattern etching

A subtractive approach can be used to apply metal to the polymer bumps. This process consists of depositing the metal over the entire surface of the substrate in the desired thickness, by evaporation or sputtering, and applying an etch mask to the bumps, then etching the field metal away. The subtractive technique described has few steps and can be done with equipment available in a typical IC manufacturing facility. However, this is a costly method, because the metal thicknesses required can result in lengthy evaporator cycles, and most of the metal deposited in the vacuum deposition process is etched away.

7.5.3   Pattern plating

An electroplating process can also be used to metallize the polymer bumps. An electrolytic plating approach requires that a thin plating bus be deposited, typically by sputtering, and then a plating mask applied to define the desired areas to be plated. After electroplating in the desired areas, the plating mask and the plating bus are removed. An additive process using electroplating could be performed using equipment that is commonly used in the PCB industry. This method is more cost-effective than pattern etching because the evaporator/sputter cycle times are short, and very little material is lost in the final etching process. This metallization approach was used to demonstrate the feasibility of the compliant bump process approach.

7.6   Fabrication Cost Analysis

Conventional solid metal bumping techniques usually include sputtering a thin metal interconnect layer onto a passivated substrate. The substrate/wafer is masked using photoresist, and metal is electroplated into the resist openings. Finally, the resist and metal interconnect are removed. This method is simple, well established, and has high yield. A typical price for gold bumps on a 4-in wafer is $150 per wafer. If compliant bumps solve the problems noted in Sec. 7.1 at a competitive cost, this technology is suitable for a wide range of applications. The analysis that follows looks at the cost of producing compliant bumps using a polymer/metal composite structure.

Two different processing sequences are analyzed: the *dry-etch process* and the *photoimagible polymer process*. The main difference between these two approaches is the method of polymer formation. The standard process uses a metal mask for dry etching of the polymer. The reduced steps process uses a photoimagible polymer to form the bumps.

Our analysis shows that compliant bumps can be fabricated at a cost that is competitive with conventional gold bumps, and that capital equipment cost is dominant at volumes below 10,000 wafers per year. For the analysis that follows, certain definitions are useful:

- *Dedicated equipment.*   Equipment purchased for the production of a certain product. All capital depreciation cost for this equipment must be absorbed by that product, even if it uses the equipment for only a small fraction of the available line time. For example, suppose that a facility needs to do polymer coating on 4000 wafers per year and that one coater can do all 4000 coatings. If production time for those 4000 wafers is 2 min per wafer, the coater will be used for only 2 percent of the working hours during the year. No other process will be run on the machine to fund the cost of the coater. Manufacturing cost with dedicated equipment is highly dependent on manufacturing volume.

- *Nondedicated equipment.*   Equipment that is shared between product lines. Each product covers capital depreciation cost only for the portion of available line time that it uses the equipment. Using the example cited above, suppose that a machine is used for 2 percent of the working hours in a given year to produce polymer coatings for compliant bumps. Then only 2 percent of the cost of ownership for the equipment is charged to compliant bump formation.

Manufacturing cost on nondedicated equipment is independent of manufacturing volume, as long as the equipment is fully used by a mix of product lines.

- *Labor.*   Labor includes only direct labor (with benefits) used to run equipment. Labor rates were assumed to be $30/h, representative of technician wages and benefits. These rates do not include maintenance, administration, or other overhead, which may account for as much as 60 percent of total production cost.

- *Materials.*   Materials that are used to form the compliant bump, the ancillary materials used during production, and a waste disposal multiplier.

7.6.1   Cost for the "dry-etch process"

The most important cost factor in the dry etch process is throughput at the dry-etch step. Etch rates of polymers are usually 1 μm/min. At that rate, wafer throughput is less than 10 wafers per hour. Low throughput leads to high capital and labor cost at the etch step.

For the dry-etch process, dedicated capital equipment costs are dominant at production volumes below 10,000 wafers per year (see Fig. 7.11 and Table 7.1). At production volumes of 100,000 wafers per year, capital equipment and labor charges are roughly even at 40 to 45 percent of the total cost per wafer. Material cost is just over 10 percent of the total cost. Compared to conventional solid gold bump costs of about $150 per 4-in wafer, and assuming that half of the $150 is

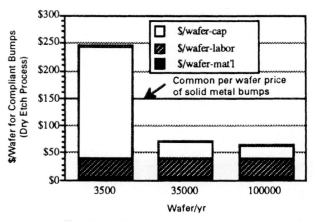

Figure 7.11   The effect of volume on fabrication cost of compliant bumps when using plasma etch to form the polymer. Labor and material costs are not affected by volume, but capital is directly affected. This figure assumes dedicated equipment.

TABLE 7.1   Cost ($ per Wafer) Breakdown* for Dedicated and Nondedicated Equipment at Various Yields and Production Volumes Using Standard Plasma Etch (Dry-Etch) Processing.

Cost center	Nondedicated equipment		Dedicated equipment	
	100% yield	80% yield	100% yield	80% yield
3500 Wafers per Year, $ per Wafer				
Labor	30.03	37.54	30.03	37.54
Material	8.28	10.35	8.28	10.35
Equipment	23.49	29.37	208.00	208.00
Total	61.80	77.26	246.31	255.89
35,000 Wafers per Year, $ per Wafer				
Labor	30.03	37.54	30.03	37.54
Material	8.28	10.35	8.28	10.35
Equipment	23.49	29.37	35.00	35.00
Total	61.80	77.26	73.00	83.00
100,000 Wafers per Year, $ per Wafer				
Labor	30.03	37.54	30.03	37.54
Material	8.28	10.35	8.28	10.35
Equipment	23.49	29.37	27.32	31.43
Total	61.80	77.26	65.63	79.32

*Costs are listed by cost center.

"overhead," the dry-etch process can achieve a similar cost of $150 per 4-in wafer with dedicated equipment at volumes of greater than 35,000 wafers per year, or with nondedicated equipment at any volume provided the equipment is fully utilized when not in use for compliant bump processing.

7.6.2   Cost for the photoimagible polymer process

The process flow used for the analysis of the photoimagible polymer method allows for a more balanced production line. The lowest wafer throughput at any step is 15 wafers per hour. Production volumes can exceed 50,000 wafers per year before any duplicate equipment needs to be purchased. The total number of process steps and associated capital equipment spending is much smaller than that used for the dry-etch process. As a result, production cost for the photoimagible polymer method is approximately 50 percent less expensive than the dry-etch process (see Fig. 7.12 and Table 7.2). Compared to conventional solid gold bump costs of about $150 per 4-in wafer, and assuming that half of the $150 is "overhead," the photoimagible process can

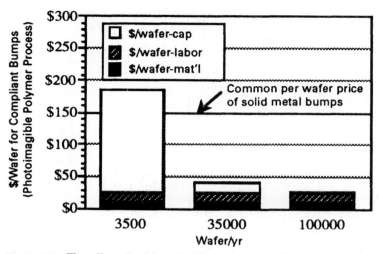

Figure 7.12 The effect of volume on fabrication cost of compliant bumps when using photoimaging to form polymer bumps. Labor and material are not affected by volume, but capital is directly affected. This figure assumes dedicated equipment.

TABLE 7.2 Cost ($ per Wafer) Breakdown* for Dedicated and Nondedicated Equipment at Various Yields and Production Volumes Using Photopolymer Processing.

Cost center	Nondedicated equipment		Dedicated equipment	
	100% yield	80% yield	100% yield	80% yield
3500 Wafers per Year, $ per Wafer				
Labor	10.95	13.69	10.95	13.69
Material	9.35	11.69	9.35	11.69
Equipment	3.88	4.85	160.20	160.20
Total	24.18	30.23	181.00	186.00
35,000 Wafers per Year, $ per Wafer				
Labor	10.95	13.69	10.95	13.69
Material	9.35	11.69	9.35	11.69
Equipment	3.88	4.85	16.02	16.02
Total	24.18	30.23	36.32	41.40
100,000 Wafers per Year, $ per Wafer				
Labor	10.95	13.69	10.95	13.69
Material	9.35	11.69	9.35	11.69
Equipment	3.88	4.85	5.01	5.50
Total	24.18	30.23	25.31	30.88

*Costs are listed by cost center.

achieve a cost of $100 per 4-in wafer at very low volumes. Further reductions in the material cost of the photoimagible polymer, combined with high-volume production of 50,000 wafers per year, will allow the cost to be further reduced, possibly to less than $50 per 4-in wafer.

7.7 Reliability

For reliability assessment, the MCC FC216 test vehicle was used. This test vehicle kit includes the chip, substrate, and TAB tape designs. The chip measures 8.9×8.9 mm and has 216 I/Os on 150-μm pitch. These 210 I/Os are connected in a dual nested daisy-chain pattern. This allows continuity checking of these connections with two resistance measurements, each reading through 105 connections. Adjacent bump electrical isolation is determined by measuring the resistance between the two daisy chains.

Six of the chip pads are used to connect a triple-track feature that can be used for surface insulation resistance and electromigration tests. Triple tracks can also be used for thermal testing by biasing two of the tracks as heating elements (for a total of 25 W per chip) and using the center track for a thermistor. The substrate used for this project was display quality borosilicate glass, sputtered with chrome/gold, and patterned with the FC216 substrate design, which complements the chip features.

True four-wire contact resistance measurements are not possible with the FC216 test vehicle. An approximation of compliant bump contact resistance is determined by comparing total daisy-chain resistance for compliant bump assemblies to that of other FC216 adhesive flip chip technologies that have also been characterized in a separate, four-wire-capable test vehicle (see Fig. 7.13).

Chips were first aligned and pressed to substrate sites without the introduction of adhesive in order to determine the minimum force application needed to establish contact at all pads. Force was increased until there was no further reduction in daisy-chain resistance. The results are summarized in Fig. 7.14.

Mechanical cycling of the bumps was performed on the flip chip bonder, using compression ranges from 20 to 70 lbf [20 to 70 pounds force; equal to approximately 20 to 70 kpsi (kilopounds per square inch) over the total bump area] for 1000 cycles. After this test, the elastic range of compression was reduced by 50 percent because of plastic deformation in the gold overcoat. There were no cracks observed in either the metallization or the polymer core. Bumps cycled 3000 times showed significant degradation of the metal overcoat (see Fig. 7.15). This mechanical cycling, however, was performed without adhesive. In a real assembly, adhesive would fully surround the

Figure 7.13 The 16-chip test vehicle is the MCC FC216.
The test vehicle in the lower right is for true four-wire
contact resistance measurements of flip chip connections.

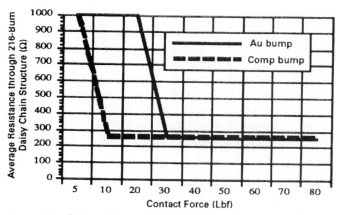

Figure 7.14 Daisy-chain resistance, measured as contact force, is
increased for both solid gold-bumped chips and compliant bumped
chips. The substrate in this test was borosilicate glass with very
thin, sputtered chrome/gold metallization. Thus, about 250 Ω in
each reading is due to bulk resistance of the substrate and chip
metallization.

bump, supporting the metal and reducing metal extrusion radially
outward as was seen in our mechanical cycling tests.

On the basis of this data, a range of initial bonding force parameters was chosen, and chips assembled to glass using a commercially
available UV-curing adhesive. Typical UV exposure times were 15 to
30 s. For the lower exposure times, a UV "postbake" was performed to

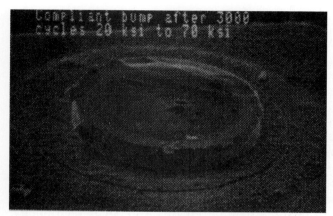

Figure 7.15 Compliant bump after 3000 mechanical cycles from 20- to 70-kpsi loading. Significant elasticity range is lost, but the bump remains conductive and somewhat compliant.

ensure total cure. As a control, assemblies were made with solid gold-bumped chips to identical substrate sites.

Compliant bump and gold bump FCOG assemblies were subjected to a high-temperature operating test. Resistance through the daisy chains was monitored while assemblies were heated slowly to 140°C. All solid gold bump assemblies failed between 65 and 80°C. Compliant bump assemblies failed between 120 and 135°C (well above the glass transition temperature of the adhesive). All assemblies that lasted to 135°C were assembled using identical force parameters. All compliant bump assemblies returned to nominal resistance when cooled to room temperature. Open circuits in the solid gold bump assemblies remained when returned to room temperature.

Gold and compliant bump FCOG assemblies were subjected to 121°C, 100% RH (relative humidity) for 2 h. At the conclusion of this test, 50 percent of the gold bump samples had failed and 12.5 percent of the compliant bump samples had failed. Subsequently, compliant bump assemblies were coated with a commercially available silicone encapsulant. These assemblies have survived 8 h of 121°C at 100% RH with no failures.

7.8 Rework

Rework was performed on several assemblies through the course of this testing. A simple fixture was made to immerse the chip to be removed in a solvent while protecting the adjacent chips from solvent exposure. Solvent soak was successful at room temperature, and can be accelerated by raising the solvent temperature to 80°C. Both the substrate and the chip emerge from this process intact and reusable. Some of the chips and substrate sites included in the tests described above had been reworked twice.

Compliant bumped chips have also been assembled to FR-4 and MCM-D substrates, using heat-curing adhesives. Two adhesives have been used to date. One provides commercial reliability levels and is easily reworkable. The other provides full MIL-STD-883 schedule D reliability, but is reworkable only in partially cured form. Flip chip assemblies with this second adhesive can be subjected to solder assembly temperature profiles with no degradation of the adhesive flip chip connections. This is important for applications where the flip chip assembly must subsequently be solder-assembled to a motherboard. For example, a low-profile ball grid array (BGA) package can be built using existing chip designs (compared to C4 or DCA technology, where the chip I/O pads must be $\geq$250-μm pitch, which usually dictates an array format).

7.9   Conclusion

Compliant bump technology has been demonstrated using a polymer core bump with a thin metal overcoat. The compliant bump process approach solves many of the problems associated with adhesive flip chip assemblies made with solid metal bumps or small conductive particles. Polymer bump formation and metallization can be done in several different ways. All approaches use equipment that is commonly purchased for TAB bumping.

Compliant bumps solve the problem of z-axis movement that is associated with UV- or heat-curing adhesives (due to mismatched CTE and moisture absorption) by elastically rebounding to maintain electrical contact. The compliant bumps also promise to reduce the detrimental effects of bowed and twisted substrates and irregular pad heights. The bumps can be manufactured with standard TAB bumping processes and equipment at a cost comparable to that of solid metal bumps.

Preliminary reliability testing has shown compliant bump flip chip assemblies to be more reliable than identical assemblies using solid gold-bumped chips. Further reliability testing is under way. Additionally, compliant bumps can be reworked either at room temperature or at slightly elevated temperature.

Acknowledgments

The authors would like to acknowledge the efforts of Robert Froehlich of Northern Telecom Labs, Candice Brown of Planar Systems, Inc., and John Lau of Hewlett-Packard for their support and assistance.

References

1. Minges, M. L., *Electronic Materials Handbook,* vol. 1, ASM International, Materials Park, Ohio, chap. 9, pp. 1046–1047, 1989.
2. Vardaman, E. J., I. Yee, and R. Crowley, *Worldwide Developments in Flip Chip Interconnect, A Multi-Client Study,* TechSearch International, Inc., chap. 4, pp. 103–111, 1994.
3. Yamauchi, A., and S. Yamada, "Chip on Glass Technology," *Proceedings of SID Conference,* pp. 25–26, 1994.
4. "Flat Panel Displays in Japan 1991," *Nikkei Business Publications,* 1990.

Anisotropic Conductive Adhesive for Fine Pitch Flip Chip Interconnections

**Hiroaki Date, Yuko Hozumi, Hideshi Tokuhira
Makoto Usui, Eiji Horikoshi, and Takehiko Sato**

8.1 Introduction

Flip chip connection is very effective for high-density LSI interconnections. Various interconnecting methods are suggested as an alternative to soldering.[1–4] A conductive adhesive consisting of conductive particles and adhesive has been proposed. This adhesive is very suitable for fine-pitch interconnections of I/O pads, has a low interconnection temperature, and has begun to find practical applications, such as in interconnecting TAB outerleads bonding and COG to an LCD panel.[4]

This adhesive was developed for use in ≥100-μm-pitch LSI (large-scale IC) electrode interconnections; below 100-μm pitch, problems arise. Conductive particles must remain between the electrodes on the LSI and those on the circuit board. When the electrode density increases, the electrodes on the LSI and the circuit board become finer. To use the adhesive for interconnecting the LSI and the circuit board, we need to increase the number of conductive particles in it. However, this produces an electrical short circuit between adjacent electrodes.

To solve this problem, we developed an anisotropic conductive microcapsule adhesive (MCA) which consists of an adhesive and a filler microcapsule filler (MCF). The MCF consists of conductive particles coated with a thin dielectric resin.

This chapter describes the characteristics of the MCA and the reliability of the interconnection between the LSI and the circuit board using the MCA by the flip chip method.

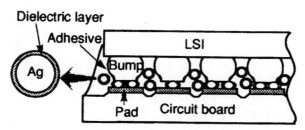

Figure 8.1 Cross section of the interconnection using MCA.

8.2   Principle of Interconnection Using MCA

Figure 8.1 shows a cross section of the interconnection using MCA. First, the LSI is mechanically pressed against the circuit board. The bumps[5] and the electrodes on the circuit board come into close contact, and the pressure is highest in this region. The high pressure destroys the MCF resin layer between the electrodes, and an electrical connection is established through the MCF. The electrical insulation between adjacent electrodes is good, however, because the MCF in these regions remains intact. Then, maintaining the pressure, the adhesive between the LSI and the circuit board is heated. Therefore, the LSI and the circuit board are set by using the contractile force of hardening.

8.3   Experimentation

We selected silver particles because of their low electrical resistance. They were produced by electrolysis, with an average diameter of 7 ±3 µm.

8.3.1   Preparation of MCF

Figure 8.2 shows the process involved in preparating the MCF. First, we produced aqueous phase by dissolving an emulsifier and amine in

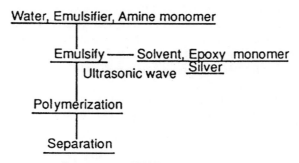

Figure 8.2 Preparation of MCF.

water. Then, we produced oil phase by adding silver particles in a dissolving epoxy monomer solvent. After the oil phase was exposed to an ultrasonic wave, the agglomerated silver powder dispersed independently. We gradually added oil phase to aqueous phase by stirring, and prepared a suspension. Heating caused a reaction between epoxy and amine to coat the silver particles. Finally, we extracted the silver particles with a resin layer from the suspension.

8.3.2 Investigation of resin layer on MCF

We analyzed the resin layer on the MCF by using SEM (scanning electron microscopy) and TGA (thermogravimetric analysis), in order to study the defects in the resin layer, the thickness of the resin layer, and how the particles dispersed. We observed the surface and a cross section of the MCF by using SEM.

When some metals are heated in air, they will easily oxidize, and will gain weight. If the resin layer of the MCF has defects such as pinholes, its weight increases by oxidation because metal at the defects is exposed. Making use of this, we investigated the defects in the resin layer of the MCF by using TGA. In this evaluation, we used copper particles with diameters almost the same as those of silver. This is because copper easily oxidizes but silver does not. The process of coating the surface of copper with resin is the same as in the MCF. We henceforth refer to copper particles coated by resin as *microcapsule copper* (MCC). We investigated the defects in the MCC by comparing the change in its weight with that of the copper particles without the resin layer.

8.3.3 Production of MCA

We prepared two types of MCA: (1) an MCA which is an adhesive consisting mainly of epoxy resin with an MCF of 0.1 vol% and (2) an MCA with an MCF of 0.5 vol%. We dispersed the MCF in the adhesive by stirring sufficiently, thus producing MCAs. MCA type 1 was used to investigate interconnecting electrical characteristics. MCA type 2 was used to examine the electrical insulation resistance of MCA and the reliability of LSI and circuit board interconnection.

8.3.4 Measurement of electrical characteristics of MCA

First, we put MCA type 2 on the surface of the electrode pattern with gaps between adjacent electrodes of 10, 20, and 40 µm. We then cured this adhesive at 175°C for 30 s, and measured the electrical insulation resistance of adjacent electrodes. For comparison, the same mea-

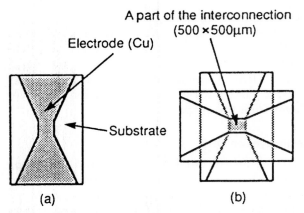

Figure 8.3 Evaluation of substrate for electrical interconnection characteristics.

surement was carried out for the conductive adhesive using silver
particles without the resin layer. We will refer to silver particles as
silver particle adhesive (SPA).

The electrical pattern on the glass epoxy substrate (Fig. 8.3*a*) was
used to realize the electrical characteristics of interconnection. First,
we coated MCA with an MCF of 0.1 vol% over a glass epoxy substrate
with a copper electrode. We then placed another substrate to this to
align each electrode as illustrated in Fig. 8.3*b*. In this case the contact area should be 500×500 μm. Two electrodes were placed under
various pressures. Adhesion between two glass epoxy substrates
cured by heating. Therefore, we could produce samples for measuring.
We diluted the MCF content to 0.1 vol% because the area of the cross-
ing electrodes was 50 times wider than the 80-μm-diameter bump
area. The same measurement was carried out for SPA and using
spherical resin particles plated with gold (Au/Res, Res: resin, diameter of particle: 7 ± 0.1 μm). Au/Res has been cited as an example of
conductive particles.[6]

Using a four-point probe method, we measured the electrical resistance of the samples taken at various pressures. We examined the *V-I*
voltage and current characteristics, changing the voltage by again
using a four-point probe method. The sample was produced under a
pressure of 4 kgf/mm^2.

8.3.5 Reliability

We tested the reliability of the interconnections by using a model chip
and a glass epoxy substrate. The model chip was a 10 mm^2 silicon
chip with a 100-μm pitch of 360 I/O pads.

Figure 8.4 Appearance of the circuit board with Si chip using MCA.

I cm

We produced gold bumps on the I/O electrodes of the silicon chip by using a wire bonder. The diameter of each bump was 80 μm. These bumps were pressed against a flat surface to make their heights uniform. This leveled bump height was 30 ± 3 μm, and the area of the top of the bump was 5000 ± 500 μm².

We coated MCA on the glass epoxy substrate. We then aligned electrodes on the silicon chip and the substrate. The chip and the substrate were pressed at 25 g per bump. Maintaining the pressure, we raised the temperature of the MCA between the chip and the substrate to 175°C, and maintained this temperature for 30 s. The silicon chip was thus settled on the substrate. For comparison, we made samples using SPA in the same way. Figure 8.4 shows the appearance of the sample.

We observed a cutaway view of the pad/bump connection using MCA by optical microscopy. We could thus investigate the stability of a part between the pad/bump and the MCF. We then measured the electrical and insulation resistances of the samples. We conducted a high-temperature, high-humidity bias test at 85°C, 85% RH, 5 V DC, for 500 h. We then remeasured the electrical insulation resistance. We finally conducted a thermal cycle test at −40°C to +125°C for 700 cycles. We measured the electrical resistance during and after this test.

8.4   Results and Discussion

8.4.1   The resin layer on the MCF

Figure 8.5 shows the surface and a cross section of the MCF. From Fig. 8.5*b* we confirmed that the MCF has a thin dielectric layer of epoxy resin whose thickness is 0.1 to 0.3 μm at the surface of the silver particles. We observed no defects such as pinholes on the resin layer of the MCF by SEM analysis, and we observed virtually no agglomerate particles.

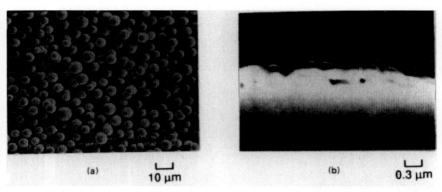

Figure 8.5   SEM photographs of (*a*) a surface and (*b*) a cross section of MCF.

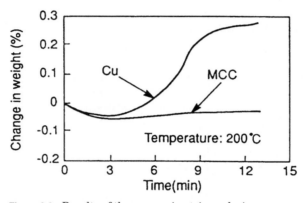

Figure 8.6   Results of thermogravimetric analysis.

Figure 8.6 shows the results of analyzing MCC and Cu (copper particles) by TGA. The weights of both samples decreased during the early stage (3 min) because of the evaporation of absorbed water. After heating for 3 min, the weight increase for Cu was greater than that of MCC. This result led us to notice that MCC was almost completely surrounded by a resin layer around a surface of Cu and also that MCC had no defects, such as pinholes, in the resin layer.

From these results, we concluded that the MCF using silver particles has a good forming state in the resin layer.

8.4.2   Electrical characteristics

Figure 8.7 shows changes in the electrical insulation of MCA and SPA with several electrode gaps after a high-temperature, high-hu-

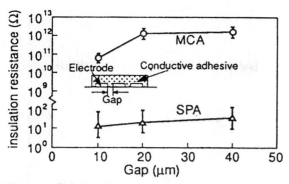

Figure 8.7 Relationship between electrode gap and insulation resistance.

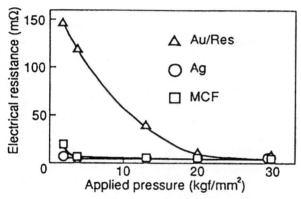

Figure 8.8 Relationship between applied pressure and electrical resistance.

midity bias test. The insulation resistance of SPA was <10^2 Ω, suggesting that adjacent electrodes at each gap might have had an electrical short circuit. Compared with SPA, MCA had better insulation resistance (>10^{10} Ω) despite a narrow electrode gap of 10 µm. With this result, we confirmed that the coating resin provided sufficient electrical insulation.

Figure 8.8 shows the relationship between applied pressure and electrical resistance for an adhesive with the Au/Res, Ag, and MCF. All adhesives indicate almost the same resistance of about 3 mΩ in regions of >20 kgf/mm². For pressures of <20 kgf/mm², MCF and Ag adhesive show still lower electrical resistance. On the contrary, the electrical resistance of the adhesive with Au/Res changes largely with applied pressure; the maximum interadhesive electrical resistance range is 1 to 50 mΩ.

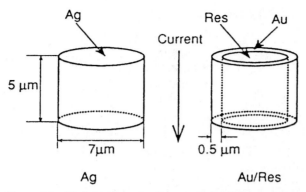

Figure 8.9   Models of conductive particles between electrodes.

As shown in Fig. 8.9, we assume that fillers of the Au/Res, Ag, and MCF between contacting electrodes of sample are columns with diameter 7 μm and height 5 μm. Resistance R can be calculated as follows. Using resistivity values of 1.6 μΩ-cm for silver and 2.3 μΩ-cm for gold, $R = 11.3$ mΩ for Au/Res, and $R = 2.1$ mΩ for Ag are obtained. Thus, the MCF and Ag have the advantage of low resistance only with respect to conductive resistance. A disadvantage of Au/Res adhesive is its high electrical resistance caused by separation of gold or cracks from particle deformation.

On the other hand, the resistance of Au/Res was almost the same as that of the MCF and Ag, in an area where the applied pressure was >20 kgf/mm². This is because Au/Res was crushed and each electrode experienced almost direct face-to-face contact when the pressure exceeds 20 kgf/mm². This is because Au/Res is much softer than MCF and Ag.

Next, we compared the MCF with Ag. The MCF resistance was 20 mΩ higher than that of Ag at 2 kgf/mm², but both fillers had almost the same electrical resistance of 3 mΩ for an area of >4 kg/mm². We assumed that the resin layer of the MCF was not completely destroyed at 2 kgf/mm². All samples, except samples with Au/Res at a pressure of >20 kgf/mm², were interconnected through the conductive particles (Fig. 8.10).

Figure 8.11 shows the V-I curve for the interconnection using the MCF and Au/Res. The maximum permissible current for the MCF is 4000 mA, which is about eight times larger than that for Au/Res. As mentioned before, the number of fillers remaining between both electrodes in this sample was almost the same as the number of fillers remaining under the 80-μm-diameter bump, so this value is applicable

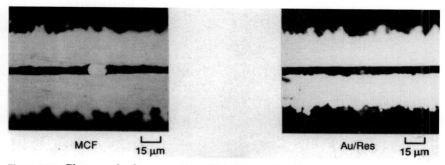

Figure 8.10 Photograph of cross section of interconnective parts.

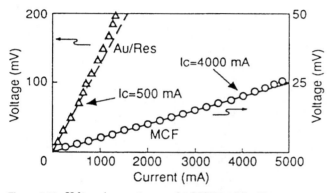

Figure 8.11 Voltage/current curve for MCF and Au/Res.

to the actual interconnection with a bump of 80 μm diameter. Another reason why the maximum permissible current of the MCF is larger than that of Au/Res is that the electrical resistance of Au/Res is higher than that of the MCF, as mentioned before, and the amount generated heat of Au/Res increases to a greater extent than does that of the MCF at the same current.

8.4.3 Reliability

Figure 8.12 shows a cross section of the pad/bump connection using MCA. We confirmed that the bumps and electrodes of the substrate were stably interconnected through the MCF. We observed that the amount of MCF between the bumps and electrodes of the substrate was sufficient to effect a stable interconnection.

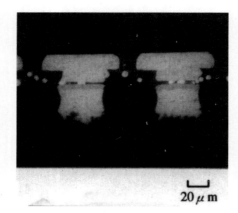

Figure 8.12 Cutaway view of pad/bump connection using MCA.

20 μ m

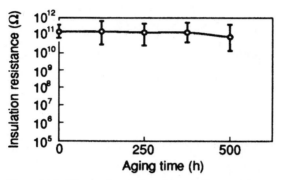

Figure 8.13 Change in electrical insulation resistance with aging time.

Figure 8.13 shows the results of a high-temperature, high-humidity bias test using MCA. MCA had a high electrical insulation resistance of over 10^{10} Ω and a low change in electrical insulation resistance after 500 h. Figure 8.14 shows the results of a thermal cycle test. Electrical interconnection resistance was as low as 3 mΩ per bump for each sample. MCA indicates that the change in electrical resistance was less than 10 percent.

These reliability characteristics would appear to be sufficient for practical applications.

8.5   Summary

We developed an MCA which consists of an adhesive and an MCF, and the MCF consists of conductive particles coated with a thin dielectric resin, as a new material to connect a bare chip with a fine-

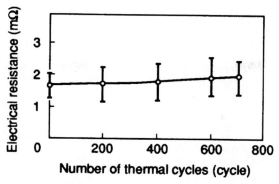

Figure 8.14 Change in electrical resistance with number of thermal cycles.

pitch I/O pattern to the circuit board face down. We investigated the electrical characteristics and the reliability of the interconnection by connecting an LSI of 100-µm pitch to a printed-circuit board. For comparison, we also investigated the electrical characteristics of silver particles without the resin layer and Au/Res. The results were as follows.

1. Anisotropic conductive adhesive
   a. The MCA consists of an epoxy-based adhesive and an MCF of 5 vol%.
   b. The MCF has a thin resin layer with 0.1 to 0.3 µm thickness around the surface of the silver particles, and hardly any defects such as pinholes.
2. Electrical characteristics
   a. The MCF particles show much higher electrical insulation resistance than do the silver particles without the resin layer.
   b. The MCF contact has a very low resistance (several milliohms). The resistance of Au/Res is 10 to 50 times higher than that of MCF contacts.
   c. The maximum permissible current for MCF is 4000 mA, which is about eight times that for Au/Res.
3. Reliability for interconnecting LSIs of 100-µm pitch
   a. The MCA has a high insulation resistance ($>10^{10}$ Ω) after 500 h of high-temperature, high-humidity bias testing under the following conditions: 85°C, 85% RH, and 5 V DC.
   b. The change in electrical resistance after 700 temperature cycles between −40°C to +125°C was less than 10%.

We conclude that this new anisotropic conductive adhesive using an MCF seems to be a promising material for fine-pitch interconnections.

References

1. Chan, D. D., et al., "An Overview and Evaluation of Anisotropically Conductive Adhesive Films for Fine Pitch Electronic Assembly," *Proceedings of Electronics Component Technology Conference,* pp. 320–326, 1993.
2. Vol, C., and M. Leray, "The 3D Interconnection—Applications for Mass Memorize and Microprocessors," *Proceedings of International Society for Hybrid Microelectronics '91,* pp. 66–68, 1991.
3. Liu, J., "Reliability of Surface-Mounted Anisotropically Conductive Adhesive Joints," *Circuit World,* pp. 4–11, 1993.
4. Matsubara, H., et al., "Bare-Chip Face Down Bonding Technology Using Conductive Particles and Light-Setting Adhesives Elastic Method," *Proceedings of International Microelectronics Conference,* pp. 81–87, 1992.
5. Kusagaya, T., H. Kira, and K. Tsunoi, "Flip Chip Mounting Using Stud Bumps and Adhesive for Encapsulation," *Proc. International Conference on Multi-Chip Modules,* pp. 238–245, 1993.
6. Suzuki, "The Application to Technology with Interconnection for LCD," *ADEE Japan '92Internepcon,* pp. 21–29, 1992.

Flip Chip Interconnection Technology Using Anisotropic Conductive Adhesive Films

Itsuo Watanabe, Naoyuki Shiozawa, Kenzo Takemura, and Tomohisa Ohta

9.1 Introduction

Anisotropic conductive adhesive films (ACFs) (under the trade name Anisolm reported in this chapter) consist of conducting particles and adhesives which provide both attachment and electrical interconnection between electrodes. ACFs are widely used for high-density interconnection between liquid-crystal display (LCD) panels and tape carrier packages (TCPs) to replace the traditional soldering or rubber connectors. In LCD applications, traditional soldering may not be as effective as ACFs in interconnecting materials between indium tin oxide (ITO) electrodes and TCP. Recently, ACFs have also been used as an alternative to soldering for interconnecting TCP input lead bonding to printed-circuit boards (PCBs).

ACF has been used for direct chip interconnections such as chip-on-glass (COG) and chip-on-board (COB) assembly in the interest of product miniaturization, multiple connectivity, and cost reduction.[1–8] Conducting particles randomly dispersed in ACF make contact with both conductor surfaces, but not with each other; this is due to the low fraction of conducting particles when chips are joined with supporting substrates such as ITO glass and PCBs under high heat and pressure. However, the risk of short circuit due to the contact between different conducting particles should be considered when ACF is used for direct chip attach (DCA) interconnections with very fine pitch (<40 μm).

In this chapter, we introduce the principle of interconnection using ACF and discuss the role of conducting particles dispersed in ACF with respect to interconnection resistance. We also describe double-layer ACF consisting of nonfilled and conducting particle–filled adhesive layers for high-density bare-chip interconnections.

9.2   Principle of Interconnection Using ACF

ACF consists of conducting particles dispersed uniformly throughout an adhesive matrix at a concentration lower than the percolation threshold.[9,10] ACF is easy to handle because it is formulated in a dry format. Figure 9.1 summarizes the procedure for interconnecting TCP to a substrate using ACF. First, ACF is placed between the TCP and the glass substrate, and the electrodes to be bonded are aligned. Second, pressure and heat are simultaneously applied to effect interconnection between electrode surfaces, where conducting particles are deformed between electrode surfaces as shown in Fig. 9.2. Bonding conditions for interconnecting a TCP to an ITO-coated glass substrate and PCB are usually 170°C, 20 kgf/cm^2 and 20 s, during which time no

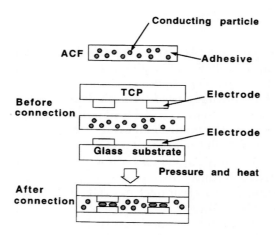

Figure 9.1   Procedure of interconnection using ACF.

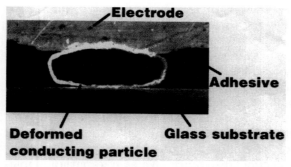

Figure 9.2   Cross section of interconnection using ACF.

adjacent electrodes are short-circuited by nondeformed conducting particles. The interconnection is then retained by the compressive force between electrode surfaces because of the curing shrink of the adhesive. When electrode surfaces are bonded as shown in Fig. 9.1, adhesive resins are squeezed out and conducting particles are trapped between conductor surfaces. Consequently, electrical conduction in ACF is restricted in the z direction (normal to the plane of adhesive films) while electrical insulation in xy plane is maintained. Therefore, ACF does not need to be patterned, unlike soldering, and therefore is more versatile. However, in flip chip technology, ACF does not self-align[11] which would allow misplaced chips to be pulled into correct position corresponding to substrate electrodes by surface tension forces of molten solder. This indicates that flip chip technology using ACF would strongly depend on the alignment ability of the equipment.

ACFs are based on a thermoplastic adhesive, a thermosetting adhesive, and a mixture of both. Thermoplastic adhesives have several advantages in terms of reworkability and pot life. On the other hand, thermosetting adhesives based on epoxy resins provide high heat resistance and excellent adhesion to a variety of substrates, although they are not thermally reversible and cannot be dissolved in common organic solvents, because they are crosslinked after bonding. Also the latter adhesives provide low interconnection resistance, because they are easy to exclude between bonding electrodes during the bonding process. Therefore, epoxy-based adhesives are suitable for flip chip interconnection technology using ACF because of their high reliability and low interconnection resistance.

Conducting particles[12] used in ACF are solid metals such as Ni[13] and InPb, metal-coated glass spheres, and metal-coated polymer spheres. Since electrical conduction in the z direction is achieved by electrical contact between conducting particles and bonding electrodes, increasing the contact area between them is very important in order to decrease the interconnection resistance.[14] Metal-coated polymer spheres are very convenient for this purpose. They can be deformed between bonding electrodes during actual bonding, and therefore this deformation increases the contact area between conducting particles and bonding electrodes. Bonding process parameters such as temperature, pressure, and curing time give significantly influence the interconnection resistance between both electrodes. Figure 9.3 shows the influence of bonding conditions on the interconnection resistance between bumps of a chip (bump size: 70×70 μm) and ITO electrodes on a glass substrate. As shown in this figure, interconnection resistance decreases with increasing temperature and pressure. In an epoxy resin used as an adhesive that can be cured, if the bonding temperature falls below 160°C, mechanical contact between bonding electrodes is

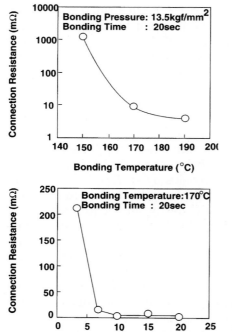

Figure 9.3 Influence of bonding condition on interconnection resistance in COG assembly using ACF.

not maintained. Also if the bonding pressure is too low, conducting particles dispersed in adhesive films cannot make good electrical contact with bonding electrodes. When the chips and the substrates are bonded for 20 s at 170°C under pressures of 9.5 kgf/mm^2 (area of a bump), an interconnection resistance of <10 mΩ is obtained.

9.3   Influence of Deformation of Conducting Particles (Metal-Coated Polymer Spheres) on Electrical Characteristics of the Interconnection Using ACF

Electrical characteristics of the interconnection using ACF are determined by the degree of the deformation of conducting particles (gold-coated polymer spheres).[14] Figure 9.4 illustrates the electrical characteristics of the interconnection and specification of electrodes on flexible printed circuits (FPCs). A gold-plated electrode on FPC was interconnected to a tin-plated electrode on another FPC using ACF by applying heat and pressure (170°C, 20 kgf/cm^2, 20 s). In order to clarify the relationship between the deformation of conducting particles and

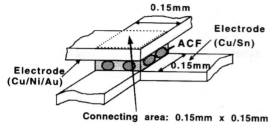

Figure 9.4 Specification of electrodes on FPC.

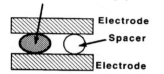

Deformed conducting particle

Figure 9.5 Schematic model for controlling the deformation of a conducting particle using undeformable spacer.

interconnection resistance, undeformable dielectric particles, termed the *spacer* in this chapter, were mixed into ACF. Figure 9.5 shows a schematic in which the deformation of conducting particles is controlled by the spacer. The deformation ratio (DR) of conducting particles was controlled by the size of the spacer and calculated by the following equation:

$$DR = (\text{diameter of a conducting particle} - \text{diameter of a spacer}) \times 10\%$$

The number of conducting particles trapped between the bonding electrodes was determined using an optical microscope. Conducting particles of 10 μm diameter and spacers of 4, 6, 8, and 10 μm in diameter were mixed into epoxy-based adhesives.

Figure 9.6 displays the electrode configuration of the interconnection used to apply a four-point probe procedure to measure the interconnection resistance. A constant current of 1 mA (I) is applied and a voltage (V) is measured between the electrodes. Interconnection resistance is calculated as $R = V/I$.

The relationship between the deformation ratio of conducting particles and the interconnection resistance between FPC electrodes using ACF is shown in Fig. 9.7. The interconnection resistance decreases as the deformation ratio of conducting particles increases (i.e., interconnection resistance and DR are inversely proportional). This result indicates that deformable gold-coated polymer spheres are suitable as conducting particles for ACF. Figure 9.8 shows the influence of the deformation of conducting particles on the current-carrying capacity

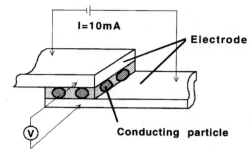

Figure 9.6 Electrode configuration of the interconnection using a four-point probe method for the measurement of interconnection resistance.

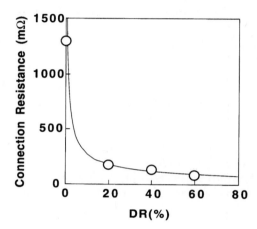

Figure 9.7 The relationship between the deformation ratio of conducting particles and interconnection resistance.

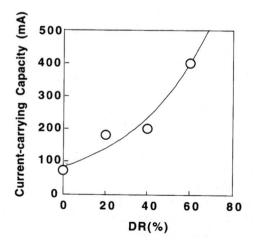

Figure 9.8 The relationship between the deformation ratio of conducting particles and current-carrying capacity.

in interconnection between FPC electrodes using ACF. Current– voltage characteristics were obtained by measuring the voltage between electrodes after applying a constant current for one minute. Current-carrying capacity was determined from the maximum current value from current–voltage characteristics following the Ohm's law. As shown in Fig. 9.8, current-carrying capacity increases with increasing deformation ratio of conducting particles in ACF (i.e., I-carrying capacity and DR are directly proportional). These results indicate that the deformation of conducting particles significantly influences the electrical characteristics of the interconnection using ACF. We also investigated the influence of the deformation ratio of conducting particles on the interconnection resistance between FPC electrodes after high-temperature, high-humidity testing (85°C, 85% RH). Low resistance and high interconnection reliability are obtained by the deformation ratio in >40 percent of instances reported.[14] This indicates that the large contact area between conducting particles and bonding electrodes provides low interconnection resistance and high reliability in interconnection using ACF, and that gold-coated plastic spheres are suitable for ACF applications.

9.4 Flip Chip Interconnection Using Double-Layer ACF

9.4.1 Concept of double-layer ACF

In direct chip interconnections using conventional single-layer ACF, conducting particles facing down to bumps flow into conductor spaces. Adhesives are squeezed out during bonding process; therefore, with decreasing interconnection conductor spacing, it becomes more difficult to maintain in-plane electrical insulation with single-layer ACF.

In order to trap a sufficient number of conducting particles to make electrical contact with both conductor surfaces but prevent electrical contact between the different conducting particles even in fine-pitch (<40-μm-pitch) applications, double-layer ACF[15] has been proposed (see Fig. 9.9). Double-layer ACF consists of nonfilled and conducting particles–filled adhesive layer. A nonfilled adhesive layer facing the bumps of the chip as shown in Fig. 9.9.

Figure 9.10 shows the influence of arrangements of double-layer ACF on the number of conducting particles trapped on a bump of a chip. The number of conducting particles on a bump of the interconnection is determined using an optical microscope. When conducting particles-filled adhesive layer in ACF faces to the bumps of the chip, the conducting particles are squeezed out with the adhesive resin during bonding and then flow into conductor spaces. Therefore, in

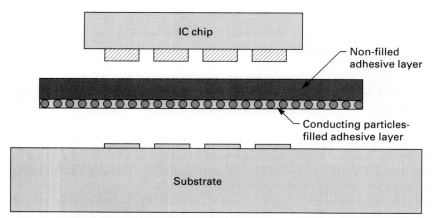

Figure 9.9   Structure of double-layer ACF.

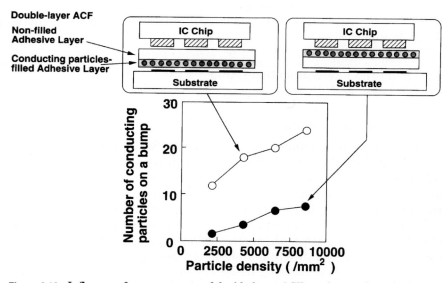

Figure 9.10   Influence of arrangements of double-layer ACF on the number of conducting particles on a bump of a chip.

this case the number of conducting particles is much less than when the nonfilled adhesive layer faces the bumps of the chip. This also indicates that the conducting particles facing the bumps of the chip in single-layer ACF flow into conductor spaces during bonding and are wasted. Figure 9.11 depicts the relationship between the number of conducting particles on a bump of a chip and conducting particle density in ACF. The number of conducting particles on a bump of a chip

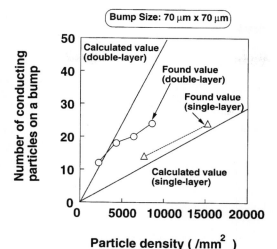

Figure 9.11 Relationship between the number of conducting particles on a bump of a chip and conducting particle density in ACF.

depends on the ACF layer structure. Both calculated and found (observed) values show that the number of conducting particles in double-layer ACF is much higher than those in single-layer ACF. This indicates that even though particle density in double-layer ACF is low, conducting particles to effect electrical contact with both conductor surfaces are trapped more effectively between conductor surfaces in double-layer ACF. Consequently, the potential risk of short circuit in double-layer ACF is less than that in single-layer ACF.

Figure 9.12 shows the influence of layer structure on probability of short circuit in conductor spaces. In the case of single-layer ACF, electrical insulation resistance higher than 10^{12} Ω is maintained down to 20-μm conductor spacings, but short circuiting takes place in <15-μm conductor spacings. On the other hand, in the case of double-layer ACF, electrical insulation resistance higher than 10^{12} Ω is maintained even in 10-μm conductor spacings. Thus, double-layer ACF appears to be suitable as a interconnect material for very fine (20-μm pitch) interconnections. Also, low interconnection resistance less than (<10 mΩ) was obtained using double-layer ACF, as described in the next section.

9.4.2   Flip chip interconnections on glass

The IC chips with gold bumps of 15 μm height on a pitch of 100 μm (bump size 70$\times$70 μm) were bonded with ITO-coated glass substrates for 20 s at 170°C under pressure (13.5 kgf/mm^2) using double-layer ACF. Figure 9.13 shows the interconnection reliability of COG assembly using double-layer ACF. Interconnection resistance between a gold

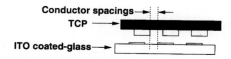

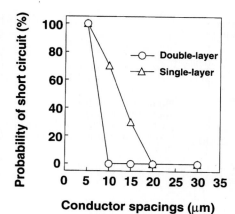

Figure 9.12 Influence of layer structure on probability of short circuit in conductor spaces.

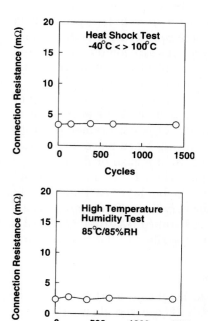

Figure 9.13 Interconnection reliability of COG assembly using double-layer ACF.

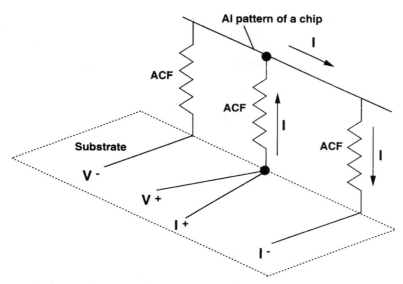

Figure 9.14 Electrode configuration employed in a four-point probe method used to measure the interconnection resistance in COG and COB assemblies.

bump of the chip and an ITO electrode is measured by the four-point probe method as shown in Fig. 9.14. The initial interconnection resistance is as low as 3 mΩ per bump and changes negligibly even after 1400 h of high-temperature, high-humidity testing (85°C, 85% RH) and 1400 cycles of heat cycle testing (-40 to 100°C). Figure 9.15 shows in-plane electrical insulation resistance of COG assembly using double-layer ACF as a function of bump spacings. As shown in Fig. 9.15, double-layer ACF provides high in-plane electrical insulation resistance down to 15-μm bump spacings in COG assembly. Such interconnection reliability and high in-plane electrical insulation resistance are sufficient for practical applications such as COG assembly.

Figure 9.16 is an optical photograph of a surface and a cross section of the pad/bump connection using double-layer ACF in COG assembly. It is observed that many conducting particles are sandwiched between substrate electrodes and bumps of the chip.

9.4.3 Flip chip interconnections on printed-circuit boards

ACFs seem promising as interconnect materials for COB assemblies because of several advantages such as low-temperature assembly, high-density interconnection, low cost, and fluxless bonding, which

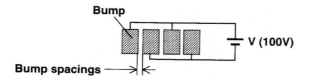

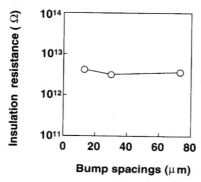

Figure 9.15 Insulation resistance of COG assembly using double-layer ACF as a function of bump spacings.

eliminates the need for cleaning. However, there is concern that interconnection resistance with ACF may be too high, because interconnection using ACF is due to mechanical contact as shown in Fig. 9.16, unlike metal bonding in soldering. Interconnection between the chip and PCB using double-layer ACF was performed by applying heat and pressure as shown in Fig. 9.17. PCBs are fabricated of FR-4 and are plated with a layer of NiAu that has been deposited on the 10-μm-height copper. Despite interconnection due to mechanical contact, interconnection resistance in COB assembly using double-layer ACF is as low as 3 mΩ. In addition, as shown in Fig. 9.18, the current–voltage characteristics measured by the four-point probe method in the COB interconnection using double-layer ACF exhibited a current-carrying capacity of 2000 mA, which is sufficient for flip chip interconnections. Yet, reliability, high frequency properties, and yield are required in applications of ACFs for solder replacement in flip chip technology.

9.5 Summary

The principle of ACF and the influence of bonding condition including temperature and pressure on interconnection resistance in the flip chip technology using ACF were described. We also demonstrated

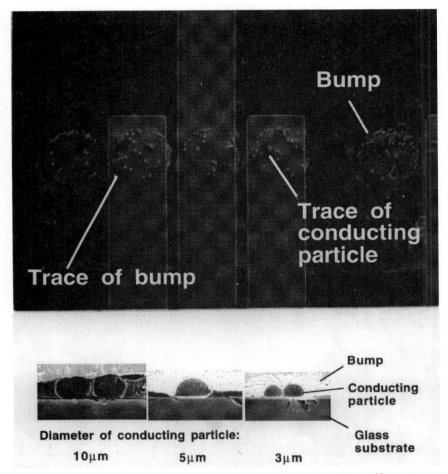

Figure 9.16 Optical photograph of a surface and cross sections of the pad/bump interconnection using ACF in COG assembly.

that the deformation of conducting particles (metal-coated polymer spheres) significantly influences the interconnection resistance and reliability between bonding electrodes.

Double-layer ACF consisting of nonfilled and conducting particles–filled adhesive layer was proposed for applications involving very fine interconnections. We also found that conducting particles of double-layer ACFs, even in lower-conducting-particle-density adhesive films, are trapped more effectively between conductor surfaces than are those of conventional single-layer ACFs in COG assembly. We also confirmed that double-layer ACFs are applicable for very fine (20-μm pitch) inter-

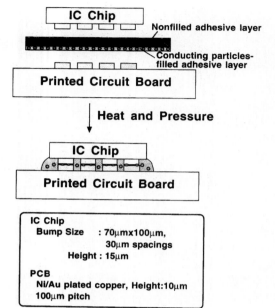

IC Chip
Bump Size   : 70μmx100μm,
                30μm spacings
         Height : 15μm

PCB
 Ni/Au plated copper, Height:10μm
 100μm pitch

Figure 9.17 COB assembly using double-layer ACF.

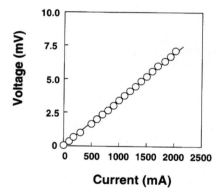

Figure 9.18 Current–voltage characteristics of COB using double-layer ACF.

connections. In addition, we demonstrated that interconnection resistance in COG and COB assemblies using double-layer ACF is as low as 3 mΩ, even though electrical conduction in the z direction of ACF is due to mechanical contact between conducting particles and bonding electrodes. Therefore, ACF is expected as a new interconnect material for solder replacement in flip chip interconnections.

Acknowledgment

We are very grateful to our coworkers K. Isaka, O. Watanabe, and K. Kojima at Shimodate Research Laboratory of Hitachi Chemical Co., Ltd.

References

1. Liu, J., "Reliability of Surface-Mounted Anisotropically Conductive Adhesive Joints," *Circuit World,* **19**(4): 4–11, 1993.
2. Basavanhally, N. R., D. D. Chang, B. H. Cranston, and S. G. Seger, "Direct Chip Interconnect with Adhesive-Connector Films," *Proceedings of 42nd IEEE Electronic Components and Technology Conference* (ECTC), pp. 487–491, 1992.
3. Takakahashi, W., K. Murakoshi, J. Kanazawa, M. Ikehata, Y. Iguchi, and T. Kanamori, "Solderless COG Technology Using Anisotropic Conductive Adhesives," *IMC 1992 Proceedings,* Yokohama, pp. 93–98, 1992.
4. Yamaguchi, Y., I. Tsukagoshi, and T. Ohta, "Anisotropic Conductive Adhesive Films," *Technical Report of IEICE, EMD92-69,* pp. 29–36, 1992.
5. Chang, D. D., J. A. Fulton, A. M. Lyons, and J. R. Nis, "Design Consideration for the Implementation of Anisotropic Conductive Adhesive Interconnection," *NEPCON West,* p. 1381, 1992.
6. Chung, K., T. Devereaux, C. Monti, M. Yan, and N. Mescia, "Z-axis Conductive Adhesives as Solder Replacement," *Proceedings of 1993 Surface Mount Technology International,* pp. 554–560, Sept. 1993.
7. Chang, D. D., P. A. Crawford, J. A. Fulton, R. McBride, M. B. Schmidt, R. E. Simtski, and C. P. Wong, "An Overview and Evaluation of Anisotropically Conductive Adhesive Films for Fine Pitch Electronic Assembly," *IEEE Transact. Components Hybrids, Manufact. Technol.,* **16**(8): 828–835, 1993.
8. van Noort, H. M., M. J. H. Kloos, and H. E. A. Schafer, "Anisotropic Conductive Adhesives for Chip on Glass and Other Flip Chip Applications," *Adhesives in Electronics '94,* Berlin: VDI/VDE, 1994.
9. Gurland, J., *Transact. Met. Soc. AIME,* **236:** 642, 1966.
10. Blythe, A. R., "Electrical Properties of Polymers," *Cambridge Solid Science Series,* Cambridge University Press, 1978, pp. 123–127.
11. Wong, C. C., "Flip Chip Connection Technology," in *Multichip Module Technologies and Alternatives,* Doane, D. A., and P. D. Franzon, eds., Van Nostrand Reinhold, New York, pp. 429–449, 1993.
12. Li, L., and J. E. Morris, "Structure and Selection Models for Anisotropic Conductive Adhesive Films," *Adhesives in Electronics '94,* Berlin: VDI/VDE, 1994.
13. Shiozawa, N., I. Tsukagoshi, A. Nakajima, and T. Itoh, "Anisotropic Conductive Adhesive Film Anisolm AC-2052 for Connecting the One Metal Electrode to Another," *Hitachi Chemical Technical Report,* no. 23, pp. 23–26, 1994.
14. Shiozawa, N., K.Isaka, and T. Ohta, "Electric Properties of Anisotropic Conductive Adhesive Films," *Adhesives in Electronics '94,* Berlin: VID/VDE, 1994.
15. Hirosawa, H., I. Tsukagoshi, H. Matsuoka, I. Watanabe, K. Takemura, N. Shiozawa, and T. Ohta, "Double-Layer Anisotropic Conductive Adhesive Films," *1995 Display Manufacturing Technology Conference, Digest of Technical Papers,* Santa Clara: Society for Information Display, vol. 2, pp. 17–18, 1995.

Anistropic Conductive Flip Chip-on-Glass Technology

Chang Hoon Lee

10.1 Introduction

Flip chip on glass (FCOG) is the cutting edge of flip chip technology. Several major Japanese LCD (liquid-crystal display) manufacturers have developed their own flip chip bonding techniques for chip-on-glass application. Although tape-automated bonding (TAB) and quad flat packaging (QFP) in surface-mount technology (SMT) form are the leading packaging methods for driving ICs, small-size LCDs such as watches, viewfinders, videogame equipment displays, or light valves of liquid-crystal projectors have been using FCOG technology for their IC-driving packaging. Recently, Citizen started mass production of large-size (5.7 to 9.4-in diagonal) VGA-grade dual-scan color super-twisted nematic (STN) LCDs using FCOG technology.

The TAB technique, which is now the most popular method for the interconnection between an LCD panel and its driving ICs, has been rapidly developed mainly for large and thin LCDs, the word processor displays, and laptop computers. LCDs increasingly require high resolution and large capacity equivalent to that of a cathode-ray tube (CRT). As a result, there is a great demand for LCDs with a very fine pitch of display patterns. However, TAB technology has a limitation on its fine-pitch outer lead bonding (OLB). By bonding IC chips on the glass substrate of the LCD panel directly, a finer-pitch interconnection between the panels and the IC can be realized.

The finest OLB pitch of TAB-packaged LCD driving ICs which is currently available on the market, from some major Japanese electronics companies such as Toshiba, Texas Instruments, Hitachi, Epson, and Sharp, is 70 μm. These ICs have been specially designed for sub-note-

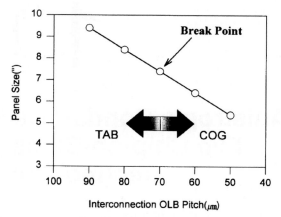

Figure 10.1 Dependence of OLB pitch on size of the dual-scan color STN LCD.

book-size dual-scan color STN LCDs. The 70-μm pitch will limit OLB TAB packaging technology for quite some time. When a finer pitch of display resolution is required, then FCOG technology is a better choice. Figure 10.1 shows an example of the limitation of the OLB pitch of TAB packaging in a VGA-grade dual-scan color STN LCD.

Several kinds of FCOG techniques have been developed by some Japanese companies. These methods can be divided into three major categories according to the intermediate materials between driving ICs (with or without bumps) and electrode pads on the glass substrate of a display by which two members are electrically interconnected.

1. The first category of the intermediate materials is a conductive paste, epoxy resin dispersed with tiny AgPd particles. This conductive paste is coated on top of Cu bumps of a driving IC with a thin overcoat layer of Au or on top of stud bumps of a driving IC,[13] in which the IC surface contacts the thin-film surface of this intermediate material.

2. The second intermediate material is a low-melting-point metal, indium alloy, formed as a low-melting-point bump on top of the gold bumps of driving ICs by using the dipping method.[5] The ICs and electrode pads of the display are interconnected at temperatures lower than 150°C.

3. The third intermediate material is conductive particles, Au-plated resin balls having elastically deformable properties, which are distributed on the Al pads of the driving IC or on the electrode pads of the glass substrate. This material is also supplied in *paste form,* as a base adhesive resin mixed with conductive particles, called *anistropic conductive adhesive* (ACA) with which the electrode pads of glass substrate are pasted; or in *film form,* as a film sheet of precured ACA, called

TABLE 10.1   Categories of FCOG Technology by Intermediate Materials

Intermediate materials	Bonding	IC	Substrate	Company
Conductive paste	AgPd	Au-plated Cu bump	ITO pads	Citizen
	Ag	Stud bump	ITO pads	Matsushita
Low-melting-point metal	In	Au bump	ITO pads	Toshiba
Conductive particle	ACF or ACA	Au bump	ITO/NiAu pads	Oki
	UV resin	Conductive particle bump	ITO pads	Sharp
	UV resin	No bump	Conductive particles' pads	Seiko Epson
	UV resin	Au bump	Opaque pads	Mitsubishi
	ACA	Au bump	Opaque pads	Samsung
	ACF	Au bump	ITO pads	Casio and Hitachi
Non	UV resin	Au bump	ITO pads	Matsushita

anistropic conductive film (ACF). There is a somewhat different film form,[12] called *vertical interconnection sheet* (VIS), which is not film-shaped precured ACA, but pure Au balls formed by a plating method on a polyimide film through the holes and aligned on a monolayer with an equally spaced matrix arrangement in vertical and horizontal directions (i.e., columns and rows). The specific process is explained below.

In addition to these three categories, there is another method called *microbump bonding technique* (MBB),[2] in which ICs with Au bumps and the electrode pads of the substrate are directly interconnected under high pressure and held by highly shrunk UV resin without the intermediate materials defined above. In this case, the Au bumps themselves serve as intermediate material. Table 10.1 summarizes this technique.

This chapter covers only the third FCOG method of using conductive particles as intermediate material for electrical interconnection between driving ICs and electrode pads on the substrate.

10.2   Conventional Flip Chip-on-Glass Technology Using Anistropic Conductive Adhesive

10.2.1   Introduction

Conventional flip chip-on-glass technology uses anistropic conductive adhesives as an intermediate material.[1] These are adhesives that are

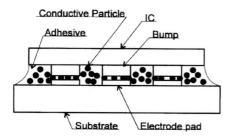

Figure 10.2 Principle of electrical interconnection with ACA.

electrically conductive across the bond line and are electrically insulating in the *xy* plane. The adhesives are available in paste form as ACAs and in film form as ACFs. Using an ACA or ACF is simple: (1) apply ACA or ACF, (2) align and place IC, and (3) cure the adhesive. ICs bonded with ACAs or ACFs must be bumped, typically with a gold bump. Although simple, the ACA/ACF flip chip bonding process has some technical shortcomings. These issues are covered below (see also Fig. 10.2).

10.2.2 Materials

ACA (anistropic conductive adhesive). The conventional ACA is a paste adhesive consisting of conductive particles dispersed in an adhesive matrix. Conductive particles can be pure metals such as gold, silver, or nickel; or metal-coated particles whose cores are plastic, quartz, or glass. The adhesive matrix can be thermosetting, thermoplastic, or UV-curable. Currently, only thermosetting and UV-curable ACAs offer the necessary reliability needed for flip chip attachment.

While isotropic conductive adhesives (ICAs) contain 30 to 40 vol% conductive filler, ACAs typically contain only 5 to 25 vol%. And, while most ICAs contain flake-shaped particles, ACAs contain spheroidal particles. Mean ACA particle sizes range from 5 to 15 μm, with smaller particles used for finer-pitch interconnections. When the IC is placed into the adhesive and applied by force, the adhesive is pressed to a monolayer of conductive particles, creating *z*-axis electrical contacts. However, with low-particle loading, there is no electrical continuity in the *xy* plane. Figure 10.2 illustrates the principle of electrical interconnection with ACA.

ACAs are available both from the United States and Japan. American suppliers include Zymet, 3M, and AI Technology. Japanese suppliers include Hitachi Chemical, Three Bond, and Sony Chemical.

ACF (anistropic conductive film). ACFs have achieved broad acceptance for TAB OLB-to-glass interconnection,[22] from larger notebook computer displays to smaller alphanumeric displays found in pagers (beepers). ACFs are not commonly used for flip chip application yet. However, some suppliers have begun producing ACFs for this application, and trials using ACF for flip chip assembly are increasing.

ACF is in film form, typically packaged in tape-on-reel format. ACFs utilizing the adhesive matrix can be thermosetting or thermoplastic. However, again, thermosetting versions are needed for flip chip applications. Recently, there have been some interesting advances in these materials.

Casio has developed an advanced ACF,[10,19,23] under the trade name Microconnector, using conductive particles with an additional insulating layer over a metal-coated plastic particle. The insulating layer is broken when heat and force are applied. The new particle reduces the likelihood of an electrical short circuit between two adjacent electrodes.

Hitachi has also developed an advanced "double-layer" ACF,[24] which comprises an adhesive layer and a particle layer. The particle layer has only one monolayer of particles. The double-layer design is reported to reduce particle flow during IC placement, increasing the number of particles on the interconnection electrodes.

Suppliers of ACFs include 3M and AI Technology in the United States and Hitachi Chemical and Sony Chemical in Japan. The use of ACFs for FCOG assembly is discussed further below.

10.2.3 OKI's example

In this section, OKI's paper[16] regarding ACA flip chip technology introduced at the IMC (International Microelectronics Conference) of June 1992 is summarized. OKI is also well known for its use of ACF material instead of ACA for flip chip technology. OKI presented a good paper regarding ACA flip chip bonding that will help us understand the overall concept of conventional ACA flip chip technique and important technical issues related to this method.

Table 10.2 specifies the materials and specimens of the samples used in their FCOG experiment. The IC has Al pads, on which fine Au

TABLE 10.2 Specifications of Test Samples

Substrate	Material	Barium borosilicate glass
	Electrode	CrAu, pitch: 120 μm
IC	Size	4.8×25×1.1 μm
	Bump	Material: Au Size: 60×120 μm Pitch: 120 μm
ACA	Base resin	UV-curable
	Conductive particle	Au-plated plastic ball Diameter: 5 μm

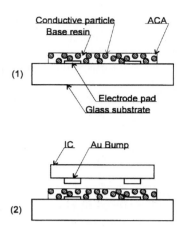

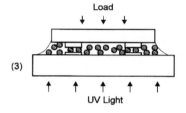

Figure 10.3 Schematic view of ACA FCOG process.

bumps are formed with 120-μm pitch. The glass substrate has an Au pattern on it. The interconnection resistance and insulation resistance between neighboring interconnections are measured. The ACA is made from a UV-curable base resin mixed with conductive particles, and the conductive particles are Au-plated resin balls and have a uniform diameter of 5 μm. Two UV adhesives are evaluated as the base resin.

Figure 10.3 is a schematic of OKI's ACA FCOG process. This process is composed of three simple steps as mentioned in Sec. 10.2.1.

1. Printing the ACA on the glass substrate

2. Aligning the IC with the glass substrate

3. Bonding the IC on the glass substrate with an appropriate load and curing the ACA by UV light exposure from under the substrate while it is under load.

Figure 10.4 shows the dependence of the number of conductive particles existing on a pad on the content of the conductive particles. The number of conductive particles existing on a bump increases as the content of conductive particles increases. These measurement values almost match the values obtained through calculation. The content of

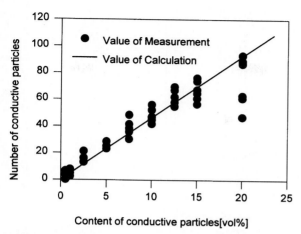

Figure 10.4 Dependence of conductive particle content on number of conductive particles existing on a pad.

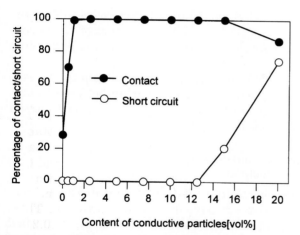

Figure 10.5 Dependence of conductive particle content on percentage of contact per short circuit.

the conductive particles seriously affects the connectability. Figure 10.5 shows the dependence of the contact and the short circuit on the content of the conductive particles. The results show that the most suitable amount of content is within the range of 2.5 to 12.5 vol%. Figure 10.6 shows the dependence of interconnection resistance on the content of the conductive particles. Interconnection resistance decreases as the volume content of conductive particles increases until reaching <7.5 vol%. However, in the region of >7.5 vol%, the interconnection

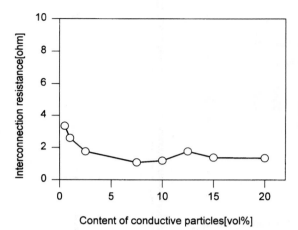

Figure 10.6 Dependence of conductive particle content on interconnection resistance.

resistance stabilizes at 1.5 to 2.0 Ω. The insulation resistance of the neighboring interconnections measured in the condition of 12.5% content of conductive particles and loaded voltage 50 V is said to be more than 10^{12} Ω.

In addition to the preceding experiments, OKI presents the stress–strain curve per conductive particle, the relationship between the resistance and the current, and the dependence of contact resistance on temperature change and reliability tests.[16]

10.2.4 Technical shortcomings and main issues of conventional FCOG method

The most serious technical shortcoming of conventional FCOG technology using anistropic conductive adhesive is that the particle distribution existing on the bumps of driving ICs cannot be exactly controlled after they are bonded on the substrate. Although the conductive particles are supposed to have a uniform distribution within the base resin, the results of the number of particles on the bumps differ from bump to bump.

Williams and Whalley[20] studied this matter quantitatively. They assumed that the particle distribution on the bumps follows a random distribution function such as binomial or Poisson equations and calculated the probability of conductivity based on the probability of particle distribution for a certain substrate size, particle loading (vol%) within base resin, particle diameter (pure Ni balls), and bonding pressure. They also calculated the probability of short circuiting between neighboring interconnections on the basis of certain assumptions.

If there is a small amount of conductive particles mixed with base resin inside, the potential for disconnection between driving ICs and electrode pads on the substrate increases. However, the more conductive particles are mixed with the base resin, the higher are the chances for short circuiting between neighboring interconnections. Of course, it is preferable to have more particles on the bumps, which results in lower contact resistance. However, it is theoretically impossible to make a monolayer with maximum packing density on the bumps because of short circuiting.

In practice, the distributions of conductive particles show different results according to the ACA dispensing methods such as pin transfer, screen printing, stencil or syringe dispensing, according to the viscosity of the base resin and also according to the bonding process. ACA also has the additional shortcoming of only short-term storability because the particles tend to settle down inside the resin as a result of weight variation between conductive particles and resin. This phenomenon is more pronounced with pure metal particles.

Conventional solid metal bumps such as Au formed for TAB inner lead bonding (ILB) purposes are not suitable for use in ACA flip chip bonding. As seen in Fig. 10.7, conventional solid metal bumps typically have a mushroom shape with a recessed top surface of >1 μm. This is because the IC pad is recessed in the passivation layer. The elastic region of the conductive particles should be under the inflection point. If the inflection point of deformable plastic ball having a metal shell is anticipated at 15 to 20 percent, an elastic region of the particles of around 10 percent would be reasonable. If the recessed depth is 1 μm and the diameter of the ball is 7 μm, the balls existing on the recessed area of a bump will not contribute to electric connections provided the conductive particles do not penetrate into the bumps. It is obvious that the bump top surface must be as flat as possible, for ACA flip chip bonding application, especially when the particle size is smaller than 10 percent of the recessed depth in the supposition given above. However, it is difficult to make such bumps.

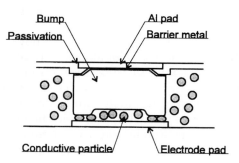

Figure 10.7 Some shortcomings of conventional ACA.

The bonding pressure imposed onto the backside of the IC must be controlled precisely according to the total number of particles on the bumps of the IC to achieve a reasonable elastic region of the particles when they are depressed. The bonding pressure influences contact resistance as well as particle deformation. High pressure results in low contact resistance. The coplanarity of the collet and the table of the bonding equipment is also very important to achieve a uniformly depressed ratio of the particles distributed over the whole area of the chip.

Bump size and alignment also influence the achievement of low contact resistance. To ensure a bigger bump size, bumps should be aligned in a staggered manner rather than in a straight line. However, this will result in a larger chip size.

It is clear that the bump configuration, the size of the conductive particles, the number of conductive particles that should be loaded into the base resin, the properties of the base resin, the method in which ACA should be dispensed, and the degree of bonding pressure imposed are the main issues in achieving the maximum condition. The size of the conductive particles must be decreased to reduce the risk of short circuit, and the amount of conductive particles must be increased to maintain the contact resistance as the pitch of interconnections narrows. Smaller particles reduce contact resistance, and many particles increase the risk of short circuit. Thus, the conventional ACA flip chip-on-glass technology is becoming less suitable for mass production, as the interconnection pitch becomes finer.

These shortcomings of the ACA method are almost the same as those of ACF flip chip bonding, except for the storage limitation because it is impossible to make ACF with a monolayer and square alignment of conductive particles. Only VIS seems to solve these problems, but it also has some technical obstacles to a wider application. This is discussed below.

10.3   Modified Flip Chip-on-Glass Technology Using Anistropic Conductive Adhesive Particles

10.3.1   Introduction

Because high temperature is required, the traditional solder bumps method, called *controlled-collapse chip connection* (C4), could not be adapted to a FCOG application. It is therefore logical that several Japanese LCD manufacturers have started to employ anistropic flip chip bonding technology using conductive particles, which is suitable for chip-on-glass application because it is simple and a low-temperature process is possible. However, the conventional ACA flip chip bonding method including ACF, as mentioned in a previous section, has some critical technical shortcomings regarding particle distribution, dispensing, storage, bump configuration, particle loading, etc. It is therefore believed that major Japanese LCD manufacturers have developed their own modified anistropic FCOG method using conductive particles to solve the preceding shortcomings, especially the risk of short circuit, of the conventional ACA method.

In this section, the modified FCOG methods using conductive particles developed by LCD manufacturers and three special materials developed by materials suppliers are explained and compared.

10.3.2 Sharp's method

Sharp has developed a new flip chip bonding method using conductive particles[7,11,14,15] called *electrical interconnection using light-setting adhesive selectively cured and conductive particles* (ELASTIC) which is very useful for chip-on-glass application. It is a new nonmetallurgical bonding method in which elastic and conductive particles are placed only on IC pads, that is, ICs with conductive particle bumps.

There are three possible classifications: a wafer process, a conductive particle mounting process, and a bonding process. Figure 10.8 shows the wafer process, which is as follows.

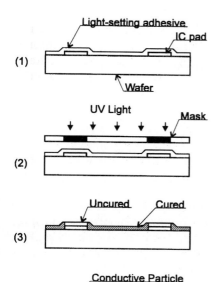

Figure 10.8 Schematic view of the conductive particle mounting process.

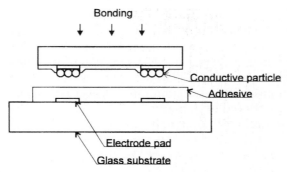

Figure 10.9 The bonding process using Sharp's FCOG method.

1. Applying an insulating light-setting adhesive onto the wafer on which the devices have been formed

2. Curing the light-setting adhesive except the electrode pad portion by irradiation with UV light through a mask

3. Mounting conductive particles above IC electrode pads which still have as yet uncured sticky portions

4. Dicing the wafer into chips

The bonding process between IC and substrate is the same as the conventional flip chip bonding, similar to OKI's method (shown in Fig. 10.3) except that the adhesive should contain pure resin without conductive particles inside. Figure 10.9 is a diagram of Sharp's method.

Sharp's method has some advantages over the conventional method. First, no short circuits will occur between neighboring interconnections, so fine-pitch interconnection is possible. Second, it is possible to obtain near-maximum or maximum monolayer packing density of conductive particles (or maximum packing density of monolayer of conductive particles) on the IC electrode pads, so the lowest contact resistance can be realized.

However, since a wafer itself should be processed before dicing to make conductive particle bumps, mounting conductive particles on the IC electrode pads will be a disadvantage. If users cannot purchase the driving ICs with conductive particle bumps, they will have to handle the wafers to do the conductive particle mounting process. Also, performing probing work for electrical die sorting (EDS) tests on the particles after the conductive particle bumps have been formed will be difficult.

10.3.3 Seiko Epson's method

Seiko Epson has developed a modified FCOG bonding method using conductive particles which employs standard aluminized driving ICs

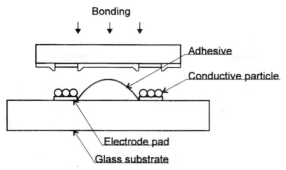

Figure 10.10 Seiko Epson's FCOG method.

instead of bumped driving ICs.[4,6] The conductive particles are printed on the electrode pads of the glass substrate and the bumpless ICs are bonded with adhesive in a face-down position. The portions on which the conductive particles are formed are the same as the Al pads of the driving IC when these members are bonded. Printed conductive particles exist only under Al pads. Figure 10.10 shows this diagram. What we can see clearly here is that the bonding result of Seiko Epson's method is exactly the same as that of Sharp's method. No one can distinguish between the Sharp and Seiko Epson's methods by analyzing the results of the bondings.

The basic technique of Seiko Epson's method is the printing process; however, they did not mention the method of printing the conductive particles in their papers.[4,6] None of the printing methods seem suitable for mass production of LCDs because the conductive particles must be printed on the exact portions (in accordance with Al pads of driving IC) of the electrode pads of the LCD glass panel assembled with two glasses. Handling the LCD glass panel to print conductive particles is also not suitable for mass production. If a reasonable, simple, and practical printing process were available, then Seiko Epson's method would eliminate all the disadvantages while retaining all the advantages of Sharp's method.

10.3.4 Mitsubishi's method

Mitsubishi's modified FCOG method[17] using conductive particles is the same as Seiko Epson's method in distributing the conductive particles on the electrode pads of the glass substrate before flip chip bonding. However, this technique uses a glass substrate having opaque electrode pads for backside imaging, and the particles remain over the entire area of the electrode pads, including the lead tracks. Typically, an ITO (indium tin oxide) pattern is used in case of LCD glass panels, on which Ni and Au are plated. Therefore this technique also uses driving

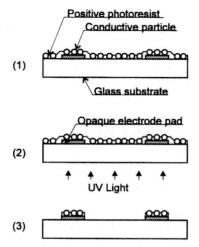

Figure 10.11 Schematic view of the backside imaging process for conductive particle distribution.

ICs with bumps. The processes can be classified into a backside imaging process for distributing the conductive particles on the electrode pads of the glass substrate and a bonding process. Figure 10.11 shows the backside imaging process, which is as follows.

1. Coating a positive photoresist in which the conductive particles are dispersed onto the glass substrate by a spin-coating method and prebaking the coated resist

2. Irradiating with UV light the glass substrate coated with the positive photoresist from the backside of the substrate, where the pattern of electrode pads acts as a photomask.

3. Removing the positive photoresist with conductive particles on the outside of the electrode pads, which are exposed to UV light, so that the conductive particles are mounted only on the electrode pads

The bonding process between IC and substrate is the same as in the methods described above. Figure 10.12 is a diagram of the bonding process. Although this method cannot load as many conductive particles on one electrode pad of glass substrate as can Sharp's or Seiko Epson's method, where almost maximum packing density of the monolayer of conductive particles on the electrode pads of either IC or substrate is possible, the backside imaging process is much more simple and also eliminates the lateral short-circuit problem.

Because many conductive particles are wasted in spin coating and developing, this method is not economical. The spin-coating process seems not to be suitable for mass production because an LCD panel with electrode pads must be spin-coated or a COG module must be

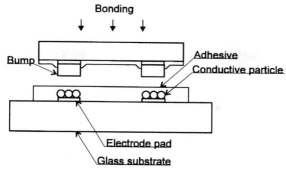

Figure 10.12 The bonding process of Mitsubishi's FCOG method.

made separately from the LCD panel and interconnected to the LCD panel later. By using the stenciling method instead of spin coating, the two shortcomings mentioned above can be overcome. However, it is difficult to form a monolayer of conductive particles by the stenciling method, and short circuiting will occur as conductive particles are squeezed between the interconnections when an IC is bonded on a substrate.

10.3.5 Samsung's method

Samsung (Samsung Display Devices) and Zymet, an American manufacturer of anisotropic conductive adhesives, have codeveloped a modified ACA FCOG bonding method using peak-shaped dielectric dams between electrode pads of the glass substrate, formed by a backside imaging process. This functions as a blocking dam which insulates the two electrical interconnections on right and left, and eliminates short circuiting by preventing any conductive particles from being positioned on top of the dam. This method requires driving ICs with bumps and glass substrates with opaque electrode pads, and uses the same conventional ACA material.

There are two classifications: backside imaging process and a bonding process. Figure 10.13 shows the backside imaging process, which is as follows:

1. Applying a negative-acting photoimagable dielectric to the glass substrate by the screen printing, stenciling, or syringe dispensing method

2. Projecting light from the backside of the glass substrate to expose the negative-acting photoimagible dielectric residing between the two opaque electrode pads, including lead tracks

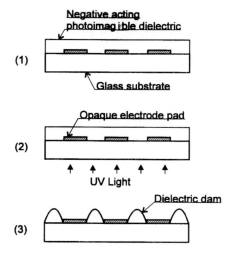

Figure 10.13 Schematic view of the backside imaging process for dielectric dam.

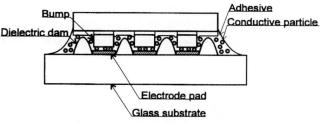

Figure 10.14 Samsung's FCOG method.

3. Removing the negative-acting photoimagible dielectric on the electrode pads, which are not exposed to UV light, thereby forming the peak-shaped dielectric dam

The bonding process is the same as in conventional flip chip bonding, but the dielectric dams themselves can also serve as an alignment guiding dam. Figure 10.14 shows a diagram of Samsung's method. The main focus of Samsung's method is building peak-shaped dielectric dams, especially dams with high wall angles. The gap between the top of a dam and the passivation layer of a driving IC should be less than the diameter of a conductive particle to prevent any lateral short circuiting. The wall angle of the dam should be made to have a peak-shaped top according to the distance from the surface of a glass substrate to the passivation layer of a driving IC and the gap between two opaque electrode pads of a glass substrate. This method can also eliminate the problems of short circuits and contact resistance of conventional ACA FCOG and provides a simple process for building dams because both stenciling and syringing for printing dielectric and backside imaging for developing dielectric are simple.

10.3.6 Casio's Microconnector

Casio has developed an advanced ACF,[10,19,23] the Microconnector, using conductive particles with an additional insulating layer over the thin metal layer plated on the plastic balls, which is broken on thermocompression bonding. This material can also be supplied in the form of ACA. When this material is used in ACA, Casio's method is exactly the same as the conventional ACA FCOG method except for the use of different conductive particles with an additional insulating layer.

The insulating layer comprises a large number of insulating micropowder particles, each of which is far finer than the plastic balls and serves to electrically insulate the outer surface of a corresponding metal layer so that this material can allow finer-pitch interconnection with much less lateral short circuiting between interconnections. This is formed by causing particles of an insulating micropowder having a low melting point (about 100 to 200°C) to adhere to the surface of the metal layer with an electrostatic effect, the Coulomb force under the presence of an electric field. The base adhesive resin should be a thermo type such as thermoplastic of thermosetting, to produce thermocompression when the bonding process is done. When the bonding heat and pressure are applied, the insulating layer which is in contact with the bump surface of the IC is broken, but the other layer remains, so it produces only z-axis electric interconnections and lateral short circuiting is prevented. This method also requires driving ICs with bumps. Figure 10.15 is a diagram of Casio's method.

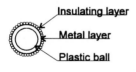

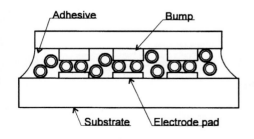

Figure 10.15 Casio's FCOG method.

Casio's method is believed to solve most of the shortcomings of the conventional ACA FCOG method while keeping the process simple. With an additional insulating layer, a fine pitch and low contact resistance without the risk of lateral short circuiting can be realized by increasing the particles per unit volume to be mixed with the base adhesive resin or film. If the insulating layer works as well as they claim and does not significantly increase the cost of the material, Casio's method is an excellent technique.

10.3.7 Hitachi's double-layer ACF

Hitachi also has developed another advanced ACF with bilayer which consists of an adhesive layer and a particle layer in monolayer.[24] Conventional ACF has one layer of adhesive material in the sheet form in which conductive particles are dispersed in a random distribution. Figure 10.16 illustrates the difference between the two materials. The conductive particles layer and the adhesive layer are formed separately and attached together later. The conductive particle layer is thin, similar to the diameter of a conductive particle, and has a high-viscosity thermosetting material in which conductive particles are arranged in monolayer, whereas the adhesive layer is thick (depending on the bump thickness) and has a low viscosity (lower than that of particle layer) pure thermosetting resin. The conductive particle layer contributes only to electrical interconnection, and the adhesive layer contributes only to attaching and holding together the two components. This material can be supplied only in ACF type, not ACA, and requires driving ICs with bumps.

Because ACF thickness is much greater than the diameter of a conductive particle, when the conventional ACF is depressed under pressure and temperature between the driving IC and the substrate, conductive particles beneath the bumps become monolayer and the

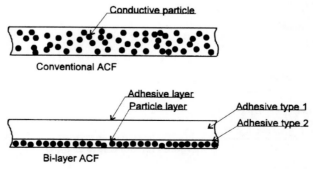

Figure 10.16 Comparison of Hitachi's double-layer ACF with conventional ACF.

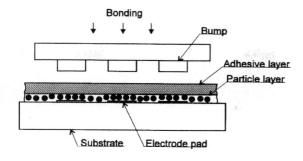

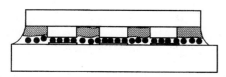

Figure 10.17 Schematic view of the bonding process of Hitachi's ACF.

other many conductive particles which did not contribute to the monolayer are squeezed between the bumps, thereby causing short circuiting. However, Hitachi's bilayer ACF already has a monolayer of conductive particles with higher viscosity and clearing points than those of the adhesive layer, so conductive particles are not squeezed and remain in the same place, thereby preventing short circuiting. Figure 10.17 shows a schematic view of the bonding process of bilayer ACF. Hitachi claims that they have successfully achieved 10-μm-pitch interconnection in laboratory experiments using their double-layer ACF. Hitachi will commercialize this newly developed bilayer ACF soon for FCOG application with 20-μm-pitch interconnection. This bilayer ACF will be a very promising FCOG material if it is cost-effective.

10.3.8 Sumitomo's VIS

Sumitomo has developed a somewhat different style of anistropic conductive film,[12] called *vertical interconnection sheet* (VIS), which consists of pure Au balls aligned in an equally spaced matrix arrangement in vertical and horizontal directions in monolayer on a polyimide film and formed by a plating method on the film through the holes on the film. This sheet is inserted later for electrical interconnection, before bonding, between a driving IC and a substrate after they are aligned with each other. Au balls are melted and attached onto Al electrode pads of a

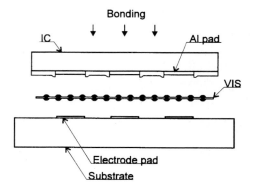

Figure 10.18 Sumitomo's FCOG VIS.

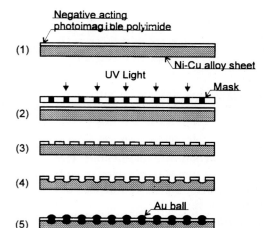

Figure 10.19 Schematic view of the VIS process.

driving IC and electrode pads of a substrate by thermocompression when the two components are pressed with high temperature and high pressure. Figure 10.18 is a diagram of Sumitomo's VIS. Sumitomo's VIS process is seen in Fig. 10.19 and consists of the following:

1. Coating a negative-acting photoimagable polyimide, approximately 4 μm, to the NiCu alloy sheet, approximately 70 μm thick

2. Irradiating UV light through a mask from above

3. Developing polyimide film, thereby forming equally spaced aligned holes in vertical and horizontal directions

4. Etching a NiCu sheet to form bowl-shaped spaces penetrating into the sheet

5. Forming Au balls by plating through the holes

6. Etching out a NiCu sheet from the polyimide film

This material also eliminates the problem of short circuiting. However, VIS is a difficult and complicated process and is also very expensive. The spacing between Au balls on VIS should be decreased for finer-pitch interconnection, so the VIS process will become more difficult. Furthermore, it is difficult to handle VIS because the sheet is weak and fragile.

10.4 Conclusions and Future Directions

Processes and materials for several FCOG assembly methods using anistropic conductive adhesives or films, along with each method's technical merits and demerits, have been discussed. If their technical shortcomings are not overcome, none of these methods can be used in mass production successfully. Developers of FCOG processes, mainly Japanese LCD manufacturers, will continue to refine their methods. Also, new and interesting materials developments such as Casio's Microconnector, and Hitachi's bilayer ACF will open new doors to further advances.

In addition to the technical issues that have been discussed in this chapter, other factors, including (1) inspection, (2) reworkability, and (3) reliability, will strongly influence which methods will ultimately be implemented. These three issues have not been covered here. However, readers can study them by reading the references listed at the end of this chapter.

TAB bonding is currently the predominant interconnection technology for LCD driving ICs and will continue to be in the foreseeable future for larger displays. With the advent of 70-μm-pitch TAB OLB, 11-in XGA (Extended Graphics Array)-grade (1024×768) dual-scan color STN LCDs and 5.5-in workstation-grade (1680×1024) color TFT (thin-film transistor) LCDs can be realized. These high resolutions satisfy consumer needs for large-size LCDs such as notebook and subnotebook computers on the market. Thus, there is little incentive to retool existing TAB assembly lines with less mature flip chip bonding processes and to implement a technology with less proven reliability in large-scale LCD applications.

However, FCOG technology is expected to play a significant role in the manufacture of small displays. Applications include displays for watches, viewfinders, videogame displays, and handheld TVs. It is also expected that newer handheld communications devices and application-specific computing tools will create significantly larger demands for small, high-resolution displays.

Acknowledgments

The author would like to thank Mr. Karl I. Loh, president of Zymet, Inc. for his contribution to the section on ACA material and Mr. Lawrence Peck, legal consultant of Samsung Display Devices' licensing department for his advice on writing this chapter.

References

1. Seikosha, U.S. Patent 4,113,981, Sept. 12, 1978.
2. Hatada, K., and H. Fujimoto, "A New LSI Bonding Technology: Micro Bump Bonding Technology," *Proceedings of the 38th IEEE Electronic Components Conference,* pp. 45–49, May 1989.
3. *Nikkei Microdevices,* no. 49, 41–65, July 1989 (in Japanese).
4. Masuda, M., K. Sakuma, E. Satoh, Y. Yamasaki, H. Miyasaka, and J. Takeuchi, "Chip on Glass Technology for Large Capacity and High Resolution LCD," *1989 IEEE/CHMT International Electronics Manufacturing Technology Symposium,* pp. 55–58, 1989.
5. Mori, M., M. Saito, A. Hongu, A. Nitsuma, and H. Ohdaira, "A New Face Down Bonding Technique Using a Low Melting Point Metal," *IEEE Transact. Components, Hybrids, Manufact. Technol.,* 13(2): 444–447, June 1990.
6. Sakuma, K., K. Nozawa, E. Sato, Y. Yamasaki, K. Hanyuda, H. Miyasaka, and J. Takeuchi, "Chip on Glass Technology with Standard Aluminized IC Chip," *Proceedings of International Symposium on Microelectronics,* ISHM, pp. 250–256, 1990.
7. Nukii, T., N. Kakimoto, H. Atarashi, H. Matsubara, K. Yamamura, and H. Hatsui, "LSI Chip Mounting Technology for Liquid Crystal Displays," *Proceedings of International Symposium on Microelectronics,* ISHM, pp. 257–262, 1990.
8. Sharp, U.S. Patent 4,963,002, Oct. 16, 1990.
9. Yoshigahara, H., Y. Sagami, T. Yamazaki, A. Aburkhart, and M. Edwards, "Anisotropic Adhesives for Advanced Surface Mount Interconnection," *Proceedings of the Technical Program, NEPCON West,* vol. 1, pp. 213–219, 1991.
10. Casio, U.S. Patent 4,999,460, March 12, 1991.
11. Sharp, U.S. Patent 5,065,505, Nov. 19, 1991.
12. Yasuo, N., and Y. Tetsuo, "Micro-film Connector for High Density Interconnection," *Electric Materials,* 28–35, Nov. 1992 (in Japanese).
13. Bessho, Y., Y. Horio, T. Tsuda, T. Ishida, and W. Sakurai, "Chip-on-Glass Mounting Technology of LSIs for LCD Module," *IMC 1990 Proceedings,* Tokyo, pp. 183–189, 1992.
14. Atarashi, H., N. Kakimoto, H. Matsubara, K. Yamamura, T. Mukii, and H. Matsui, "Chip-on-Glass Technology Using Conductive Particles and Light-Setting Adhesives," *IMC 1990 Proceedings,* Tokyo, pp. 190–195, 1990.
15. Matsubara, H., H. Atarashi, K. Yamamura, N. Kakimoto, K. Naitoh, and T. Nukii, "Bare-Chip Face-Down Bonding Technology Using Conductive Particles and Light-Setting Adhesives: ELASTIC Method," *IMC 1992 Proceedings,* Yokohama, pp. 81–87, 1992.
16. Takahashi, W., K. Murakoshi, J. Kanazawa, M. Ikehata, Y. Iguchi, and T. Kanamori, "Solderless COG Technology Using Anisotropic Conductive Adhesive," *IMC 1992 Proceedings,* Yokohama, pp. 93–98, 1992.
17. Otsuki, H., T. Kato, F. Matsukawa, M. Nunoshita, and H. Takasago, "Chip-on-Glass Packaging Technology Using Conductive Particles," *IMC 1992 Proceedings,* Yokohama, pp. 99–103, 1992.
18. Ogunjimi, A. O., O. Boyle, D. C. Whalley, and D. J. Williams, "A Review of the Impact of Conductive Adhesive Technology on Interconnection," *J. Electron. Manufact.,* 2: 109–118, 1992.
19. Casio, U.S. Patent 5,123,986, June 23, 1992.

20. Williams, D. J., and D. C. Whalley, "The Effects of Conducting Particle Distribution on the Behaviour of Anisotropic Conducting Adhesives: Non-uniform Conductivity and Shorting between Connections," *J. Electron. Manufact.* **3:** 85–94, 1993.
21. Vardaman, E. J., and R. Crowley, "Emerging Flip Chip Use," *Adv. Packag.,* Fall 1993.
22. Andoh, H., Y. Yanada, and Y. Fukuda, "Fine Connection Technology Using Anisotropic Conductive Film," *HIBRIDS* **8**(6): 19, 1993 (in Japanese).
23. Casio, U.S. Patent 5,180,888, Jan. 19, 1993.
24. Hitachi Chemical, *Connecting Materials for COG,* technical report, Sept. 1994.

Wire-Bonding Flip Chip Technology for Multichip Modules

Stacey T. Baba and William D. Carlomagno

11.1 Introduction

As advances are made in multichip module technology, it is important to address the issue of electrical interconnection of the chips to the circuit board or substrate. The three primary methods of electrically interconnecting bare chips today are wire bonding, TAB, and flip chip.[1,2] Each of these interconnect technologies offers specific advantages, but each has certain disadvantages as well. A new proprietary assembly process, *Bonded Interconnect Pin* (BIP) technology developed by Raychem Corporation, combines the simplicity of wire bonding, the pretest capability of TAB, and the packaging density of flip chip. BIP technology enables large numbers of chips to be interconnected to a wide variety of circuit boards and makes it possible to access and take advantage of the very high routing density of high-density interconnect (HDI) microcircuits.

The BIP process utilizes thermosonic ball bonding and reflow soldering to create electrical interconnections between the chip and the circuit board. A typical BIP assembly is shown in Fig. 11.1. Any chip that can be thermosonically ball-bonded can be electrically interconnected using BIP. A BIP Chip (BIP and BIP Chip are trademarks of the Raychem Corporation) is assembled face-down on the HDI microcircuit minimizing the path length of electrical signals and optimizing routing and packaging densities. The BIP length can be varied to accommodate the thermal coefficient of expansion mismatch associated with conventional flip chip structures.[3,4] BIP assemblies are repairable using reflow soldering techniques. In addition, BIP can be

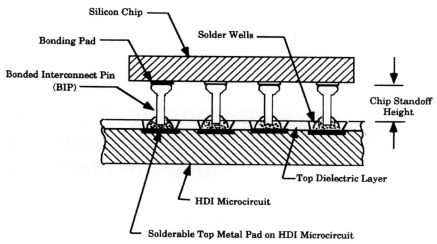

Figure 11.1   The BIP concept showing a basic assembly structure.

applied to individual die; thus availability of the wafer (a sensitive problem for semiconductor manufacturers) is not an issue.

11.2   Fundamental Studies of BIP Test Assemblies

As part of the development of BIP technology, a fundamental studies program was established to characterize the behavior of BIP/HDI assemblies subjected to environmental tests for microelectronic packages. This program was not intended to serve as a reliability program. Instead, the focus of this program was to generate preliminary data and to detect catastrophic failures in either the prototype design or material selection. In addition, the tests for this program were conducted beyond military specifications in order to fully assess the capabilities of this new technology.

11.2.1   BIP fabrication

BIPs were fabricated from bulk hard-drawn 99.9% gold bonding wire, 2.0 mil in diameter, having an elongation of 0.5 to 1.0 percent. A modified thermosonic ball bonder equipped with a proprietary bonding capillary was used to transform the bulk wire into individual interconnect pins. The process is illustrated in Fig. 11.2. As in conventional ball bonding, the bulk wire was fed through the capillary and a ball was formed at the exposed end using an electric flameoff. The wire was then welded to the interconnect pad on a chip preheated in the workholder of the bonder. To form a BIP, the wire was clamped,

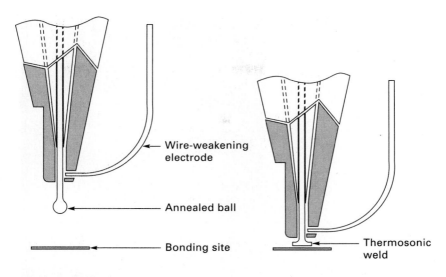

Wire-weakening electrode

Annealed ball

Bonding site

Wire is fed through capillary,
EFO arcs wire forming ball

Thermosonic weld

Capillary lowers and
welds ball to interconnect
pad on chip

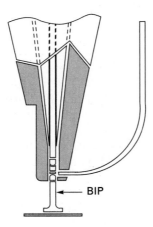

BIP

Wire is clamped, drawn up
in tension and is weakened
by heating; wire breaks
and pin is formed

Figure 11.2   Illustration of the BIP process.

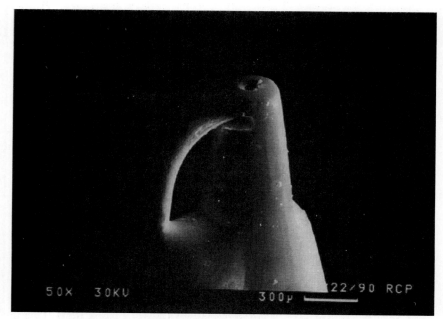

Figure 11.3 SEM micrograph of modified capillary.

drawn up in tension, and locally weakened by discharging an arc from a small electrode (buried in the sidewall of the capillary) to the wire. The capillary is shown in Fig. 11.3. This caused a tensile break to complete the formation of the pin. The BIP process allows for 100 percent nondestructive testing of each pin. If a flaw is present in the wire or a poor weld is made, a failure will occur during the wire tensioning step in the process. After BIPs were formed, they were planarized using a shear plate. A BIP Chip after shearing is shown in Fig. 11.4.

Selection of the BIP diameter and length was based on chip size, chip pad size, handling, and thermal stress considerations. To optimize handling of BIP Chip and assemblies, the maximum diameter wire for the chip pad was used. Figure 11.5 shows the nominal BIP length versus chip size. For this study, BIP diameter and lengths of 2.0 and 20 mils, respectively, were used.

11.2.2 Test chip configuration and construction

A test chip was designed to simulate the bonding conditions of commercial integrated circuits and to provide multiple bonding sites for BIP. The test chips were fabricated on silicon with AlSiCu bonding pads using standard semiconductor processing techniques. Two chip

Figure 11.4   SEM micrograph of a BIP Chip.

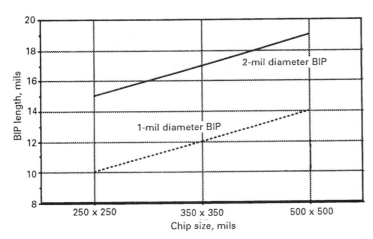

Figure 11.5   Nominal BIP length versus chip size.

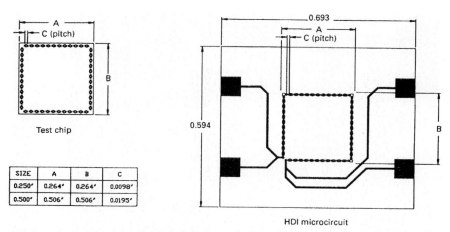

SIZE	A	B	C
0.250′	0.264′	0.264′	0.0098′
0.500′	0.506′	0.506′	0.0195′

Figure 11.6 Layout of test chips and HDI microcircuits for 250×250-mil chip and 500×500-mil chip assembly.

sizes, 250×250 mil and 500×500 mil, each having 96 bonding pads were used in this study. Each bonding pad was 120×120 μm. The pitch between pads was 9.84 mils for the smaller chip and 19.68 mils for the larger chip. Layouts of the chips are shown in Fig. 11.6.

11.2.3 HDI Microcircuit configuration and construction

The HDI microcircuits (fabricated by Raychem) were designed to mate with 250×250-mil and 500×500-mil BIP test chips. When assembled, the BIP test chip and HDI microcircuit formed a daisy chain of 96 BIP sites. Conductive traces from each end of the daisy chain led to pads on the top surface of the HDI for resistance measurements of the interconnect assembly. To accommodate BIPs, the HDI substrates were fabricated with a thick dielectric top layer. Blind vias were processed in the top dielectric layer, creating solderable interconnect sites for the BIPs. After processing the blind vias, the substrates were tested for electrical continuity and isolation, visually inspected, and diced into individual microcircuits. The HDI microcircuit board layout is shown in Fig. 11.6.

11.2.4 Preparation of the HDI microcircuit for BIP assembly

A 80/20 AuSn solder was selected for the BIP assembly. AuSn is a hard solder and was selected because of its matched thermal conductivity to the bond wire and its resistance to thermal fatigue, due to its elastic rather than plastic deformation.[5,6] The solder, in paste form, was squeegeed into the interconnect sites or "solder wells." After the

Figure 11.7   SEM micrograph of solder wells.

solder wells were visually inspected, the solder was reflowed in a furnace in a nitrogen environment. The eutectic temperature of the solder was 280°C. The microcircuits were then cleaned and visually inspected. A typical solder well with and without solder is shown in
Figs. 11.7 and 11.8.

11.2.5   BIP/HDI assembly

Prior to assembly, a thin layer of RMA flux was wiped over the interconnect sites on the HDI microcircuit. To complete the assembly
process, a BIP chip was placed on the HDI microcircuit and was then
reflow-soldered into position using a nitrogen-purged furnace as
shown in Fig. 11.9. After reflow, the assemblies were cleaned and inspected for proper solder fillets and electrical continuity. Figure 11.10
shows the layout of a typical test assembly. Figure 11.11 is an SEM of
an actual test assembly.

11.3   Test Plan and Evaluation

The BIP assemblies were subjected to environmental tests selected
for the purpose of screening critical aspects of the interconnect technology, specifically:

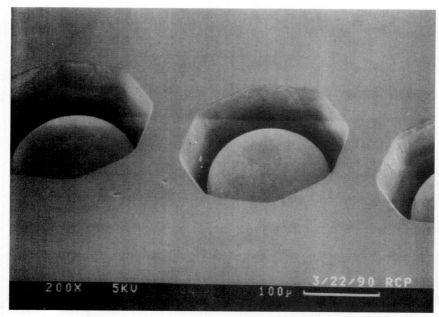

Figure 11.8 SEM micrograph of solder wells with solder (after reflow).

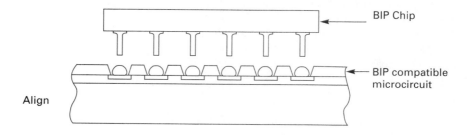

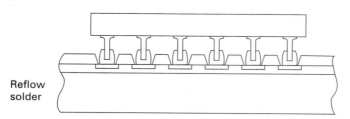

Figure 11.9 BIP assembly process.

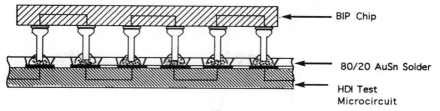

Figure 11.10   Layout of test assembly.

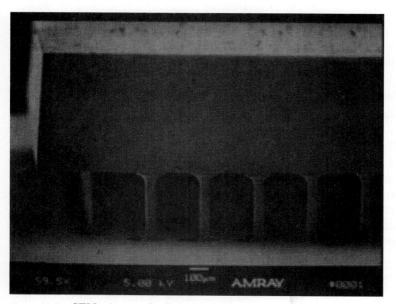

Figure 11.11   SEM micrograph of test assembly.

1. The resultant metallurgical properties of the solder joint
2. The mechanical stress on the assembly due to differential expansion during thermal cycling and shock
3. The robustness of the assembly

The tests selected for the fundamental studies program are listed below. A summary of these fundamental studies is shown in Table 11.1. The Military Standard 883 was used for the test specifications.[7] It should be noted that not all size assemblies were subjected to test. We selected what we considered to be worst-case scenarios. As an example, only the test assemblies with the larger die were subjected to

TABLE 11.1   Summary of the Fundamental Studies Test Plan

	Chip size	
Tests	250×250 mil	500×500 mil
Temperature cycle		X
Thermal shock		X
Stabilization bake	X	
Mechanical shock	X	X
Random frequency vibration	X	X
Random vibration	X	X
Substrate attach strength	X	

mechanical tests. It was assumed that if these designs passed, the smaller die designs would also pass.

11.3.1   Temperature cycling

The objective of this test was to assess the robustness of the BIP structure to exposures at extremes of high and low temperatures and to the effect of alternate exposures to these extremes. The procedure used was per MIL-STD-883C, method 1010, condition C. Under these conditions the samples were cycled from −65 to 150°C. The readouts were taken at 0 cycles and at 100 cycles and 500 cycles. The test criteria used to determine changes was variation in the resistance of the assemblies. The sample size for this test was three assemblies or 288 solder sites.

11.3.2   Thermal shock

This test was used to assess the robustness of the device to sudden exposure to extreme changes in temperature. Unlike the temperature cycling test, during the thermal shock test, the samples are moved quickly (within a 10-s period) between the hot and cold extremes of −65 to 150°C. This test was conducted per MIL-STD-883C, method 1011, condition C. Measurements were taken at 0, 10, 20, 50, 75, 100, 125, 150, and 200 cycles. The resistance of the assemblies was measured to note any changes. Three assemblies were used in this test, consisting of 288 solder sites.

11.3.3   Stabilization bake
(high-temperature storage)

The objective of this test was to determine the effect of storage at elevated temperatures on the BIP assembly and to accelerate the growth and effect of intermetallics in the solder. The method used was per MIL-STD-883C, method 1008. The assemblies were stored at 150°C for up to 1000 h. The resistance of the assemblies was measured. In addition, SEM cross sections were taken to observe changes in the solder composition.

11.3.4   Mechanical shock

This test was conducted to assess the robustness of the BIP/HDI design to moderately severe shocks as a result of rough handling, transportation, or field operations. The procedure used was conducted per MIL-STD-883C, method 2002. During this test, the assemblies were subjected to several conditions. There were mechanical shocks at 500g for duration of 1 ms, shocks at 1000g for a duration of 0.75 ms, and at 1500g at a duration of 0.50 ms. The test criteria used to determine changes were measurements of the resistance of the assemblies after each condition. In addition, a visual examination was used to note the orientation of the pins. Both large and small HDI/BIP assemblies were used.

11.3.5   Random frequency vibration

This test was used to assess the effect of vibration on BIP assemblies in the specified frequency range. This test and the selected ranges were based on MIL-STD-883C, method 2007, conditions A, B, and C. In this test, the vibration frequency varies logarithmically between 20 and 2000 Hz. The peak acceleration varies from 20g for condition A, 50g for condition B, and 70g for condition C. The samples were measured prior to test and then after each condition. The test criteria used to measure changes in the assembly was resistance measurements and visual inspection of the BIP.

11.3.6   Random vibration

The objective of this test was to assess the ability of BIP assemblies to withstand the dynamic stress exerted by random vibration applied between upper and lower frequency limits to simulate various service-field environments. This test was conducted per MIL-STD-883C, method 2026, condition E, test condition II. During this test the test assembly was subjected to a random vibration with a spectral power density of 0.2 and an overall root-mean-square G of 18.7. One assembly per size (96 solder sites) was tested. Resistance measurements were taken after the test along with a visual examination of the assembly.

11.3.7   Substrate attach strength

The purpose of this test was to determine the strength of the BIP and associated interfaces and to note any changes of strength due to high-temperature storage. This test was based on MIL-STD-883C, method 2027. BIP assemblies were fabricated, stored at 150°C in air, and placed in tension. The force required was recorded as well as the failure mode. Two assemblies were tested at the initial time, after 250 h, after 500 h, and after 1000 h.

11.4   Test Results

The results for temperature cycling, thermal shock, and stabilization back are shown in Figs. 11.12, 11.13, and 11.14, respectively. The data indicate that there is no significant change in the resistance of the as-

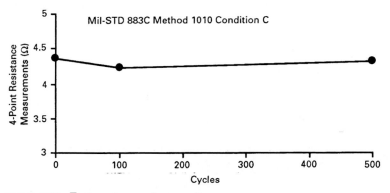

Figure 11.12   Temperature cycling results.

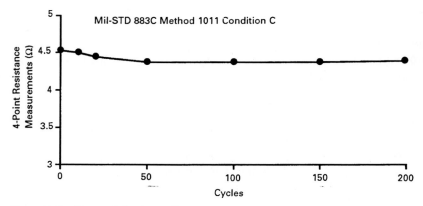

Figure 11.13   Thermal shock results.

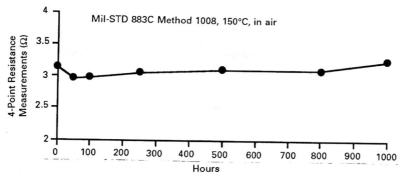

Figure 11.14 Stabilization bake results.

semblies after 500 h of thermal cycling, 200 cycles of thermal shock, or 1000 h at 150°C. There was some concern that during stabilization bake, possible intermetallic growth could occur in the AuSn solder, causing embrittlement. Although some intermetallic growth was observed (cross sections of samples as assembled at 500 h and at 1000 h are shown in Fig. 11.15), results from the substrate attach strength test (as reported below) show that the failure occurred in the wire and not in the solder well, indicating the ability of the solder to withstand elastic deformation.

The results for mechanical shock are shown in Table 11.2, and the results for random frequency vibration and random vibration are shown in Table 11.3. No change in the resistance of the test assemblies was observed. Some BIP deformation was observed. The pins were bent during the force of the test. However, no complete failure of any pin was observed.

The results for substrate attach strength are shown in Fig. 11.16. Failure occurred within the BIP wire. No solder pullout was observed. A decrease in the substrate attach strength was observed after 270 h. However, in a separate test, not reported in this work, it was this phenomenon that was also observed in raw stock bonding wire.[8] Therefore, this observed decrease in strength is not due to BIP formation or assembly but is an inherent characteristic of the bond wire used.

11.5 Summary

The initial results from the fundamental studies of BIP did not indicate any catastrophic failures in either the prototype design or materials.

Further development of the BIP assembly process included the design and fabrication of an electrically functional multichip DRAM module. The module, configured in a SIP format, utilized a multilayer HDI substrate, NAS edge clips, MT 1259 [256K (256-kbyte)] DRAM

X Section of BIP Solder Well As Assembled (535X)

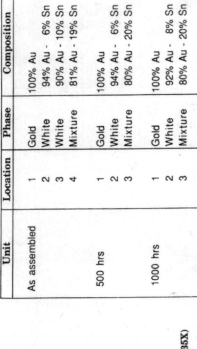

X Section of BIP Solder Well; 500 hrs at 150°C (535X)

Unit	Location	Phase	Composition
As assembled	1	Gold	100% Au
	2	White	94% Au - 6% Sn
	3	White	90% Au - 10% Sn
	4	Mixture	81% Au - 19% Sn
500 hrs	1	Gold	100% Au
	2	White	94% Au - 6% Sn
	3	Mixture	80% Au - 20% Sn
1000 hrs	1	Gold	100% Au
	2	White	92% Au - 8% Sn
	3	Mixture	80% Au - 20% Sn

X Section of BIP Solder Well; 1000 hrs at 150°C (535X)

Figure 11.15 SEM micrograph of intermetallic structure after stabilization bake.

TABLE 11.2 Mechanical Shock Results

MIL-STD 883C, method 2002

Condition	Peak acceleration level g	Duration, ms	Results*
A	500	1.00	Pass
A (modified)	1000	0.75	Pass
B	1500	0.50	Pass

Pass criteria: no change in resistance >10 percent.

TABLE 11.3 Random Frequency Vibration and Random Vibration Results

MIL-STD 883C, method 2007

Condition	Peak acceleration level g	Results*
A	20	Pass
B	50	Pass
C	70	Pass

*Pass criteria: no change in resistance >10 percent.

MIL-STD 883C, method 2026, test condition II

Condition	Result*
E	Pass

*No change in resistance >10 percent.

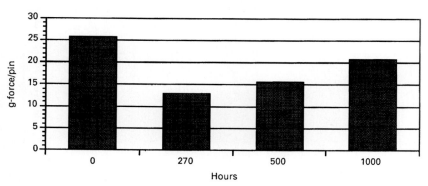

Figure 11.16 Substrate attach strength results.

die, and surface-mount chip capacitors. The die were reflow-soldered to the substrate using 80/20 AuSn solder and electrically tested for short circuits and open circuits. All nonfunctional die were removed from the substrate during a repair operation.

This functional 12×256K DRAM module—having length, height, and thickness dimensions of 2.000, 0.560, and 0.125 in, respectively—demonstrates the superior packaging density achievable when combining conventional wire-bondable die and the BIP assembly process.

The results from the fundamental studies and the preliminary evaluations conducted on the DRAM BIP module will develop a better understanding of this unique technology and its capability and future in the assembly of microelectronic circuits.

References

1. Neugebauer, C., R. O. Carlson, R. A. Fillion, and T. R. Haller, "Multichip Module Designs for High Performance Applications," in *Multichip Modules,* Johnson, R. W., K. F. Teng, and J. W. Balde, eds., IEEE Press, pp. 46–52, 1991.
2. Early, L. C., "A Series of Demonstrators to Assess Technologies for Silicon Hybrid Multichip Modules," ibid. (Ref. 1), pp. 139–143.
3. Harada, S., and R. Satoh, "Mechanical Characteristics of 96.5Sn/3.5Ag Solder in Micro Bonding," *IEEE 1990 Proceedings of the 40th Electronic Components and Technology Conference,* IEEE Press, pp. 510–517, May 1990.
4. Vaynman, S., "Fatigue Life Prediction of Solder Material: Effect of Ramp Time, Hold Time, and Temperature," ibid. (Ref. 3), pp. 505–509.
5. Matijasevic, G., C. Y. Wang, and C. C. Lee, "Void Free Bonding of Large Silicon Dice Using Gold-Tin Alloys," *IEEE Transact. Components Hybrids Manufact. Technol.,* **13**(4), 1128–1134, Dec. 1990.
6. Wright, C., "The Effect of Solid State Reactions upon Solder Lap Shear Strength," *IEEE Transact. Parts, Hybrids Packag.,* **PHP-13** (3), 202–207, Sept. 1977.
7. Military Standard, *Test Methods and Procedures for Microelectronics,* MIL-STD 883C, Aug. 25, 1983.
8. Cummings, D., and F. Guerrero, private communication, 1990.

Flip Chip Mounting Using Stud Bumps and Adhesive for Encapsulation

Kazuhisa Tsunoi, Toshihiro Kusagaya, and Hidehiko Kira

12.1 Introduction

A semiconductor recently introduced on the market is directly installed on a glass epoxy or glass polyimide resin printed-circuit board using a bare-chip and solderless flip chip mounting method. The purpose of this technology is to ensure low-cost and high-density mounting of electronic parts for large-scale mainframes as well as personal computers.

Advances in semiconductor process and fabrication technology continue to boost chip integration sales in terms of gate count and function, but the accompanying increase in pin count is beginning to pose new problems for chip carrier, packaging, and interconnect design. Fine-pitch packaging to accommodate more pins creates extra production, test, and reliability problems for the manufacturer and customer, whereas the solution of increased carrier and package dimensions contradicts the trend toward smaller, lightweight products.

Chip-on-board (COB) techniques use variations of tape-automated bonding (TAB), flip chip, and conventional wire-bonding techniques to mount bare chips directly on the printed-circuit board (PCB) or substrater, eliminating most pin-related problems entirely. This chapter describes a solderless high-density COB mounting technique based on flip chip technology.

12.2 Current Semiconductor Mounting Processes and Problems

12.2.1 Introduction

Commercially available packaged integrated circuits may have as many as 500 pins; a total pin count of 1000 is expected within the next few years. If we consider a 15-mm-square chip having 300 I/O pads mounted on 55-mm-square, 0.65-mm-pitch and 35-mm-square, 0.4-mm-pitch chip carriers, the mounting efficiencies are only 14 and 6, respectively. By adding in the additional printing wiring substrate area required to mount these packages on a board, we can immediately see expensive board real estate going to waste. This chip-to-carrier, carrier-to substrate interconnect wastage has been the stimulus behind COB technology development.

12.2.2 Bare-chip assembly techniques

Three typical COB techniques are compared as follows (see also Fig. 12.1):

1. *Tape carrier package* (TCP) (Fig. 12.1a), whereby bare chips are supplied on a tape carrier on which the I/O pads of the individual chips are prebonded to a flexible lead frame structure which is transferred and bonded to the substrate

2. *Wire bonding* (Fig. 12.1b), whereby fine wires bond the bare chip I/O pads to the printed wiring substrate in much the same way as a chip is connected to the lead frame in a packaged chip

3. *Flip chip mounting* (Fig. 12.1c), whereby bare chip I/O pads are bumped and the chips are mounted face down on the substrate.

The chip-to-substrate interconnects of both TCP and wire bonding take up a certain amount of board space, not to mention access area for tooling; hence, the total area is larger than the chip itself. In contrast, flip chip COB techniques mount chips face down on the sub-

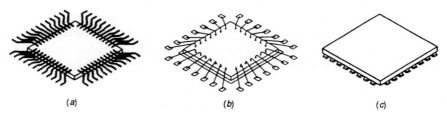

(a)   (b)   (c)

Figure 12.1   Bare-chip assembly techniques: (a) TCP; (b) wire bonding; (c) flip chip.

strate, which is an advantage in terms of mounting density. Bare chips intended for flip chip mounting, however, must undergo formation of protective films such as Ni and Au on chip bonding pads, at the water stage to protect the pads against corrosion during solder bumping. This step rules out off-the-shelf bare-chip purchase. The solderless flip chip COB technique we developed solved this problem.

12.3   Solderless Flip Chip Mounting

12.3.1   Introduction

Flip chip mounting is notorious for poor chip-to-substrate connection reliability due to widely differing coefficients of thermal expansion of the mating surfaces. The fourfold difference in coefficient between the glass epoxy substrate and the silicon chip produces shear stresses at the interconnect as temperature varies; the larger the chip, the greater the stress.

Using finite element analysis, we examined and compared two structures: one in which the chip-to-substrate interstice is filled with a thermosetting adhesive, and the other in which the interstice is left unfilled.

12.3.2   Structure

As the cross section of the structure (Fig. 12.2) shows, a stud bump is formed on each chip I/O pad and covered with a conductive paste. The chip–printed wiring interstice is filled with enough thermosetting adhesive to ensure that the sides of the chip are covered.

A precalculated quantity of a thermosetting adhesive at a predetermined viscosity is silk-screened onto the substrate to minimize adhesive flow after printing and facilitate mounting of chips. The epoxy-based adhesive has a glass transition temperature T_g of 119°C and a coefficient of thermal expansion (α) of 6.2×10^{-5}°C^{-1}.

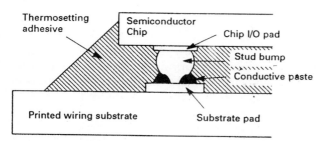

Figure 12.2   Solderless flip chip–substrate interconnect.

Connection stud bumps to the substrate pads is based on a shrinkage stress greater than the thermal stress of the thermosetting adhesive produced from heating the chip, and the two kinds of stress produced by elastic strain on the stud bump. The conductive paste compensates for variations in stud-bump height, making the connection more secure. Reliability appraisal later showed freedom from faults at the joints.

Because chips are purchased unpackaged, uranium and thorium traces, both sources of α rays, were found to be below measurable limits. Further, the content of chlorine and fluorine impurity ions, which cause joint corrosion, was below 10 ppm.

Still another advantage of this structure is that all steps—thermosetting adhesive application conductive paste application and pre-bumped chip mounting on the substrate—are done at the same time. This is instrumental in minimizing the maximum principal stress, acting on the edges of the chip as a result of differences in thermal expansion coefficients of the chip and substrate, as well as limiting the maximum shear strain on the bump. The findings of a stress analysis are presented below.

12.3.3 Stress analysis

The stud bumps and the substrate I/O pads were isolated for stress analysis. A model joint comprised one bump and was made two-dimensional because both the structure and the manner in which the stress is applied are symmetric. Beginning with a initial temperature of 170°C, local stresses and displacements were calculated using the ABAQUS finite element analysis program at temperatures of ≤25°C.

Allowance was made for the characteristics of temperature dependence of Young's modulus and the coefficient of thermal expansion of the thermosetting adhesive and printed wiring substrate.

Two different structures, one with interstitial thermosetting adhesive (Fig. 12.3b) and the other without (Fig. 12.3a), were examined by finite element analysis and compared to verify stress displacement.

The displacement of the chip in the adhesive-filled structure was smaller because the adhesive restricted movement (Fig. 12.3). The adhesive-filled structure's local stresses were reduced by about 25 percent. This was evident from reliability appraisal. This adhesive-filled structure makes flip chip mounting using commercially available bare chips possible, including Au stud bumping by a wire bonder, and solderless flip chip mounting by interconnecting the stud bump and the electrodes on the printed wiring substrate with a conductive adhesive and a thermosetting adhesive.

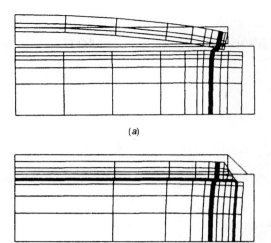

(a)

(b)

Figure 12.3 Results of finite element analysis: structure without (a) and with (b) thermosetting interstitial filler.

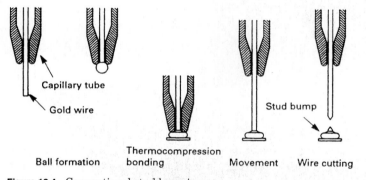

Figure 12.4 Conventional stud bumping.

12.3.4 Stud-bump formation

The process of stud-bump formation is the same as that of wire bonding IC die to lead frames. It is both economical and reliable. Figure 12.4 shows the stud-bumping process.

First, as in conventional wire bonding, a gold wire is passed through the hole in a capillary tube and subjected to an electrical discharge to form a gold ball, which is bonded to the chip I/O pads by thermocompression and ultrasonic energy. Next, the capillary tube clamping the wire is withdrawn and the wire is cut. This process forms stud bumps

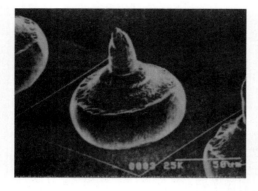

Figure 12.5 Stud bumping by an alternate method.

Figure 12.6 Stud bump after height leveling.

of uniform height and diameter. The Young's modulus of conventional bonding wire is too low to produce uniform bump height, so we substituted wire having a higher modulus. Figure 12.5 shows a stud bump formed by using a gold wire with a high Young's modulus.

12.3.5   Stud-bump leveling

In next step, stud bumps are pressed against a flat surface to make their height uniform (Fig. 12.6).

A pressure of about 30 g is applied per bump.

Making bump height and shape uniform in this way optimizes bump contact with the substrate when the chip is mounted and helps control the rate of transfer of the conductive paste.

12.3.6   Conductive paste transfer

The next step is to apply a conductive paste to the stud bumps. This is done by squeezing a conductive paste film of a predetermined thick-

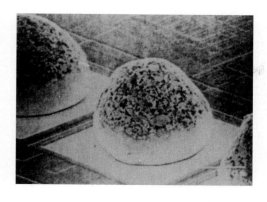

Figure 12.7 Stud bump with conductive paste coating.

ness onto a jig having a surface roughness not exceeding 0.2 μm (such as a sheet of glass), then pressing the stud-bump surface against the film to transfer the paste. The rate of transfer is controlled by varying the film thickness. This technique confines the spread of the conductive paste to within the outer diameter of each bump, and prevents short circuits to adjacent bumps (Fig. 12.7).

The chip mounting sequence begins by feeding thermosetting adhesive onto the printed wiring substrate by a silk screen or dispenser process. Only enough adhesive is supplied to form sufficient adhesive fillets on the chip sides to ensure proper seal and strength. Next, the chip and the printed wiring substrate are adjusted to be parallel to within ±3 μm, followed by alignment of the chips to the printed wiring substrate pads by image recognition. This is done with a half-mirror placed between the chip and the substrate, using a camera to adjust planar position, height, and angles of rotation (x, y, z, θ). Mounting precision is within ±5 μm. Figure 12.8 shows the bonder-alignment mechanism.

When alignment is complete, the chip is heated and pressurized from the reverse side to set the adhesive and thus make the printed wiring substrate pad and chip bump connections. The paste is allowed to set fully to complete the mounting process.

12.3.7 Repair technology

Because chips are tested only by DC testing via a wafer probe and by static function testing, chips may malfunction after mounting on the substrate. This makes technology for replacing faulty chips essential.

To facilitate repair, the conductive paste is thermoset at the mounting temperature but becomes plastic at higher temperatures (Fig. 12.9).

To make a repair, the chip and printed wiring substrate are heated and pressurized from the reverse side in the same way as they were

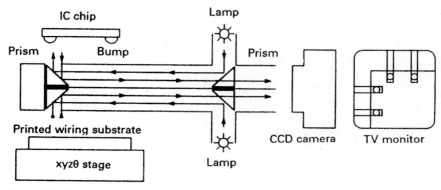

Figure 12.8 Bonder-alignment mechanism.

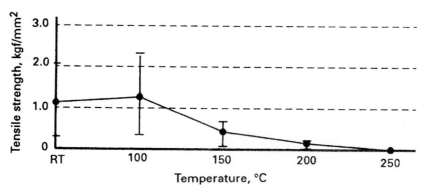

Figure 12.9 Adhesive bonding characteristics.

bonded. When the adhesive softens, a slight shear force is applied to the side of the chip to separate it from the substrate. After separation, residual adhesive on the substrate is polished off with a hand-held buffing tool. We have also confirmed that excimer laser works quite well for this purpose. The substrate is then recoated with adhesive and a new chip mounted. This completes the repair operation.

12.3.8 Connection reliability

We conducted a temperature cycling test of 500 cycles between −40 and +105°C to ensure connection integrity.

The maximum contact resistance between stud bumps and substrate pads did not exceed 10 mΩ (Fig. 12.10).

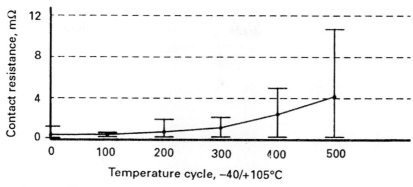

Figure 12.10 Contact resistance.

12.3.9 Reliability testing

The chip used for reliability testing was a gate array with a chip measuring 10 mm^2 and I/O pads on 120-μm centers: In total, 25 to 32 samples were subjected to the following three tests:

1. High-temperature operation test (32 samples)
2. High-temperature, high-humidity bias test (25 samples)
3. Temperature cycle test + autoclave test (25 samples)

Table 12.1 summarizes the results of the reliability test. Evaluations were done by dynamic testing. The samples had a defect rate of 0 percent, without the migration anticipated with the conductive paste.

12.3.10 Feasibility study

As a practical test, we built and evaluated a 1-Mbyte IC memory card using a glass epoxy substrate (Fig. 12.11).

TABLE 12.1   Reliability Test Results

Test conditions	Rate of function error, %
High-temperature operation test: 1000 h, 100°C, 5.5 V	0
High-temperature, high-humidity bias test: 3000 h, 85°C, 85% RH, 5.0 V	0
Temperature cycle test: 500 cycles, −40°C to +105°C	0
Autoclave test: 144 h, 105°C, 100%, 1.2 atm	0

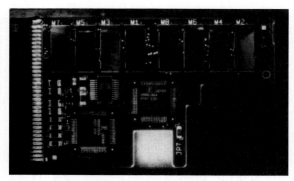

Figure 12.11 Memory card.

The prototype consists of a copper foil pattern on a glass epoxy resin substrate and 1-Mbit SRAM (static random-access memory) devices. The I/O pads of the 32-pin SRAM devices are placed on 240-μm centers. Eight devices could be mounted on one side of the board—twice the previous number—which formerly mounted four TSOP packages on each side.

12.4 Summary

A solderless chip-on-board (COB) technique that enables highly integrated bare die to be mounted on printed wiring boards has been proposed. A conductive paste eases thermal and mechanical stresses resulting from differences in die and board coefficients of thermal expansion. This process is inexpensive and reliable, and supports repair.

Reference

1. Rao, R. T., and E. J. Rymaszewski, *Microelectronics Packaging Handbook,* Nikkei BP, 1991.

Wire Interconnect Technology An Ultra-High-Density Flip Chip–Substrate Connection Method

Larry Moresco, David G. Love, Bill Chou, and Ven Holalkere

13.1 Introduction

Over 25 years ago engineers at IBM[1–3] recognized the need for a low-cost, high-density connection technique for integrated circuits. For maximum density, considering the topology, they chose a surface to surface area array connection. The solder bump or C4 technology was developed. This reliable,[4,5] connection technique has virtually gone unchanged over the past 30 years except for the type of substrate used to connect the integrated circuits together. The use of solder bumps in the electronics industry is gradually becoming more common today to meet the demand for high numbers of connections between chip and substrate for both signals and power supply.

In 1991 Fujitsu Computer Packaging Technologies, Inc. (FCPT) was formed to explore and develop technologies to interconnect modern VLSI circuits for high-speed computers. Although solder bumps provide a tremendous improvement over peripheral connection techniques such as wire bonds or TAB (tape automated bonding), they are also not without limitations.

Wire interconnect technology (WIT) was invented at FCPT.[6] This technology provides new opportunities for the IC designer unavailable from any other IC connection method. Structurally, WIT (shown in Fig. 13.1) is a metal post (approximately 10 μm in diameter, 50 μm long) requiring a small (25-μm) pad to solder attach. These posts can connect chip to substrate on a 50-μm area array pitch. This amounts

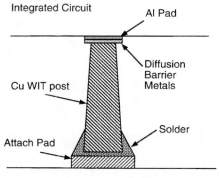

Integrated Circuit

Al Pad

Diffusion
Barrier
Metals

Cu WIT post

Solder

Attach Pad

Figure 13.1 WIT structure in cross section.

Polymer Based Interconnect

to over 1 order of magnitude improvement over existing solder-bump connections. The metal post is grown on either the IC surface or the substrate. During thermal cycling the WIT metal post is in bending (instead of shearing). The number of cycles to fatigue are improved over the solder-bump connection method. The volume of solder and the distance of the solder from the active surface of the IC result in at least 3 orders of magnitude improvement with regard to α-particle radiation dosage on the active circuits at the IC surface. With a pad diameter of approximately 10 μm at the base of a WIT (on the chip surface), this technology reduces the connection capacitance by an order of magnitude over any other method of chip attach available today.

13.2 WIT Fabrication and Assembly

The present method of creating WIT posts is plating. The WIT post is plated on either the substrate or on the IC. High-aspect-ratio vias in a thick photoresist layer are patterned followed by a fine-grain plating process. The resulting copper metal post has nearly the same properties of annealed pure copper. The decision as to which structure has the WIT post attached is based on yield and associated manufacturing cost of the integrated circuits and the substrate to which they are attached. To allow multiple replacement of WIT attached chips, a thin layer of nickel is coated over the copper WIT posts. The nickel acts as a diffusion barrier between the solder and copper materials. This barrier allows multiple solder attach and replacement cycles without degradation of the solder joint materials. The WIT post can also be embedded in dielectric layer such as polyimide (resulting in a via connection between metal layers) if needed. This processing technique can be applied to fabricate a high-density multilevel interconnection structure.

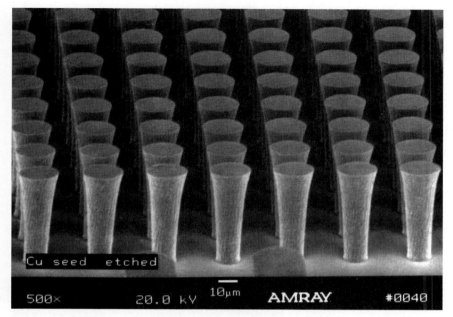

Figure 13.2 WIT "forest" on silicon with 30-μm area array pitch.

Figure 13.2 shows a "forest" of WITs grown on a pitch of 30 μm. These particular WITs are approximately 10 μm in diameter and 47 μm tall. To ensure proper assembly, the WIT post must be fabricated to within very tight height tolerances. The current manufacturing process is capable of fabricating a 45-μm-tall WIT within ±2.5 percent over the area of a 20-mm² VLSI surface. This process maintains a ±5 percent variation in WIT post height across a 6-in wafer. FCPT has developed a plating method that yields WIT height uniformity across a 6-in wafer of ±5 percent (± 3σ).

Figure 13.3 is a SEM photo of a cross section of a WIT assembly after carefully potting it in epoxy and polishing the sample to approximately the center of a WIT. As of this writing, WIT assemblies have been successfully reworked three times and subsequently power cycle reliability tested with no degradation in reliability results. A fluxless process is preferred to keep the WIT posts from trapping flux because of the tight pitch and tolerances required.

13.3 Reliability Testing

A full battery of environmental tests must be used appropriate to the product specifications. One of the most difficult tests of flip chip (C4)

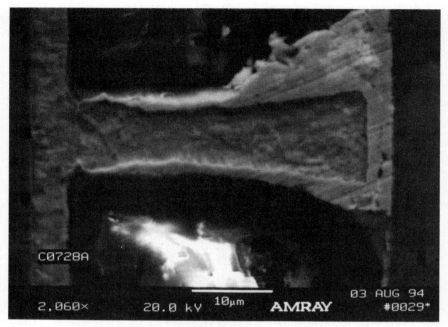

Figure 13.3   SEM photo of an assembled WIT.

reliability is considered to be temperature cycling. For any IC contact (and especially flip chip attach methods) power cycling reliability fatigue life must be determined. During on/off cycles, there will be mismatch in temperature between chip and substrate. A rapid change in power dissipation will result in thermal temperature gradients. These temperature gradients can result in a mechanical displacement of one end of a WIT post relative to the other end of the post. The same thing occurs with solder bumps except the WIT post can bend and not shear plastically. The displacement [Eq. (13.1)] can be calculated as follows:

$$\Delta I = (T_{\text{Si}} \times \text{TCE}_{\text{Si}} - T_{\text{sub}} \times \text{TCE}_{\text{sub}}) \text{DNP} \qquad (13.1)$$

where ΔI = displacement
T_{Si} = temperature of silicon
TCE_{Si} = thermal coefficient of expansion of silicon
T_{sub} = temperature of substrate
TCE_{sub} = thermal coefficient of expansion of substrate
DNP = distance from neutral point (center of IC to farthest corner pad center)

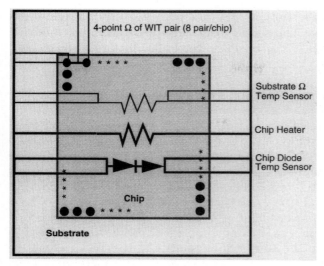

Figure 13.4 Simplified chip-on-substrate diagram.

One must solve the transient heat-transfer problem of the complete structure including heatsinks attached to the substrate or IC to determine the magnitude of this problematic transient displacement. This mismatch is time-dependent, unlike the more obvious mismatch in TCE between chip material and the substrate matrix of materials at steady-state temperature conditions for both materials.

Figure 13.4 is a simple electrical schematic of the power cycling test structure. This test substrate to chip interconnect incorporated the following features:

- *On-chip.*   Heater (polysilicon sheet, powered at four corners of the chip). Five temperature-sensing diodes at four corners and center of chip. WIT attach test pairs for resistance measurements at maximum DNP (distance from neutral point); WIT attach pads with solder on the perimeter of the chip (484 total).

- *Test substrate.*   Temperature sensors made of copper traces, matching the locations of temperature sensors on the chip, measured with a four-point milliohm meter probe; WITs matching the attach pads on the chip.

With this apparatus it is possible to measure (in real time) WIT resistance changes and the temperature difference between chip and substrate. Using this setup it is possible to precisely calculate the strain and the resistivity changes affected by the strain on the WIT assembly structures. The importance of real-time measurement should be noted here. The strain displacements are produced during the first few seconds of the on/off power cycle. Observation of these strain disturbances to the resistance of the joint is an excellent indicator of incipient joint failure. It is desirable to understand the resistance changes during the start/stop power cycle as well as any changes found after relaxation begins.

In one of the first test setups the WIT post was deposited on a rigid substrate first. The chip contained a small solder pad that was reflowed to attach the WIT post as shown in Fig. 13.5.

In order to determine the power cycle reliability of the WIT attach method, an accelerated test condition was developed. For the particular design of concern, a WIT joint at the maximum DNP on a 16-mm² chip experiences a displacement of approximately 0.5 μm. Therefore, a test setup condition was created that developed a displacement of 1.0 μm to accelerate the stress failure mode of the WIT post joints shown in Fig. 13.5.

A controlled-atmosphere thermal chamber (Fig. 13.6) was used to simulate environments under which a WIT joint may be exposed. In the controlled-atmosphere enclosure a PGA package and socket are used to allow for quick replacement of the combined test chip and substrate modules. The cavity of the PGA (where an IC would be normally die-attached) is machined to remove material and create another cavity. A substrate is placed on the ledge produced by milling out this new smaller substrate cavity. The substrate is secured with a small

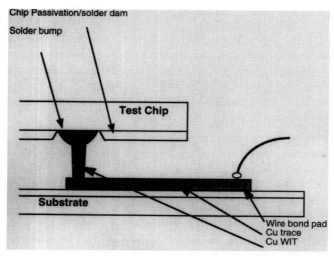

Figure 13.5 Chip-on-substrate cross-sectional view of WIT attach.

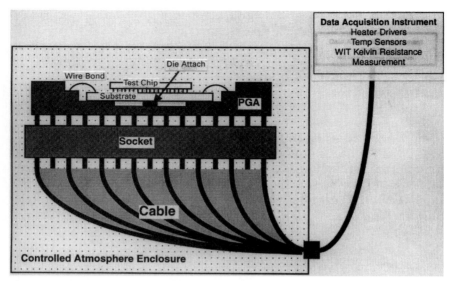

Figure 13.6 Test structures for real-time data sampling.

droplet of polyimide die attach adhesive at its center. The perimeter pads of the substrate are wire bond–attached to the PGA. The purpose of this process is to produce minimal mechanical coupling between the PGA package and the test substrate. Any artificial stresses which may cause the substrate to bend or strain would affect the measurement of the true fatigue life of the WIT attach joints.

The test deliberately causes the WIT to strain under rapid temperature cycles between the simulated integrated circuit and a rigid substrate. For this specific test setup a temperature difference of 50°C between the test chip and substrate produces a 1-μm displacement from top to bottom across the WIT post. The rigid substrate of the accelerated reliability test causes the WIT post solder joint to undergo strains more than 2 orders of magnitude greater than would be seen by a WIT attached between a polymer-based substrate and a silicon chip. During this test, the chip is repeatedly heated to 120°C while the substrate remains at 70°C and then both chip and substrate return to 50°C.

At this time the N_{50} (number of cycles for 50 percent failure) for these accelerated reliability tests is greater than 1800 of these power cycle test conditions. According to projections based on relative strain using three-dimensional finite element analysis of the accelerated test structure and the actual application structure, the number of power cycles to first failure in the actual application will be in excess of 10,000. Further testing is being performed as of this writing to determine the life expectancy in a real application.

13.4   Interconnect Technical Characteristics

As was stated earlier, the reasons for developing a new attach method are as follows: (1) local contact density from the surface of an IC to the substrate, (2) reduced interconnect pad capacitance, and (3) power cycling reliability for large-format VLSI circuits. This third requirement has recently become pronounced because of the environmental requirements of powerdown for computers. This low power state for computers reduces the electrical power consumption for computers when they are idle. This savings of power results in extra on/off power cycles for computer circuits. The result will be more strain cycles for IC contact joints such as solder bumps and WIT attach posts.

13.4.1   Electrical characteristics

The electrical characteristics of the WIT interconnect, similar to those of any other interconnect, depend on the physical relationship between the various connections to power, ground, and signal contacts. For the parameters presented here the standard connection cell shown in Fig. 13.7 is used for the analysis. Imagine that this figure is a standard cell that is repeated between the surface of an integrated circuit and its matching substrate. The pattern is distributed uniformly with one DC contact for every signal contact to the surface of

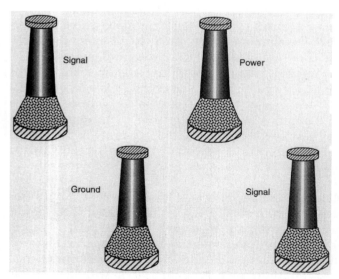

Figure 13.7 Pattern used for electrical parameter approximation.

the IC. For the purposes of this electrical analysis, the dimensions of the WIT pattern cell were as follows:

WIT height	45 µm
WIT diameter	12 µm
WIT pitch	50 µm

The resulting electrical lumped equivalent circuit for the length of the WIT post elements of this repeated structure are

Resistance	8.4 mΩ
Capacitance	10 fF
Inductance	30 pH

13.4.2 Mechanical stress

An investigation into the mechanical stress state is performed to determine the stress and strain conditions of the various WIT applications. This allows one to extrapolate between test and real applications of the WIT structures.

Introduction. In order to properly analyze the stress state of the WIT post in various configurations, a full three-dimensional elastic through plastic strain analysis was performed. This analysis was used to project the stress and strain resulting in accelerated testing and in real applications. Initially, a linear static thermal stress analysis was performed on the WIT assembly as shown in Fig. 13.8. The stresses obtained from

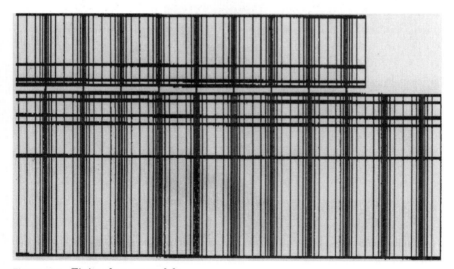

Figure 13.8 Finite element model.

this analysis were beyond the elastic range of the WIT (plated copper) and solder (63/37 PbSn). Hence, a detailed elastoplastic analysis of the WIT assembly was performed to evaluate the plastic strains in the critical areas of the assembly. As one would expect, these occur within the WIT and solder joint connecting the chip to the substrate.

The "WIT assembly" consists of chip, WITs, solder, copper layers, polyimide layers, and the base substrate plate as shown in Fig. 13.9. This structure is used to approximate the real application of the WIT attach joint. The number of interconnects (WITs) between the chip and the substrate is 484 for these simulation experiments. Figure 13.9 also contains some of the critical dimensions used for the simulations of the assembly.

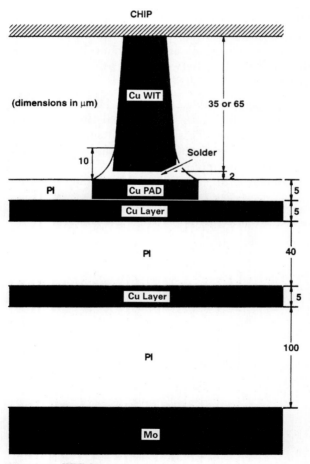

Figure 13.9 WIT dimensions.

Analysis and discussion. Analyzing the complete assembly using the finite element method is prohibitive in terms of computer cost and time. It is possible to capture the major stress and strain details of the critical areas of the assembly by analyzing a diagonal section of WITs within the assembly as shown in Fig. 13.10. It should also be noted that the minimum pitch for a WIT joint is not used because of the size of such a model and the minimal improvements that a full model would offer for these initial simulation results. Referred to as the "strip" model, the purpose of this model is to investigate the upper and lower bounds on the mechanical stress and strain conditions of the WIT assembly. Variable dimensions and temperature extremes are used to expose the stress and strain extremes. The WIT and solder assembly stresses were evaluated for a temperature differential of 100 degrees. ABAQUS,[7] a commercially available nonlinear finite element program, is used to perform the analysis. The mechanical properties used in the analysis are listed in Table 13.1. The two WIT lengths considered in the analysis were 35 and 65 μm.

The bulk mechanical properties were used in the analysis. It is important to note that the flow stress increases dramatically in thin metal films in the micrometer range of thickness based on the film thickness and grain size.[8] Since this is an elastoplastic analysis, it is

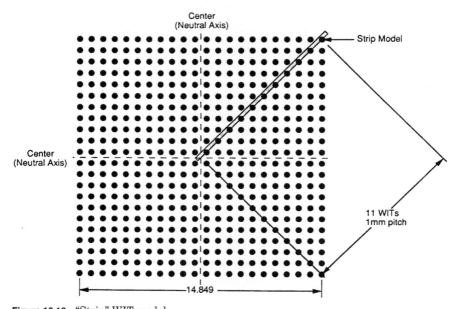

Figure 13.10 "Strip" WIT model.

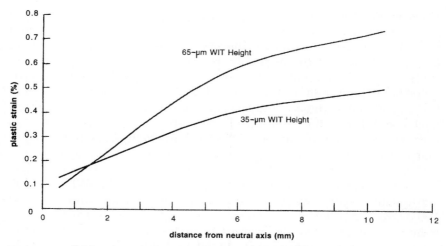

Figure 13.12   Solder plastic strain versus distance from the neutral axis.

decreases, the beam changes from a "long" beam to a "short" beam. In a "short" beam, shear effects become more predominant than bending. Hence, the critically stressed point on the solder is shifted toward the base of the solder joint, where it is absorbed by the larger solder volume. The maximum plastic strain on the solder decreases as the WIT height decreases. This mechanism is illustrated in Fig. 13.13. The resulting bending moment caused by the taller WIT increases the strain on part of the solder joint. Looked at another way, one can see that as the WIT becomes shorter, it places a greater shear load on the bond pad joint affecting a larger area and therefore reducing the strain within the solder material. Recently the thermomechanical fatigue of 63/37 SnPb solder has been studied under strain-controlled conditions. A piecewise Coffin-Manson relation has been used to determine the life prediction of high-lead solder and is given as follows:[9]

$$\Delta\epsilon_p = 114\, N_f^{-0.72} \qquad (13.2)$$

where $\Delta\epsilon_p$ is the solder plastic strain (in percent) and N_f is the number of cycles to failure.

The preceding analyses predicted high plastic strain zones in the solder as well as at the WIT/substrate interface. There was a dramatic reduction in plastic strain in the corner WIT (copper) when its height was increased from 35 to 65 μm as seen in Fig. 13.11. Equation (13.2) predicts that a 50 percent reduction in strain will result in a 2.5 times increase in the number of cycles to solder failure. Using this type of

TABLE 13.1   Material Properties

Material	Elastic modulus, kg/mm²	Poisson's ratio	Yield stress, kg/mm²	TCE, ppm/°C
Silicon	16,346	0.205		3.0
63/37 PbSn	3,164	0.4	3.16	25.0
Copper	12,655	0.34	7.23	16.1
Molybdenum	33,044	0.3		5.53
Polyimide	714	0.32	61.0	3.0

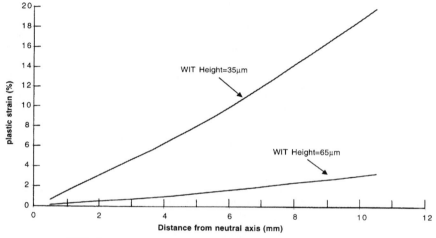

Figure 13.11   WIT plastic strain versus distance from neutral axis.

more pertinent to evaluate the plastic strain for the WIT farthest from the neutral axis at the corner of the chip and the solder connecting the WIT to the chip.

Figure 13.11 illustrates the variation of plastic strain in the WIT as a function of DNP (distance from the neutral point). This means that the bending moment applied to the copper post increases with the DNP. Hence, the corner WIT experiences the combination of maximum bending moment and shear force. As the WIT height increases, since the differential expansion between the chip and the substrate remains the same, the bending stress on the WIT decreases. One can see the reduction in strain due to an increase in WIT post height. The copper post has a bending moment applied to it by the thermal displacement. The longer the post, the lower the plastic strain resulting from this thermal displacement.

Figure 13.12 shows the variation of plastic strain in the solder connecting the WIT to the chip as a function of DNP. As the WIT height

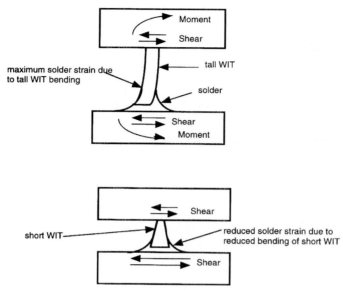

Figure 13.13   Bending and shear mechanisms of WIT and solder.

analysis helps determine the design dimensions of a WIT based on chip size, substrate structure, solder, and materials properties.

13.4.3   Heat transfer

With a WIT post average diameter of 12 μm, significant heat can be conducted into the substrate. The effective thermal resistance of this WIT post is 1000°C/W. This thermal resistance is dependent on the diameter of the WIT post. Since copper is nearly 10 times more thermally conductive than high-lead solder, the thermal resistance will drop dramatically as the WIT post diameter is increased. The increased diameter increases the stress on the solder join (another design tradeoff for product-specific optimization).

This thermal path allows one to carry heat away from the surface of an active integrated circuit through the substrate to either an air- or liquid-cooled heatsink. For comparison, a solder bump on the order of 75 μm in diameter and 65 μm tall would have a thermal resistance of 400°C/W.

The thermal resistance of this shape of WIT translates into a VLSI circuit with 1000 contacts having a chip-to-substrate top metal thermal resistance of 1°C/W. One must then design good thermal paths into the substrate to remove heat from the backside of a polymer–metal interconnect substrate.

13.4.4 α-Particle radiation

Another advantage of the WIT attach method is the lower potential for α-particle reliability problems. WIT attacks this problem from two fronts. First, the mass of solder present in the bond is reduced by nearly 700 times because of the small size of the solder joint. Second, the proximity of the solder joint is nearly 50 times farther away from active circuits when the WIT is grown on the integrated circuit and soldered to the interconnect substrate. Therefore, the radiation dosage on the active surface of the integrated circuit can be reduced with WIT assembly technology by 700 (with WIT, solder attached to the integrated circuit) to nearly 1,000,000 times (with WIT, solder attached to the substrate).

13.4.5 Routing effects

In order to take full advantage of WIT, one needs a substrate interconnect technology with enough routing paths to successfully route away from this fine-pitch connection structure. Figure 13.14 shows a repeatable bond pad "cell" for two types of IC attach pads. The perimeter pad cells on the left and the area array pad cells on the right of Fig. 13.14 are called "cells" because they are fundamental patterns that repeat as they are stepped around or across the surface of the integrated circuit. All four cell arrangements have a repeatable cell of four pads. The upper two cells contain three signal pads for each DC pad. The lower two cell patterns have three DC contact pads for every signal pad. This reason for this difference will be explained later in the text.

If one were interested in showing the overall superiority of an area array contact method over a perimeter-based contact method, a dia-

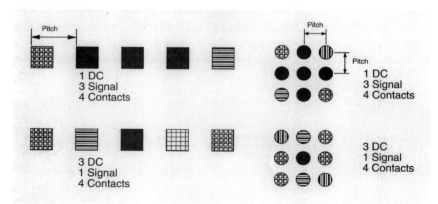

Figure 13.14   Perimeter bond pad cell versus area array pad cell.

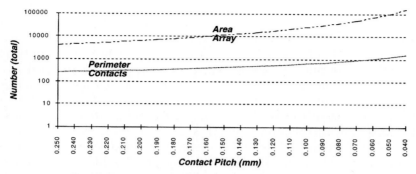

Figure 13.15   Perimeter and area array contacts versus contact pitch.

gram similar to Fig. 13.15 for a 16-mm die would be shown. Obviously an area array connection method is more than an order of magnitude better than a perimeter contact method.

Today what is becoming more important is the number of contacts possible from a given area within the IC surface. One objective is to connect as many signals as possible from the surface of the IC. Another objective is to keep the voltage levels on the surface of the IC as electrically quiet or near DC as possible. Both of these objectives require more connections to the surface. In Fig. 13.14 there are four patterns or cells for repeatable contacts. Two cell patterns have three signals for every DC contact, and two have three DC contacts for every signal. The difference between these patterns is the possible frequency of operation. Gates on the surface of the IC either communicate to other gates on the surface of the same IC or to other ICs within the interconnect system. In either case these gates need connections to DC voltages, both power and ground levels. The more contacts available to DC voltage levels, the higher the frequency of operation.

As with all of these figures, the IC is a square 16 mm on a side (e.g., see Fig. 13.16). Figure 13.17 shows the size of a square area on the

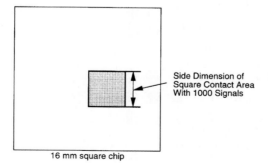

Side Dimension of
Square Contact Area
With 1000 Signals

16 mm square chip

Figure 13.16   1000-signal area within a 16-mm integrated circuit.

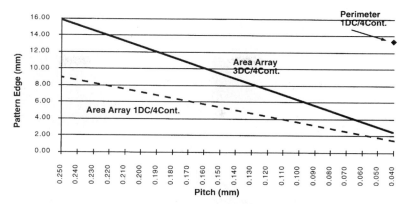

Figure 13.17 Square area dimension for 1000 signals.

surface of an IC required to achieve 1000 signal contacts to the substrate beneath the IC. The highest frequency of operation case would
be the area array three DC contacts for one signal contact pattern
(the solid line in Fig. 13.17). At a 100-μm pitch this pattern would require a 6- by 6-mm area on the surface of the IC. The only pattern
physically possible for the perimeter contact method would be one DC
contact for every three signal contacts and a contact pitch of 40 μm
(very difficult, even for fine-line TAB). In the case of the perimeter
contact method the graph in Fig. 13.17 shows a 13-mm length of
perimeter contacts on all four sides of this 16-mm IC. Using the 50-
μm WIT contact method and three DC contacts for every signal, one
could achieve the 1000 signal contacts within a 3- by 3-mm area.

Signal trapping (shown graphically in Fig. 13.18) occurs when there
are obstructions within routing layers of an interconnect substrate.
The square in the upper left corner of Fig. 13.18 depicts an IC floor-
plan. One-eighth of the chip is removed for further explanation.
Depending on the pattern of DC and signal contacts between the IC
and the substrate, certain signal lines attempting to be routed out beneath the IC to the next IC may be blocked.

One way blockage occurs is when one cannot route a line between
the vias connecting the chip to layers below the chip. Blockage can
also occur by not having enough line-to-line or line-to-via pad space
available. Either of these situations would make it impossible to take
full advantage of a smaller area array pitch without adding more
routing layers within the substrate.

Placing more routing layers in the interconnect substrate is possible
with added cost. The constraints of higher performance, number of connections between chip and substrate and cost once again collide, allowing the designer to seek an optimum.

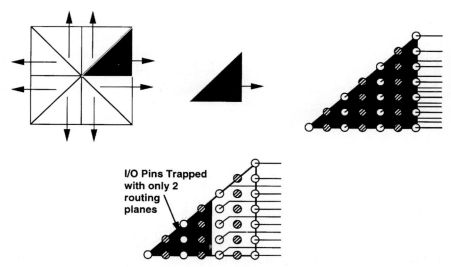

Figure 13.18   Signal trapping limitations (line pitch and layers).

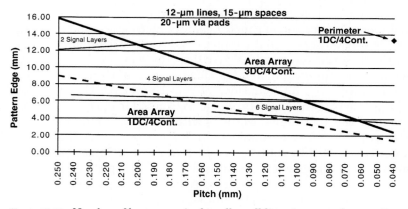

Figure 13.19   Number of layers required to allow all lines to escape from a given signal/DC cell pattern (for layers with 12-μm lines, 15-μm spaces, and 20-μm via pads).

Figure 13.19 shows the beginnings of the optimizing graph for a given interconnect substrate technology [27-μm line pitch (12-μm line + 15-μm space) and 35-μm via pitch (20-μm via + 15-μm space)], cost, structure, and area array pattern. This graph will change for each set of interconnect technologies reviewed. Using this set of technologies, one can see an example by choosing a 100-μm area array pitch and a high-speed design (three DC contacts for every signal). It can be seen from that graph that a four-signal layer substrate would be needed to route 1000 signals beneath an IC utilizing a 6- by 6-mm area on the surface of the IC. Again, if one were to stay with the 100-μm area array pitch and add two more signal layers in the interconnect substrate beneath the substrate, the net effect would be to reduce the size of the square contact area beneath the IC to 4 mm on a side.

Every design will not have 1000 signals, 100-μm area array contacts, or three DC connections for every signal required. The basic structure of graphs and optima remains the same. New graphs are required for each case in question.

13.5   Conclusions and Future Directions

The WIT attach method is a new technology with the potential for making significant advances for IC assembly. This technology provides at least 10 times more contacts compared to any available IC connection technology available today. As can be seen from the routing constraints, one needs to chose the interconnect substrate wisely to take advantage of WIT area density contact capability.

Initial reliability information is predicting at least three times longer life than solder bumps when these joints are exposed to similar power cycling stresses. As the size of circuits within a VLSI circuit decreases, the smaller pitch provided by WIT helps shorten the fanout distances between on-chip and off-chip driver circuits and reduce the capacitance load for these circuits, allowing them to perform at higher frequencies. With the use of WIT connections, one can reduce α-particle radiation effects by at least 1000 times to 1,000,000 times depending on which structure the WIT is grown (substrate or IC, respectively).

As this technology matures, we expect to be able to customize the WIT height to the required reliability for a given chip size and power dissipation. Realistically, a minimum WIT pitch will probably be 50 μm.

Acknowledgments

The authors wish to acknowledge the work of the many people who have made this idea begin to become a reality: Solomon Beilin, David Horine, Mike Lee, Connie Wong, Patricia Boucher, Vivek Mansingh, Carlo Grilletto, Dave Kudzuma, Paul Reilly, and Krysztof Kozak.

References

1. Miller, L. M., U.S. Patents 3,429,040, 1967; 3,459,133, 1969.
2. Norris, K. C., and A. H. Landzberg, "Reliability of Controlled Collapse Chip Connections," *IBM J. Research Devel.*, **13**:266, 1969.

3. Goldman, L. S., "Geometric Optimization of Controlled Collapse Interconnections," *IBM J. Res. Devel.,* **13**(3):251–265, May 1969.
4. Vaynman, S., "Effect of Strain Rate on Fatigue of Low-Tin Lead-Base Solder," *IEEE Transact. CHMT,***12**(4), 469–472, Dec. 1989.
5. Lawson, L., "A Model of Thermomechanical Fatigue in a Lead-Base Alloy," *J. Mater. Research,* **8**(4), 745, Apr. 1993.
6. Love, D. G., L. L. Moresco, W. T. Chou, D. A. Horine, C. M. Wong, and S. I. Beilin, "Wire Interconnect Structures for Connecting an Integrated Circuit to a Substrate," U.S. Patent 5,334,804, Aug. 2, 1994.
7. *ABAQUS User's Manual,* vols. 1 and 2, version 5.2, H. K. & S. Inc., Pawtucket, R.I., 1992.
8. Nix, W. D., R. Venkatraman, J. F. Turlo, R. Vinci, S. Bader, and J. C. Bravman, *Plasticity and Flow Stresses in Aluminum Thin Films on Silicon,* Dept. Material Science, Stanford Univ., Tencor Workshop on Mechanical Behavior of Microelectronic Thin Films, May 17, 1994.
9. Guo, Q., et al., *Thermomechanical Fatigue Life Prediction of 63Sn /37Pb Solder,* Dept. Mechanical Engineering, Northwestern Univ.; reprinted in *J. Electron. Packag.,* 145–151, June 1992.

14

A Compliant Chip-Size Packaging Technology

Thomas DiStefano and Joseph Fjelstad

14.1 Introduction

Virtually unheard of just a few years ago, plastic ball grid array (PBGA) packaging of semiconductors has made extraordinary strides and seems poised to replace many, if not all, of the plastic, peripherally leaded component packaging strategies. Presently, the benefits of PBGA packaging are being broadly extolled for their ability to reverse the trend toward increasingly finer-lead-pitch packaging. That trend has taken the industry to the limits of peripherally lead chip packaging, where extremely delicate leads have resulted in nearly unmountable device packages.

Now this area array approach, with all its attractive advantages, is working its way down to the size of the chip itself as a new class of IC chip packaging technologies labeled *chip-scale packaging* (CSP) makes their entrance.

While the benefits of area array mounting concepts for IC chips have been well known and successfully exploited using ceramic substrates by such major OEMs such as IBM for many years (most notably through its now-famous C4 process), broader implementation of the technology has lagged because of a number of factors. Among the roadblocks to attainment of broader use of PBGA packaging have been such matters as a limited availability of packages, lack of assembly processing knowledge, and concern about the mounting process, especially by quality assurance departments relative to their inability to make visual assessment of solder joint quality under the die. Most of these concerns have been addressed by recent research, and the results have been most encouraging. Experiences to date have been highly positive.

Improved assembly techniques and processes have resulted in low-ppm defect levels in solder joints.[1] These steady advances in PBGA

Figure 14.1   An example of a 188-pin compliant μBGA with a mating test socket.

technology continue to broaden further its acceptance as processing methods mature and more reliability data are gathered. Even with the gains offered by standard PBGA not fully exploited yet, the trend toward increased miniaturization continues, being paced by the desire of designers to capture the cost and performance advantages it offers. The compliant micro–ball grid array (μBGA)* (shown in Figure 14.1) may well represent the future of chip-scale packages, allowing product designers to achieve the highest levels of performance and miniaturization at competitive cost.

The recent introduction of a compliant μBGA technology appears to have laid an appropriate foundation on which future generations of electronic packages can be built. Figure 14.2 graphically depicts how the new technology bridges the gap between flip chip and standard surface-mount packaging technology. There are several attractive features of the new compliant μBGA package:

1. The package is virtually the same size as the die, providing the same high density and performance as with unpackaged die.

2. The package is testable; thus module yield after final assembly is at least as high as with bare-die KDG. In fact, because the die is protected from damage after testing better than in alternate methods, it is likely to increase final module yield appreciably.

---

*Trademarked by Tessera.

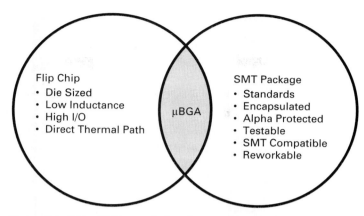

Figure 14.2 The μBGA was designed to combine the best elements of standard SMT with the best elements of flip chip.

3. Electrical performance, thermal management and manufacturability of the package are excellent.

The traces "fan in" from the die pads (unlike standard TAB, which fans out) to an area array of metal bumps under the chip. This gives a package the same size as the die which can be attached to the substrate by PBGA surface-mount technology (SMT). Thus, without sacrificing valuable real estate, equivalent numbers of interconnections, at pitches coarser than can be reliably achieved using ultra-fine-pitch technology, become available to the designer and/or user.

A key factor promoting early acceptance of the BGA is the reduction in form factor for high-pin-count ICs. By converting from perimeter leads to area array contacts, the physical size of high-pin-count packages can be reduced significantly. The size advantage of the BGA increases at high I/O because the surface area increases linearly with package I/O rather than with the square of I/O as is the case for the QFP. This is clearly illustrated in Fig. 14.3, which plots package footprints as a function of I/O for various packaging formats and assembly pitches. μBGA technology, like all area array packages, is much more extendable to high-I/O-count devices than are the peripheral packages such as QFP and standard TAB. Note that the footprints shown for the μBGA are for a fully populated 0.5-mm bump grid. In reality, pitches can be made larger if desired, since the limiting factor is the size of the chip. Figure 14.4 provides a size comparison between a 100 lead TQFP at 0.5-mm pitch and a 169-I/O BGA at 1.5-mm pitch with a 100-I/O μBGA at 0.5-mm pitch.

As alluded to earlier, now and for the future, one can expect that electronics packaging of all types will ultimately be reduced to the

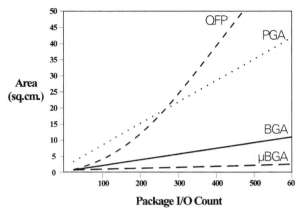

Figure 14.3 A comparison of differing packaging technologies' area use efficacy.

Figure 14.4 A size comparison between a 100-lead TQFP at 0.5-mm pitch and a 169-I/O BGA at 1.5-mm pitch with a 100-I/O μBGA at 0.5-mm pitch.

TABLE 14.1   Key Properties of Alternative Packaging Technologies

	PGA	PQFP	BGA	TBGA	μBGA
I/O count	208	208	225	224	313
Lead pitch, mm	2.5	0.5	1.27	1.27	0.5
Footprint, mm^2	1140	785	670	530	252
Height, mm	3.55	3.37	2.3	1.5	0.85
Inductance, nH	3–7	6–7	3–5	1.3–5.5	0.5–2.1
Capacitance, pF	4–10	0.5–1.0	1.0	0.4–2.4	0.05–0.2
Θ_{jc}, °C/W	2–3	0.5–0.6	10	1.5	0.2–2.0
Package die area ratio	11	8	7	5	1
Density, mm^3/lead	19.45	12.72	6.85	3.55	0.68

pursuit of optimal volumetric packaging of all types of components, with bare flip chip representing the ultimate chip interconnection method. This matter of sizing has implications that can, if properly addressed, positively impact the cost performance requirements of most, if not all, future electronics packaging. Table 14.1 illustrates how μBGA technology addresses this concept.

14.2   Construction of Compliant Micro–Ball Grid Arrays

The compliant μBGA[2–4] is a new method for packaging semiconductors and accordingly requires the creation of new processes for manufacture. The finished product itself is a chip-scale or chip-size package which appears ready to simultaneously solve the full spectrum of problems faced by the electronics packaging industry both today and for the foreseeable future.

Based on flexible circuit technology, the μBGA enables the fabrication of a chip-size package that, like standard BGA technology, reverses the increasing trend toward fine-pitch assembly. It is, in fact, the logical extension of the BGA concept taken to its natural lower limit of a chip-size package without imposing new design rules on the semiconductor manufacturer.

Shown in cross section in Fig. 14.5, the package—basically, a small polyimide flexible circuit—is attached to the chip with an elastomer several mils thick, and leads are bonded to the die pads. The elastomeric film provides a compliant layer between the semiconductor device and the substrate to which it will be mounted. This compliant layer, combined with the specially developed S-shaped bond lead ribbon, effectively decouples the device from the strain of thermal expansion induced by CTE mismatches with the ceramic, metal, or polymer

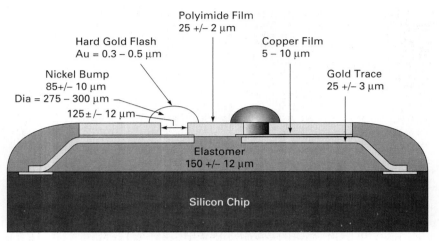

Figure 14.5   Construction of a compliant μBGA.

substrate on which the device is mounted. Such thermal excursions occur both during assembly and during the life of the product. Alternative, and more traditional approaches to chip-scale assembly (such as C4 flip chip) require the use of polymer underfill to reduce strain. This method, while executable, is one of the least attractive features of the C4 technology from an assembly standpoint.

In summary, although originally conceived of as a means for solving the known good die (KGD) problem for multichip modules (MCMs), μBGA technology seems well suited to serve the broader needs of the electronics industry. The package is both small and testable and combines the best features of TAB, wire-bond and C4 techniques while overcoming many of their individual limitations. Table 14.2 qualitatively compares the relative merits of the different technologies. Thus the μBGA approach should allow the advantages of flip chip to be enjoyed by the broadest possible range of potential users of flip chip technology without the concerns of having to handle, assemble, and underfill bare IC die.

14.2.1   Base material

Polyimide film is the most commonly used material for the μBGA. This is due to its excellent balance of physical and electrical properties. One property of significant importance is the base material's CTE, which closely matches that of commonly used organic based laminates. Certain films such as Kapton H are well suited to helping achieve the delicate balancing of properties required to create the μBGA package. The coefficient of thermal expansion of this material

TABLE 14.2   Comparison of Relative Attributes of Different IC Packaging Approaches

Attribute	Chip on board	TAB	Flip chip	μBGA
Assembly infrastructure	[open]	[solid]	X	[solid]
Cost to implement	[solid]	[open]	X	[solid]
Thermal management	X	X	[solid]	[solid]
Miniaturization capability	[open]	X	[solid]	[solid]
Testability of device	X	[solid]	X	[solid]
Die bumping required	N/A			N/A
Nonrecurring engineering	[solid]	X	X	[open]
Underfill required	N/A	N/A		N/A
Die protection offered	X	X	X	[solid]
High I/O capability	X	[open]	[solid]	[solid]
Handling concerns	X	[open]	X	[solid]
Electrical performance	X	[open]	[solid]	[solid]

[solid] good, [open] fair, X poor.

closely matches that of copper and FR-4 laminates, allowing it to nearly match the expansion rate of all the different materials used in construction with thermal changes. A further advantage of these materials is that they bond well to the elastomer, which is the stress decoupler between the IC die and the substrate. This is more completely described in Sec. 14.2.4.

14.2.2   Metal ribbon leads

The metal leads on the μBGA package are electroformed by pattern-plating the polyimide flex circuit material with 99.99% pure, soft gold. The rationale for taking such a seemingly expensive approach can be found in the need to endure a small amount of flexing during thermal cycling (as much as 0.0019 in of movement can be anticipated with a 20-mm^2 die during soldering; see Fig. 14.6). Gold, unlike other metals, does not significantly work-harden, making it a logical choice. The contribution to cost of the gold is not sufficiently significant in the early life of a new product where reliability is more often a concern. The leads on the package fan inwardly to a bump array that serve to make interconnection to the next-level package. In manufacturing μBGA product, the leads of the flex circuit carrier package are thermosonically bonded to the die after the flex circuit has been mounted on the die face. The gold leads employed can be bonded to either aluminum or gold bond pad metallurgy on the die without modification. As the wires are thermosonically bonded to the die, they are gently formed in an S shape to further enhance their compliance and flexural

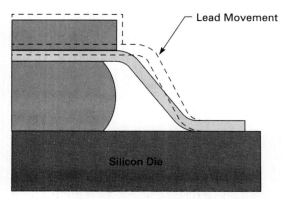

Figure 14.6 The compliant layer can expand by as much as 40 μm during soldering as illustrated here.

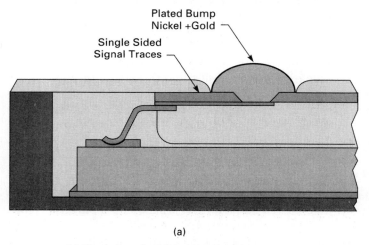

(a)

Figure 14.7 (*a*) Single-metal μBGA with plated post.

durability. Still, with the anticipated drive to lower cost, less expensive metals such as copper are already being readied for evaluation.

14.2.3 Bump/ball metallurgy

First versions of the bumps for the compliant μBGA have been made of electroformed nickel which is subsequently plated with a gold flash (see Fig. 14.7*a*). The height of these electroformed bumps is nominally 0.085 mm. Alternative metallurgies are also under investigation; for instance, copper plating with gold or tin may prove a less expensive alternative approach.

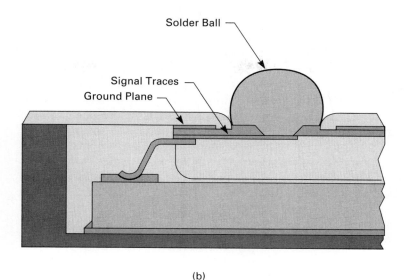

(b)

(c)

Figure 14.7 (*Continued*) (*b*) Two-metal ground-plane μBGA with solder bumps; (*c*) ground-plane μBGA with solid-core solder balls.

Solder, both plated and in the ball form, is also under investigation, and techniques are being developed to allow early implementation. Presently because of the compliance of the device, it is postulated that standard eutectic solder may serve to create reliable interconnections that can withstand thermal cycling. Solid-core balls look especially promising for this purpose. Because of the x,y,z compliance of the µBGA contacts, solid-core solder balls can be used for bumps, as shown in Fig. 14.7c. Thermally induced shear strain is taken up across the elastomer pad rather than the solder ball. The copper core ball, used on the original IBM flip chip,[5] provides dimensional control over the bump diameter and height, which are less dependent on pad size, solder paste, and surface conditions. The solid-core ball has the interesting feature that it allows a solder fillet to form at the junction of the ball to the pad for a range of solder volume, pad diameter, and ball diameter. The solder fillet acts to relieve stress at the solder-to-pad junction where brittle intermetallics often form. In addition, the solid-core ball may be easier to socket since it forms a high-aspect-ratio bump of controlled diameter. The copper core solder ball is a promising candidate for the smaller (0.25- to 0.4-mm) bumps needed for fine-pitch BGAs.

14.2.4 Elastomer

The elastomer was carefully chosen for the µBGA package. Primary among attributes sought were high purity, ready availability, and high temperature stability. The final product was selected from among several potential candidates. The one chosen was a silicone-based material which is filled with silica. In manufacture, elastomer pads are stenciled onto the flex circuit and subsequently bonded to the face of the silicon chip. The elastomer has two major functions, both of which are related to compliancy. First, it serves as a compliant pad, providing compressibility for testing the package in a socket. The second, more important, purpose is that it serves to decouple stresses that arise from the mismatching coefficients of thermal expansion between the silicon chip and the substrate to which the package is attached.

This key feature of attraction, that is, the decoupling of the die surface from the mating substrate, successfully mitigates shear forces encountered during thermal excursions. Such shear forces are capable of fatally damaging die without this feature. An additional advantage of the materials used to create the µBGA package is that they are not prone to water accumulation; thus moisture-related problems such as "popcorning" are avoided.

A final, and not insignificant, benefit of the elastomer is that it serves as a barrier to α-particle emissions from the radioactive decay of lead within the solder. This naturally occurring, low-level emission is capable of causing "soft" errors on the chip during operation. Traditional C4 flip chip demands that stringent design rules where active devices are spaced ≥150 µm from solder materials be followed to mitigate potential problems of this type.[6]

Bumped Flex Circuit

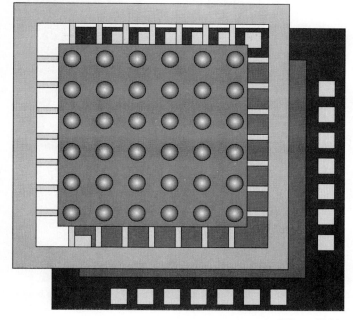

Silicon IC Die

Figure 14.8 Flex is aligned to IC die prior to bonding.

14.2.5   Micro-BGA assembly

The assembly of the micro–ball grid arrays is unlike the assembly of TAB. The primary difference is that the ribbon leads are supported to the moment of assembly. Figure 14.8 shows a representation of how the bump flex circuit is presented to the die for assembly. In actuality, the bumped flex is first bonded to the surface of the die using the elastomer pad and adhesive material. The gold ribbon leads are then formed and bonded using a specially developed technique. During the bonding process, each lead is moved laterally out of the way of the un-bonded leads, in a sequential Zipper Bond* motion that opens up the additional space to allow the bond to be made easily. The Zipper Bond technique allows for a misalignment of the flex relative to the die, since the leads are moved into proper position during the bonding process. Figure 14.9 shows an example the wire bonds made to the

---

*Trademarked by Tessera.

Figure 14.9 A photomicrograph of the gold leads bonded to the silicon chip prior to encapsulation.

die lands. Because the lead is repositioned by the tool to align with the bond pad, the initial alignment accuracy is not critical in determining capability for fine pitch. The bonding process can be used to bond ribbon leads to pads anywhere on the face of the die. In the standard process, a bond window is opened over the pad, wherever it lies, and the ribbon is formed and thermosonically bonded to the pad. Leads have been bonded to perimeter pads, center pads, and bumped pads. The critical dimensions limiting the bond pad pitch are the bonder tool dimension (feature T in Fig. 14.10), the flare in the bonded ribbon lead (feature F in Fig. 14.10) and the relative positional accuracy of the bonder for placement of adjacent leads. Bonds have been made in the laboratory with a 50-μm-wide tool and a 25-μm-wide ribbon lead. The bond leads are 99.99% gold ribbons 20 to 25 μm thick. Bonding at a 55-μm pitch has been demonstrated on a linear row of pads using a 50-μm tool to bond 20-μm ribbon leads. According to the simplified model in Fig. 14.10, the practical limit to the bond pitch is projected to be 45 μm for the near future.

Following lead bonding, the peripherally bonded leads are encapsulated with a silicon elastomer to protect them from physical damage while still allowing compliance. The encapsulant is also a semiconductor-grade silicon elastomer that will allow the compliance of the leads to be taken advantage of during thermal excursions. The encapsulant

Time Frame	Pitch (P)	Pad Width (W)	Tool (T)	Tab (S)	Flare (F)	Position Accuracy	Alignment
1994	100 μm	60 μm	100 μm	38 μm	48 μm	± 8 μm	± 25 μm
1995	80 μm	50 μm	85 μm	32 μm	40 μm	± 6 μm	± 20 μm
1996	75 μm	45 μm	75 μm	28 μm	35 μm	± 5 μm	± 15 μm
1997	60 μm	40 μm	65 μm	25 μm	31 μm	± 4 μm	± 15 μm
1998	55 μm	35 μm	55 μm	22 μm	27 μm	± 3 μm	± 15 μm
1999	45 μm	35 μm	50 μm	18 μm	23 μm	± 3 μm	± 15 μm

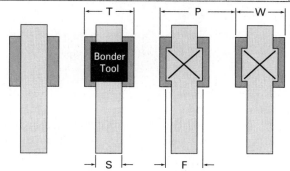

Figure 14.10 Bond pitch roadmap for μBGA technology.

for the lead is also a silicon elastomer that will allow the natural compliance of the leads to be taken advantage of during thermal cycling.

14.2.6 Package enhancements

In its most basic, yet highest-performance form, the silicon die is typically presented to the assembly with the back of the die exposed. This creates the smallest possible form factor for the assembler. However, the package can be further protected by addition of a handling ring or metal can (the "canned" or thermal spreader form is shown in Fig. 14.7a) for applications requiring an extremely robust package. This adds 0.5 to 1.0 mm on each side and 0.25 mm (or more) in height depending on the materials used and protection method chosen.

14.2.7 Cost issues

The advantages of compliant μBGA technology are many; however, a recurring concern for most potential users is cost. While this emerging technology is still very young, it is anticipated that it will be fully competitive with all existing packaging formats. By way of example, a

Figure 14.11   Yield comparison of μBGA with TAB on 70-mm tape.

single frame of 70-mm TAB tape can contain as many as 48 sites for memory-size devices (see Fig. 14.11). Thus it must ultimately be many times cheaper than the estimated $500,000 (or more) it can cost to convert an existing design in silicon from peripheral bond pads to area array solder bumps for flip chip. Such conversion technologies as the compliant μBGA can help avoid these expenses.

14.2.8   Alternate forms

While the single-layer μBGA is expected to be the most commonly used form, there are a number of alternate forms that can be created to further enhance the performance of the technology. Among those are fanin/fanout and two-metal-layer versions. In addition, there is the possibility of using the technology for multichip applications. In such applications, the devices could be bonded as with current practice. Planarity issues would be of significant importance as varying die height could affect the bonding process.

14.3   Designing for Use of Micro–Ball Grid Arrays

In designing for μBGA use, it is necessary to consider both the chip layout and the substrate. This was very well pointed out in an article by Mawer and Rhyner, who stated that: "When it comes to the routability of BGA footprints, consideration of pin functional assignments at the package level is critical. If the signal-pin assignments are made too deeply within the BGA matrix, board level escape using conventional pc-board technology becomes difficult for large matrices."[7]

In the case of the IC die, the number of inward escapes to the bump array on the flex circuit is influenced by some interrelated factors. Key among them are IC bond pad pitch and die size. Currently, in order to wire μBGAs using readily available, low-cost PWB substrates, it is best to design for a coarse bump pitch and to limit the number of rows of bumps. Initial applications for the BGA are using 1.27-mm pitch, and three to four rows of bumps to make wiring easier. At this coarse pitch, the μBGA must be fanned out or fanned in/out to an array larger than the die itself. In the future, small form factors and cost reduction will drive the development of fine via pitch laminates needed for the high-I/O chip-scale packages.

By way of demonstration, the μBGA is easily configured as a chip-scale package with the bumps arrayed on an elastomeric layer over the face of the die. At the BGA standard bump pitches of 1.0, 1.27, and 1.5 mm, the I/O of the chip-scale packaging is limited by the number of bumps that will fit under the shadow of the die, as illustrated by the 1.0-mm bump pitch curve in Figure 14.12. Thus it is apparent that chip-scale packaging such as the μBGA will accommodate memory chips and other low-pin-count parts at standard BGA grid pitches.

For microprocessors and other high-pin-count ICs, a bump pitch of <1.0 mm is necessary to wire the I/O on a chip-scale μBGA. Virtually all the present-day processor chips can be wired in a CSP with a pad pitch of 0.5 mm. Future processor chips are becoming pad limited for I/O approaching 300 to 500, as indicated by the dotted line in Fig. 14.12. The rising I/O is driving a demand for a smaller pad pitch on perimeter-padded ICs. For example, a 1-μm reduction in pitch on a pad-limited die translates into a 2 percent saving in silicon area. By reducing the pad pitch to about 50 μm, all the μBGA bumps on a 0.5-mm chip-scale package can be wired to perimeter pads. Above 1000- to 1400-I/O capability, a transition from perimeter pads to area array pads on the die is anticipated.

The μBGA roadmap extends the I/O capability of the package up to about 4000 by wiring to full array of pads on the die. A reduction in μBGA bump pitch to 0.25 mm is necessary above an I/O of 1400 in order to fit the I/O under the die of the chip-size package. The area array μBGA retains the compliant bump and elastomeric encapsulation used on the standard μBGA package, while allowing bonding over active devices to the array of pads.

As stated earlier, the I/O capability of the μBGA has been extended to 55 μm in the laboratory by techniques that allow for better control of the fine-pitch bonding process. A reduction of pad pitch to 50 μm enables full wiring capability for a 0.5-mm-pitch μBGA chip-scale package up to 20-mm die sizes. Referring back to the roadmap for implementation of fine-pitch bonding, shown in Fig. 14.10, it is antici-

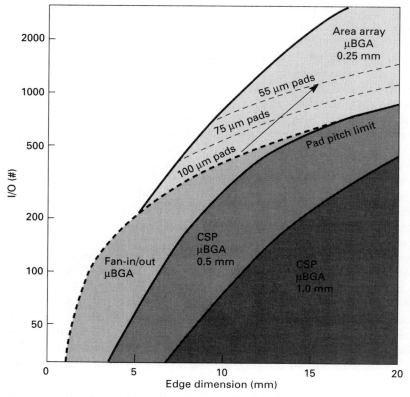

Figure 14.12 Roadmap of chip I/O for the μBGA for the chip-scale package and the fanin/fanout configuration.

pated that there will be a 55-μm bond capability on linear rows of pads in production by 1998.

14.3.1 Grid pitch and layout

The grid or bump pitch for μBGAs is definable by the user, but typical pitches will more commonly be based on 0.5-mm pitch specified in the International Electrotechnical Commission (IEC) Publication 97, *Grid Systems for Printed Circuits*. This pitch allows for natural multiples such as 1.0 and 1.5 mm to also be used in design. When using interstitial points, it is possible to get intermediate grids that should be able to serve the needs of virtually any electronics package. The pertinent citation from the IEC publication is as follows:

5.0   Preferred Grid System
5.1   For positioning connections on a printed circuit board, a grid with a

nominal spacing in the two directions of 0.5mm shall be used.
5.2   Where a grid with a nominal spacing of 0.5mm is not adequate, a
grid with a nominal spacing in the two directions of 0.05 mm shall be
used.

This seemingly simple standard will ultimately prove to be extremely important because it facilitates implementation of common
grid systems that will be necessary for future high-performance systems. This was recognized by the IEEE's Computer Society Technical
Committee on Packaging. The committee appointed a special task
force in the fall of 1990 "to seek early consensus on the need to stan-
dardize MCM sizes and to propose some possible sizes." The task
force recommended 0.5 mm center lines for peripherally leaded packages and PGA standards for area array packages.[8] This will be discussed in more detail later in this chapter.

14.4   Performance Characteristics

The materials and structure of the μBGA offer improved performance
across the full range of factors deemed important for electronic products, including thermal, mechanical and electrical products. Until very
recently all μBGA designs were single-metal-layer constructions; however, with the addition of ground-plane (two-metal-layer) capability, it
should be possible to even better address the needs of present and future very high-frequency applications. The small size of the μBGA
package allows extremely dense packing of die on a substrate, so that
delays and capacitance are also reduced on the system level. Still the
performance limits of the package are intrinsically linked to the performance of the substrate. Table 14.3 shows the wiring density capability
based on the via/bump pitch of substrates for μBGA applications.

14.4.1   Substrate description

One key area to be addressed in order to take full advantage of the
μBGA is in the area of substrate technology. New approaches to design and manufacture of interconnection substrates must be identified and implemented. Figure 14.13 illustrates one potential future
based on relatively simple design rules and "Manhattan style" routing of signal traces.

No exotic construction steps are required for such structures. Simplified processing is possible with less manufacturing steps. Designed
with a "sea of vias" or Sheldahl's "via grid" approach, where all grid
intersection points are occupied by an interconnection via, high-performance multilayer structures can be created from simple two-sided
circuits which are subsequently bonded together to produce a finished

TABLE 14.3   Sample Design Rules for Substrates for μBGA Technology

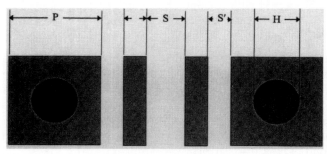

Feature	0.5-mm grid (2 lines between pads)	1.0-mm grid (2 lines between pads)	1.5-mm grid (2 lines between pads)
Line width (L)	50 μm (0.002 in)	125 μm (0.005 in)	200 μm (0.008 in)
Space (line to line) (S)	75 μm (0.003 in)	175 μm (0.007 in)	300 μm (0.012 in)
Space (line to pad) (S')	50 μm (0.002 in)	125 μm (0.005 in)	200 μm (0.008 in)
Hole size (H)	125 μm (0.005 in)	200 μm (0.008 in)	250 μm (0.010 in)
Pad size (P)	225 μm (0.009 in)	325 μm (0.013 in)	400 μm (0.016 in)
Wire rout-ability	40 cm/cm² (100 in/in²)	20 cm/cm² (50 in/in²)	13 cm/cm² (33 in/in²)

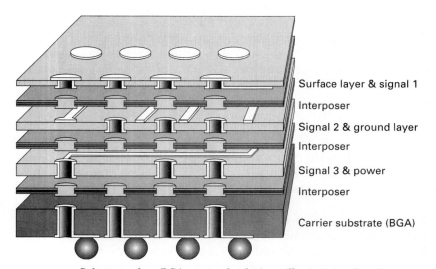

Surface layer & signal 1

Interposer

Signal 2 & ground layer

Interposer

Signal 3 & power

Interposer

Carrier substrate (BGA)

Figure 14.13 Substrates for μBGA-type technologies will require such techniques as illustrated to economize on the use of available real estate. In this illustration, the reverse blind via approach will eliminate the need for wire redistribution on the surface.

TABLE 14.4   Substrate* Electrical Parameters

Impedance	50–100 Ω (±10%)
Conductor resistance	0.5 Ω/in
Skin depth at 100 MHz	6.6 mm
Skin depth at 1 GHz	2.1 mm
Self-inductance	8.8 nH/in
Mutual inductance (adjacent trace)	0.7 nH/in
Capacitance	2.9 pF/in
Mutual capacitance (adjacent trace)	0.23 pF/in
Ground-plane power	100s of amperes
Inductance to decoupling capacitance	1 nH
Ground-plane separation	0.002 in
Ground-plane separation	0.004 in
Interposer via resistance	10 mΩ (max)

*Properties of substrate in stripline configuration [SPICE (simulation program for IC emphasis) data].

Figure 14.14 A substrate for μBGA applications with processor and memory chip.

circuit using these via layers. Table 14.4 presents modeling data on the electrical performance of such structures.

A high first pass yield is due to the ability to inspect and test (and repair) individual layers before lamination and, finally, electrical testing simplified by lack of fine-pitch contact areas. Figure 14.14 shows what such a system looks like in finished form.

14.4.2 Electrical performance

The electrical performance of the μBGA is enhanced by the short on-package trace length. The average trace length between the bonding leads and the bumps on the μBGA flex circuit is approximately 1.25 mm. This results in greatly reduced on-package parasitics. The measured loop inductance ranges from 0.5 to 2.1 nH, and mutual inductance ranges from 0.1 to 0.5 nH. Trace capacitance is 0.05 to 0.2 pF. When the package is coupled with a "Manhattan" routed substrate such as is shown in Fig. 14.12, it becomes possible to create package and interconnection systems with the highest possible performance.[9] Such substrates eliminate redistribution wiring, which can sap the energy of a high-performance electronic system.

14.4.3 Thermal performance

As stated earlier regarding mechanical issues, the thermal mismatch between the die and the substrate material is absorbed within the μBGA package. Since the μBGA is a flip chip package, efficient thermal dissipation can be achieved by heatsinking directly to the exposed backside of the die. Recent work performed in collaboration with Sun Microsystems[10] has shown that a finned heatsink attached to the die with thermal grease gives a Θ_{ja} (junction to ambient thermal resistance) of 2.2°C/W with airflow at 0.8 m/s (150 ft/min). The Θ_{jc} (junction to case) for this configuration is 0.2°C/W. With an optimized heatsink design, it is expected that it will be possible to reduce the total thermal resistance by a factor of ~2. Table 14.5 presents thermal conductivity data for alternative approaches to heatsinking μBGA packages.

14.5 Reliability Testing and Burn-in of Compliant Micro–Ball Grid Arrays

The ultimate reliability of any product is intrinsically linked to a few key elements of its construction, specifically, the design, materials, and processes used and the end-use environment. Careful analysis of the properties of these elements in combination and their potential interactions will help assess the probability of the product's failure. A seemingly minor fault in one category can result in the failure of the finished product. While the data on μBGA technology are still preliminary, they hold out great promise for the technology. Qualification testing is currently under way. In addition to the package qualification, testing is under way to determine the reliability of the solder joints formed on the surface of a printed wiring board mounted with compliant μBGA.

Burn-in of the compliant μBGA, on the other hand, is accomplished in a fashion similar to existing semiconductor chip packaging technologies. The purpose of testing is to accelerate the rate of infant fail-

TABLE 14.5   Comparison of Thermal Conductivity for Alternative Approaches to Heatsinking

Material	Thermal conductivity, W/m °C	Pad thickness, mm	Junction to ambient thermal resistance, °C/W	Junction to case, °C/W
Silicone rubber	0.4	0.20	5.8	3.8
Conductive elastomer	2.2	0.20	4.2	2.2
BN-doped elastomer	6.0	0.15	2.8	0.8
Thermal grease	0.8	0.05	2.2	0.2

TABLE 14.6   Testing Requirements for μBGA Packages

Test	Conditions	Duration	Sample/fails
Thermal shock	−65 to 150°C Method A110	100 cycles	76/0
Thermal cycling	−65 to +150°C	1000 cycles	45/0
Pressure pot	121°C, 100% RH, 2 atm	168 h	45/0
HAST	130°C, 85% RH 5.5 V bias	96 h	45/0
High temperature storage	150°C	1000 h	45/0
Mechanical shock	600 g, 2.5 ms method 2002	6-axis	32/0
Solderability	JEDEC 22-B	—	3/0
Physical dimensions	JEDEC 22-B	—	
Mark perma-nence	Method 2015	—	

ures resulting from random defects in IC die manufacture. This screening process weeds out the marginal product and provides assurance that the product reliability will meet customer expectations.

14.5.1   Testing requirements

As just indicated, the μBGA must meet the standard requirements for testing of traditional semiconductor chip packages. The testing program for such devices is fairly rigorous and includes a number of different environmental stress tests designed to weed out marginal packaging approaches. The minimum testing requirements for packages IC die are listed in Table 14.6.

14.5.2   Assembly-level testing of μBGAs

Initial data on the μBGA package are very encouraging.[11] Although only a small number of packages have been tested, no failures have yet been recorded through testing. Qualification test of the BGA package assembled to FR-4 substrates is under way. At the time of this writing they have already endured 800 cycles of 0 to 100°C without failure, with a target of 1000 cycles. Additional testing of assembled boards with μBGA packages technology planned include power cycling of assembled boards at 27 W from 25 to 90°C for 500 cycles and vibration testing of assemblies at 5.2 g random for 48 h. Other testing will be performed, as required, to meet the varied needs of end users of the package.

14.5.3   Electrical testing approaches

With respect to electrical testing of μBGAs, the relative scale of feature sizes limits the number of "off the shelf" solutions. However, there are in existence a number of approaches to testing of fine pitch devices that have been successfully implemented and employed for flip chip. Many of these are viewed as being extensible to the μBGA technology. While the "standard" μBGA grid pitch is commonly given as 0.5 mm, the roadmap presented in Fig. 9.12 clearly illustrates that pitches as great as 1.0 mm are easily obtained within the "shadow of the die." This grid pitch, while not considered easy, is clearly more amenable to address by a wider variety of test approaches. Still, seekers of highest performance systems will wish to minimize signal path length to maximize performance.

COBRA test system.   COBRA probe technology is so named because of the characteristic cobra-like curve that the test probes assume in the fixture (see Fig. 14.15). The test method was developed by IBM for

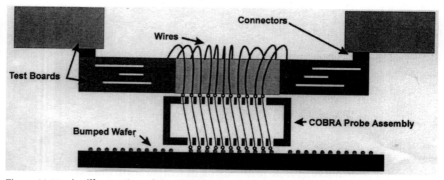

Figure 14.15   An illustration of the COBRA test fixture concept.

testing substrates. The hardware consists of two basic components, a probe assembly and a redistribution circuit. The probe assembly makes contact with the device under test, and the circuit redistributor fans the circuits out to the tester load board. This test technology is area array capable and quite suitable for grid pitches down to 0.25 mm. The fixtures used for testing chips, while very effective, tend to be quite expensive.

Planar contacts. Because of the compliance of the μBGA, it is possible, using planarized ceramic substrates, to test the die by registering and clamping the die directly to the circuitized ceramic base. A compliance of 25 μm has been measured for the μBGA bumps with an applied force requirement of 1 to 1.2 g of force per micrometer. As a test system approach, this method is a much less expensive alternative to the COBRA probe method but requires special care in clamping to make certain that the die is not damaged during the clamping operation. The primary concern is that if forces are applied unevenly, the μBGA device can be fractured. The use of a metal can or thermal spreader will mitigate some of the potential damage but the direct thermal path through the back of the silicon die is a feature that many potential users find attractive, and thus methods that minimize the fracture risk are a high priority.

Dendritic contacts. The use of denritic contacts is another method of testing very fine grid pitch devices. Developed by IBM, these test substrates are created by means of a special palladium-plating method that deposits the metal in a highly roughened form to serve as contacts for electrical test and device burn-in. The method, referred to as dendritic plating, was devised to allow for temporary attachment of solder bumped, flip chip die to substrates for test. The bumped die are clamped onto the surface of the substrate and the palladium dendrites break through surface oxides and penetrate the solder bumps. The bumps are deformed during the testing operation by the piercing effect of the dendrites. Thus after test when the chip is removed the solder is reflowed again to restore the solder balls to their original shape and facilitate the final assembly process. According to published reports, the carriers can survive more than 20 reuses.

14.6   Assembly of Micro–Ball Grid Arrays

The assembly of μBGA parallels that of standard BGAs. Much of the current assembly infrastructure in use today for standard surface mount technology is capable of performing the assembly tasks required for the μBGA. This is one key distinction between the compliant μBGA and traditional flip chip. As was explained earlier in this chapter, this is due in large part to the built-in compliance of the package that is provided by the elastomer used in the construction of the μBGA. Because of this, the use of underfill is not required after soldering to protect the device against cracking during thermal cycling. Most users of flip chip commonly use underfill to relieve the strain that occurs during thermal excursions due to the mismatch in CTE of the silicon and the substrate. Without this underfill, the solder joints will prematurely crack and fail.[12] Even with underfill, there is a warpage of the device's combined structure which acts in much the same manner as a bimetal strip due to the large mismatch in CTE between the two materials.

14.6.1 Solder paste application

Traditional solder paste stenciling has been used with success to apply solder paste to the substrate for solder attachment and joining at the assembly level. The solder paste stencil can be prepared by chemically etching the array pattern of the device into 0.10-mm stainless steel sheet stock. The size of the opening is adjusted to provide a predetermined quantity of solder paste onto the lands. The flux used in the preparation of the solder paste is preferably a "no-clean" type. This is consistent with the requirements of soldered flip chip.

14.6.2 μBGA device placement

Data from assembly equipment manufacturers indicate that the μBGA package should be able to be assembled by using any of a small number of different pick and place machines. Vision-based pick and place systems with pattern recognition algorithms have proven suitable, as have split prism systems. In an alternate approach, the data for device placement can be downloaded from the CAD-generated files used to manufacture the substrate. These files can provide the exact coordinate location as well as the orientation for each device. The x-y coordinates can be located from a single "datum point." The μBGA packages can be picked up from waffle packs or from tape reels. Early test runs used fiducials on the board to calculate the center and rotation of the substrate, and the four corner bumps on the die were used to calculate center and rotation of the μBGA.

14.6.3 Reflow soldering processes

Following solder paste application and device placement, reflow soldering is used to make permanent solder joints that serve both the mechanical and electrical interconnection needs of the device. Convection oven and vapor phase soldering methods are preferred. Hot air-gas convection type systems are capable of gradual and uniform heating required for good solder joint formation. In the reflow soldering process, temperatures in the range of 218°C are required. Infrared (IR) systems are of some concern due to the high absorption of IR radiation possible with the exposed silicon. Warpage of the die is a possibility during the assembly process with IR reflow.

14.6.4   Conductive adhesives for assembly

The use of conductive adhesives to assemble μBGAs to next level substrates is also currently being evaluated. While data are limited, it appears that this method may prove quite suitable for certain types of applications where temperature extremes do not excessively challenge the interconnection joint such as might be encountered in automotive underhood applications. Based on data gathered by conductive adhesive interconnection system developers, this appears very promising and could represent an environmentally benign or "green" approach to interconnection of chip scale packages of all types. Anisotropic adhesives have already been successfully used in the interconnection of die to the glass panels used in flat panel displays for a number of years.

14.6.5   Inspection of assembled μBGA devices

Post solder inspection methods for assemblies which have μBGA devices attached are similar in nature to those used in the inspection of flip chip assemblies. While the peripheral solder joints of leaded devices can be viewed and inspected by eye, the inspection of μBGA solder joints, which reside under the device, is limited to x-ray techniques. (See Fig. 14.16.) This is still a major concern to a number of companies. The inability of an inspector to directly view all of the solder joints of a device represents a significant barrier to acceptance of all area array technologies by certain factions of industry. Meanwhile, the experiences of long-time users of area array interconnection technology are highly positive. It is anticipated that with increased encounters, all electronics assemblers will become comfortable with their ability to reliably interconnect area array devices.

14.7   Conclusion

Chip size packaging, such as the μBGA, offers a compelling response to ongoing industry concerns and requirements for future electronics. The concerns over the problems associated with obtaining known good die (KGD) for use in electronic modules will be especially well served by this new family of devices. It has been rationally argued that the KGD concern has been misdirected. The argument is that known good die are pervasively available; however, they are all in

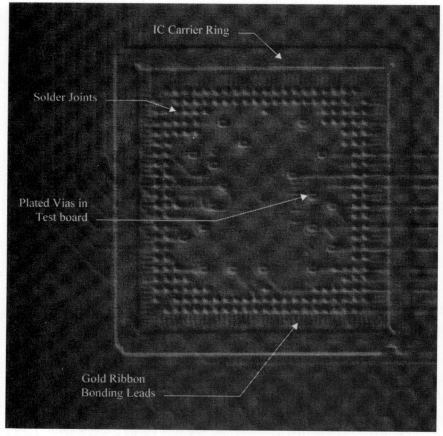

IC Carrier Ring

Solder Joints

Plated Vias in
Test board

Gold Ribbon
Bonding Leads

Figure 14.16 An image-enhanced x ray of a μBGA flip-soldered to a test board. (*Photo courtesy of Digiray Corp., San Ramon, Calif.*)

packaged form. The opposing argument is that packaged die cannot offer the kind of performance that can be achieved using bare die, especially flip chip. This argument continues that bare die are cheaper than die obtained in the packaged form. Although this argument may hold true for certain low-end product which use mature IC manufacturing technology and peripherally leaded die for chip and wire assemblies very cost effectively, it may still take some time to prove true for higher I/O count flip chip bare die. Today, semiconductor manufacturers who do offer tested known good die, commonly offer these die at price parity with the lowest cost packaged version of the same die. Thus, not only is price advantage lost, but it is the board assembler who is now tasked with the packaging of the die as a final step of assembly. This is true whether the assembler uses chip and wire (which requires a glob top encapsulant) or flip chip (where an underfill is required to compensate for mismatches in CTE between the substrate and the silicon die by forcing a bimetal type warpage locally[13]). Moreover, for traditional flip chip applications, there is often the unaccounted-for cost of adding the additional metal layers to the die face to redistribute peripheral I/O over the face of the die. This act is arguably a packaging step in and of itself, exclusive of the underfilling step. Most importantly there is the fact that die for flip chip tend not to be readily available to the full spectrum of users. To date, it remains in the domain of the largest of OEMs, who are fully vertically integrated, thus the broader market is excluded.

On the other hand, compliant chip size packages, such as the μBGA, offer solutions that are available to the broader market. These new package types seek to offer the user/assembler most of the major advantages of bare die but without the concern of procuring, interconnecting, and ultimately packaging, in situ, the known good die purchased from the IC manufacturer. These newer forms of packaged die are designed to be dealt with in the same fashion as today's standard packaged parts in that they will be packaged in standard footprint formats, tested, burned-in, and retested prior to their being committed to assembly. These important attributes should facilitate the acceptance of this important new branch of IC packaging technology.

References

1. Johnson, R., et al., "A Feasibility Study of BGA Packaging," *NEPCON East,* pp. 413–425, 1993.
2. Khandros, I., and T. DiStefano, "Semiconductor Chip Assemblies with Fan-in Leads," U.S. Patent 5,148,265, Sept. 1992.
3. Khandros, I., and T. DiStefano, "Semiconductor Chip Assemblies Having Interposer and Flexible Lead," U.S. Patent 5,148,266, Sept. 1992.
4. Khandros, I., and T. DiStefano, "Semiconductor Chip Assemblies with Fan-in Leads," U.S. Patent 5,258,330, Nov. 1993.
5. Totta, P., and R. Sopher, "SLT Device Metallurgy and Its Monolithic Extensions," *IBM J. Research Devel.,* **13:** 226, 1969.
6. *C4 Product Design Manual,* vol. 1, *Chip and Wafer Design,* pp. 2–21.
7. Mawer, A., and K. Rhyner, "Reliable BGAs Take on Extra Routes," *EE Times,* p. 98, Sept. 26, 1994.
8. Doane, D., and P. Franzon, *Multichip Module Technologies and Alternatives,* Van Nostrand Reinhold, 1993.
9. Martinez, M., et al., "The TCC/MCM: nBGA on a Laminated Substrate," *International Conference and Exhibition on Multichip Modules Proceedings,* 1994.
10. Gilleo, K., and M. Loo, "Area Array Chip Carrier: SMT Package for Known Good Die," *ISHM '93 Proceedings,* pp. 318–323, 1993.
11. Loo, M., M. Krebser, A. Hassanzadeh, and G. Shook, "Surface Mount Study of Fine Pitch Micro Ball Grid Array," *Proceedings of the 1994 International Electronics Packaging Conference,* EPS, pp. 555–561, Sept. 15–28, 1994.
12. Ries, M., et al, "Attachment of Solder Ball Connect (SBC) Packages to Circuit Cards," *IBM J. Research Devel.,* **37**(5):597–608, Sept. 1993.
13. Togasaki, T., "Reliability of a New Flip Chip Interconnection for CCD Imaging Devices," *IMC Proceedings,* pp. 513–518, Yokohama, 1992.

Flip Chip Assembly Using Gold, Gold-Tin, and Nickel-Gold Metallurgy

Elke Zakel and Herbert Reichl

15.1 Introduction

Flip chip assembly offers advantages over other methods of bare-die mounting such as wire bonding and TAB technology. It requires the lowest footprint area on the substrate and allows a significant reduction in package height. Because of the face-down assembly, the whole surface of the IC can be used for different area array configurations allowing the highest density and number of I/Os.

Another advantage of the area array configuration is that, compared to using standard ICs for wire bonding with peripheral pads, pitch and the pad size can be greater. As wire bonding and TAB technology using peripheral pads move to pitches below 100 μm[1–5] a comparable area configuration for flip chip is possible at a >300-μm pitch. Because most chips are currently designed for wire bonding or TAB, an area configuration can be achieved only with an additional redistribution layer. This additional step in the bumping process increases the cost and reduces the electrical performance of the circuit by the additional parasitic lines. Therefore, many users prefer to avoid this step by performing flip chip with peripheral pads.

Another key point for the use of flip chip technology is the bumping process. The standard process today is based on 95/5 lead/tin (SnPb) solder bumps developed for IBM's C4 process.[6–8]

Flip chip technology on printed wiring boards using surface laminate circuits (SLCs) with eutectic SnPb as final pad metallization and chips with the classic C4 bump metallurgy, consisting of 95/5 PbSn, has been introduced by IBM Japan. This process requires the application of flux for the soldering operation.[9–11] The development of underfill materials which compensate for the mismatch in the coefficient of thermal expansion (CTE) between the polymeric substrate and the silicon chip is of essential importance for the reliability and applicability of flip chip technology on these substrate materials.[12,13] The performance of the underfill material is rated by the extent of match between the underfill CTE and bump CTE.[14] On the other hand, a good adhesion between the underfill material, the substrate, and the chip surface is essential for the stress compensating function of the underfill. Flux residues can be critical for the adhesion of the underfill and therefore can reduce the reliability performance.[15] For these reasons a fluxless process for flip chip assembly is desirable. The availability of simple and flexible bumping processes for flip chip bonding is a key issue for a wide-scale application.

The flip chip approaches presented in this chapter are based on bump metallurgies possessing the following characteristics:

1. Good availability based on an existing infrastructure of companies offering commercial bumping services. This is fulfilled by electroplated Au bumps already widely used in TAB technology.
2. Good flexibility based on a software-driven process with low demand on equipment. This is fulfilled by the mechanical stud bumps, which require only a modified wire bonder. The following four factors contribute to bump metallurgy flexibility:
   a. For different chips the pad positions are software-programmable, therefore avoiding the need for additional time-consuming and costly mask processes.
   b. Stud bumping can be performed on whole wafers, substrates, and even on single chips, thus allowing a maximum of flexibility.
   c. The bump height can be influenced by using stacked bumps or by the bumping of the chip and substrate.
   d. Gold and different lead-tin alloys (98/2 PbSn and near-eutectic) are used as bump metallurgies today. In principle, the process can be extended to copper, palladium, or other solder types.
3. Fluxless soldering can be performed with the gold-tin metallurgy. Two approaches are presented:
   a. Use of electroplated AuSn solder bumps derived from the electroplated Au bumps used in TAB technology; this process does not require solder deposition on the substrate site and can be applied for self-alignment.
   b. Use of electroplated gold bumps and solder on the substrate pads.
4. Low-cost processes. The electroless bumping based on nickel-gold has the highest potential for cost reduction because of the selectivity of the electroless deposition, which requires no sputtering and photolithography. The process has high flexibility because no phototooling is needed and single-chip bumping is possible.

In summary, the presented bump processes and metallurgies are based on

- Electroplated Au bumps
- Electroplated AnSn bumps
- Mechanical stud bumps of Au and 98/2 PbSn
- Electroless deposited NiAu bumps

It will be shown that the four bump types allow the flip chip processes on various substrates covering the full range of multichip module types: MCM-L, MCM-D, and MCM-C.

The applied substrate materials are

- Printed wiring boards
- Flexible polymeric circuits
- Thick-film alumina substrates
- Thin-film aluminum and silicon substrates
- Green tape ceramic multilayer

The presented bump metallurgies allow the full gamut of existing flip chip mounting processes—from soldering to thermocompression bonding, and adhesive joining techniques.

For soldering, fluxless soldering and the self-alignment effect of the AuSn metallurgy are especially emphasized; however, these flip chip processes and metallurgies are quite new in the industry. Therefore, understanding the established or proven metallurgies, possible interactions during thermal aging and cycling, and the impact on the degradation of flip chip interconnections is important for implementation of the processes mentioned above.

For each application specific requirements of material selection, availability and cost, compatibility with other process steps (e.g., SMT assembly) and reliability requirements must be carefully analyzed for the selection of an adapted flip chip process. Because the presented metallurgies and processes are quite new in flip chip assembly, the data available today are often limited to special applications and requirements. With the further development of FC (flip chip) technology, data on a wider range will become available, facilitating process selection for new applications.

This trend is evident in the development of conference topics within the last 2 years. Widely attended conferences in the United States such as ECTC 93/94, IMC 93/94, IEMT 94, and NEPCON West 93/94 have implemented tutorials and sessions on flip chip technology. Other conferences in 1995, including the Flip Chip, TAB, and BGA Symposium

in San Jose (Calif.) or the Workshop on Area Array Packaging
Technologies in Germany are focusing on the new technologies.

15.2   Bumping Processes for Gold, Gold-Tin, and Nickel-Gold Metallurgy

Bump deposition processes are based on electroplating, electroless
plating, or mechanical procedures. Therefore, the process flow, the required equipment, and the specifications of the bumps obtained are
different. This section presents a detailed description of the bumping
processes. The main characteristics of the bumps, the process steps,
the requirements of the wafer, and bumping process limitations are
summarized in Table 15.1. From this table it is evident that the five
proposed bumping processes cover a full range of possible FC assembly methods, including thermocompression (TC) bonding, soldering,
and adhesive joining.

Especially with adhesive processes, FC assembly on nearly all types
of substrates is possible; however, additional data on the reliability requirements (operating temperature, environmental conditions) must
be taken into account in order to select the appropriate process. These
specifications are not implemented in Table 15.1. A more detailed description of special process developments is given in Sec. 15.4.

15.2.1   Electroplating of gold bumps

Electroplated gold bumps are commonly used in TAB technology
throughout the industry, especially in Japan, the United States, and
Europe. The largest market for TAB products today is in Japan,
where TAB LCD driver ICs are used. In the United States and
Europe a large number of companies have been working on the development of TAB technology in the past years. Therefore, the infrastructure of Au bumping is being studied and developed, and a number of companies are offering commercial electroplating services.

An overview of recent markets and trends in TAB technology is
given by Khadpe,[16,17] and a detailed description of the electroplating
process is given by Angelucci.[18] A series of investigations concerning
the bumping was performed by TU-Berlin. Figure 15.1 shows the
schematic process flow, which consists of sputtering the titanium-
tungsten (Ti:W) diffusion barrier and the gold-plating base. Then the
typical 20- to 30-μm photoresist is applied and photoexposed. For FC
assembly the bump height can be increased to 50 μm. After the development of the photoresist, gold is electroplated in the resist openings.
The photoresist is then removed and the plating base is selectively
etched. In order to obtain the desired hardness for the subsequent
bonding process, an annealing of the gold is performed.

TABLE 15.1 Characteristics of the Bumping Process

			Part 1		
Bumping process	Electroplated Au	Electroplated AuSn	Electroless NiAu	Stud Au	Stud 97/3 PbSn
Process steps	1. Sputter Ti: W + Au 2. Spin photoresist 3. Expose photoresist 4. Develop photoresist 5. Electroplate Au 6. Remove photoresist 7. Etch plating-base 8. Anneal Au	1. Sputter Ti: W + Au 2. Spin photoresist 3. Expose photoresist 4. Develop photoresist 5. Electroplate Au 6. Electroplate Sn 7. Remove photoresist 8. Etch plating base 9. Reflow AuSn	1. Pad pretreatment 2. Zincate 3. Zincate etch* 4. Zincate* 5. Electroless Ni 6. Electroless Au	1. Program pad position 2. Program bond parameters 3. Perform stud bumping	1. Program pad position 2. Program bond parameters 3. Install inert gas nozzle 4. Perform stud bumping
Process requirements					
Possible pad metallizations	Al (Si wafers) Au (GaAs,InP), MCM-D	Al (Si wafers) Au (GaAs,InP), MCM-D	Al pad Au pad†	Al (Si wafers) Au (GaAs,InP), MCM-D, MCM-C	Sealed Al pad with wettable metal, e.g., electroless NiAu
Pad design/wafer design	Passivation overlap for Al pads, optimum 10 μm	Passivation overlap for Al pads Nonwettable solder dam around the bump for Au pads	Water scribelines must be passivated Passivation around the pad for Au pads	Pads for wire bonding	Electroless NiAu and stud Au (e.g.)

TABLE 15.1 Characteristics of the Bumping Process (Continued)

Bumping process	Electroplated Au	Electroplated AuSn	Electroless NiAu	Stud Au	Stud 97/3 PbSn
Minimal pitch	20 μm (bump height 30 μm)	20 μm (bump height 30 μm)	150 μm (bump height 30 μm)	70 μm (bump height 45 μm)	150 μm† (Bump height 70 μm)
Bump shape	Straight wall	Straight wall	Mushroom	Stud → Coining → Cylindrical	Stud → Reflow → Spherical
Bump shape determined by	Photoresist opening	Photoresist opening	Bond pad design	Bonding/coining tool, bond parameters	Bonding tool (not reflowed) bond pad, solder volume (reflowed)
Additional hardware required for process	Photomask for bumps	Photomask for bumps	No additional tools	No additional tools	No additional tools
Suitability for					
Wafer bumping	+++	+++	++	+	+
Substrate bumping	+	+	(+) requires special substrate layout and metallurgy	++	(+) requires special substrate layout and metallurgy
Single-chip bumping	— (not possible)	— (not possible)	+	++	+
Cost for high volume	xx	xxx	x	xx	xx
Cost for low volume	xxx	xxx	x	x	xx
Sealing of the bond pad	Hermetical	Hermetical	Near-hermetical	No sealing (e.g., wire bond)	Near-hermetical, due to electroless NiAu underbump metallization

Part 2

Bumping process	Electro-plated Au	Electro-plated AuSn	Electroless NiAu	Stud Au	Stud 97/3 PbSn
Thermocompression bonding	++	—	—	+++	—
Typical bond temperature	300–400°C	—	—	300–400°C	—
Bond force	100 cN/pad	—	—	100 cN/pad	—
Substrates	1, 2 (3), (4)	—	—	1, 2, 3, 4	—
Soldering with use of flux	(+) solder from substrate	++	(+) solder from substrate	— (flux residues cause Al corrosion)	++
Typical soldering temperatures	200–350°C	300–350°C	200–250°C	—	250–400°C
Bond force	10–30 cN/pad	10–50 cN/pad	10–30 cN/pad	10–30 cN/pad	5–10 cN/pad
Substrates	1, 2, 3, 4, 5, 6, 7	1, 2, 3, 4, 7	1, 2, 3, 4, 5, 6, 7,	1, 2, 3, 4, 5, 7	1, 2, 3, 4, (5), (6), 7 eutectic SnPb solder from substrate
Fluxless soldering	(+) solder from substrate	+++	—	(+) solder from substrate	—
Typical soldering temperatures	200–350°C	300–350°C	—	200–350°C	—
Bond force	10–30 cN/pad	10–50 cN/pad	—	10–30 cN/pad	—
Substrates	1, 2, 3, 4, 5, 6, 7	1, 2, 3, 4, 7	—	1, 2, 3, 4, 5, 7	—
Soldering using self-alignment	—	+++	(+) solder from substrate	—	++
Typical soldering temperatures	—	300–350°C	250–400°C	—	250–400°C
Bond force	—	0 cN/pad	0 cN/pad	—	0 cN/pad
Substrates	—	1, 2, 3, 4, (7) substrate fixture	1, 2, 3, 4, (5), (6), (7) substrate fixture	—	1, 2, 3, 4, (5), (6), (7) eutectic SnPb solder on substrate

421

TABLE 15.1 Characteristics of the Bumping Process (Continued)

Bumping process	Electro-plated Au	Electro-plated AuSn	Electroless NiAu	Stud Au	Stud 97/3 PbSn
FC with isotropic conductive adhesives	(+) no cost or process advantages	—	+++	++	(+) no cost or process advantages
Typical assembly temperature	100–200°C	—	100–200°C	100–200°C	100–200°C
Typical bond force	3–20 cN/pad	—	3–20 cN/pad	3–50 cN/pad	3–20 cN/pad
Substrates	1, 2, 3, 4, 5, 6, 7, 8	—	1, 2, 3, 4, 5, 6, 7, 8	1, 2, 3, 4, 5, 6, 7, 8	1, 2, 3, 4, 5, 6, 7, 8
FC using anisotropic conductive adhesives	+	—	+++	++	(+) no cost or process advantages
Typical bonding temperature	150–200°C	—	150–200°C	150–200°C	150–200°C
Typical bond force	50–100 cN/pad	—	50–100 cN/pad	50–100 cN/pad	10–50 cN/pad
Substrates	1, 2, 3, 4, 5, 6, 7, 8	—	1, 2, 3, 4, 5, 6, 7, 8	1, 2, 3, 4, 5, 6, 7, 8	1, 2, 3, 4, 5, 6, 7, 8
FC using nonconductive adhesives	+	—	(+) very high substrate planarity required	+++	—
Typical bonding temperature	100–200°C	—	100–200°C	100–200°C	—
Typical bond force	50–100 cN/pad	—	50–100 cN/pad	50–100 cN/pad	—
Substrates	1, 2, 3, 4, 5, 6, 7, 8	—	1, 2, 6, 7, 8	1, 2, 3, 4, 5, 6, 7, 8	—

Suitability: +++ excellent; + possible; — not possible; ++ good; (+) possible with restrictions.
Cost: xxx very high; xx high; (blank) low.
Substrates: 1—thin-film ceramic; 2—silicon substrate; 3—thick-film ceramic; 4—green tape multilayer; 5—printed-circuit board (e.g., FR-4); 6—flexible circuit type 1 (polyimide with adhesive); 7—flexible circuit type 2 (adhesiveless polyimide; e.g., ESPANEX); 8—LCD—glass substrate.
*Double-zincate treatment.
†The minimal pitch for solder wire bumping is given by the availability of different solder wire diameters. Tanaka is offering a minimal diameter of 30 μm.

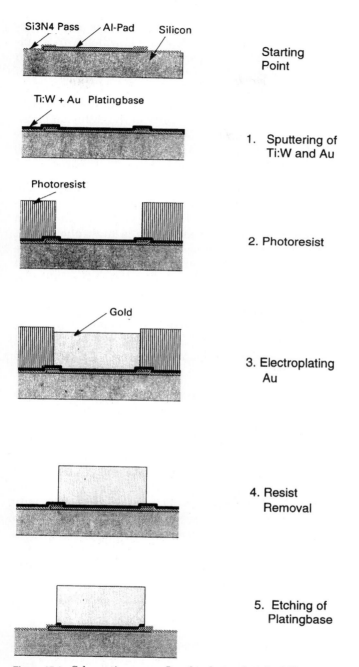

Si3N4 Pass   Al-Pad   Silicon

Starting
Point

Ti:W + Au Platingbase

1. Sputtering of
    Ti:W and Au

Photoresist

2. Photoresist

Gold

3. Electroplating
   Au

4. Resist
   Removal

5. Etching of
    Platingbase

Figure 15.1 Schematic process flow for electroplated gold bumps.

Sputtering.   The Ti:W layer has a double function; it serves as an adhesion promoter to the Al pad and as a diffusion barrier, which avoids aluminum–gold interdiffusion. Titanium nitrides and oxinitrides are often used in combination with tungsten in order to increase the effectivity of the diffusion barrier.[19,20] Optimization of the process with regard to internal mechanical stress and the improvement of the diffusion barrier quality is essential.[21] A typical test for the quality of the diffusion barrier consists of annealing at 400°C for one hour. After this the bump is selectively removed using a gold etchant. An optimized layer shows no Al–Au interdiffusion in the optical inspection of the pad integrity.[22–25]

Photoresist process.   The *photoresist* determines the shape of the bumps. The process presented by Engelmann et al.[26] allows straight-wall bumps of $10\times10~\mu m^2$ and a height of 25 μm in 20-μm pitch.[26] In order to achieve well-defined opening geometries, a high photoresist thickness is required; the liquid photoresist is applied by multiple spin in steps followed by a special baking procedure. The photoresist is then exposed with a mask aligner in vacuum contact. Typically liquid-positive photoresists are applied for Au bumping.

In addition to the requirement for a well-defined opening in the resist with vertical sidewalls, after the resist is developed it must be compatible with the subsequent Au plating solution. Figure 15.2 shows an example of openings in the photoresist for 10-, 25-, 50-, and 100-μm bumps.

Electroplating.   Before electroplating, possible photoresist residues on the bottom of the resist openings are removed by plasma etching. If not removed, these residues can cause a reduced adhesion of the bumps, an increased contact resistance and an inhomogeneous growth of the electroplated gold layer. The bump equipment consists typically of a fountain or cup plater. The fountain plater principle is based on the vertical flow of the electrolyte to the wafer, which is fixed with pins on a cup and floats over the electrolyte, and is applied to obtain optimal homogeneity of bump height over the wafer. Typically, height tolerances of <1 μm over a chip and <3 μm over the wafer are required in today's TAB applications. There are similar requirements for the use of this bump type in flip chip technology. The Au-plating solutions are typically sulfite-based.[28,31] Important criteria determining the choice of electrolyte are

- Optimal current density—this is important for the throughput and the shape of the bump.
- Compatibility with the applied photoresist system.
- Obtained height distribution and bump shape at the maximal current density.
- Gold surface morphology (fine crystal sizes).

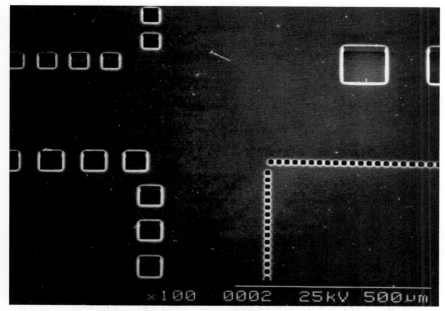

Figure 15.2 Openings in a photoresist for straight-wall bumps in 20-, 50-, 100-, and 200-μm pitch. (*Courtesy of TU-Berlin.*)

- Au hardness after electroplating (this can range from 50 to 140 HV)
- Annealing behavior in order to achieve a defined and reproducible hardness of typically 50 to 70 HV in a temperature range below 400°C.

After the electroplating, the photoresist is removed in a stripping solution, and the plating base of Au and Ti:W are selectively etched (dry- or wet-etching techniques are applied). The wet etching of the Au-plating base causes a slight roughening of the bump surface, whereas dry etching results in a smoother surface morphology. Figure 15.3 shows the bumps from a test mask with different pitches of 200, 100, 50, and 20 μm. The tiny bumps with a 10×10-μm pad size have a height of 25 μm. A further development of the photoresist process at TU-Berlin shows that bump heights of ≤80 μm can be achieved.[22,32] Therefore, the process can be used not only for bumping but also for implementing additional structures such as microcoils for wireless communications on the chip surface.[39] The pad design for TAB technology typically requires a square structure. For flip chip application an octagonal or round shape is more suited in order to minimize the mechanical stress of the FC interconnections. Figure 15.4 shows an example of a pad design rule and a FC bump.

Figure 15.3 Electroplated gold bumps from a test mask in 20-, 50-, 100-, and 200-μm pitch.

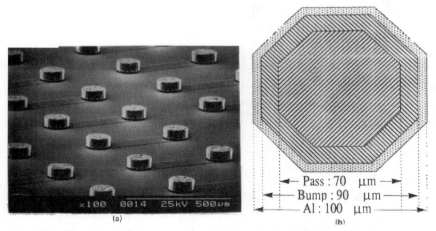

Figure 15.4 Design rule for an octagonal bondpad for flip chip assembly: (*a*) octagonal gold bump for FC-assembly; (*b*) design rule for an octagonal bond pad.

Cost aspects.   The Au-bumping process can be performed on wafer level only. Depending on the wafer size, additional equipment may be required. Upscaling of a process to wafer sizes of ≥8 in from an existing process using 4- or 6-in wafers requires costly new investments, especially for the mask aligner. Additional to this, minor changes in the existing equipment (sputtering, electroplating) must be done.

For the bumping mask, the exact positions and sizes of bond pads and test pads must be obtained from the wafer manufacturer. The cost for the design and manufacture of the bumping mask must be added to the total bumping costs. Because the facility performing the bumping will add nonrecurring engineering costs for each wafers, the process is cost-effective only for high-volume production.

15.2.2   Electroplating of gold-tin solder bumps

This process is based on the electroplating of gold bumps with an additional step, in which a tin cap is deposited. Figure 15.5 is a schematic flowchart consisting of cleaning of the Al pad by backsputtering and sputtering of the plating base of Ti:W and Au. The photoresist is applied, exposed, and developed. After this, gold followed by tin are electroplated and the photoresist is removed. Finally the plating base is selectively removed.

Typically prior to FC assembly the tin cap is reflowed in order to obtain the eutectic 80/20 AuSn solder, which permits a fluxless FC assembly.

This process was initially developed by TU-Berlin for reducing the bonding pressure during the inner lead bonding of TAB devices.[34–37] The use of this bump type is also reported elsewhere.[38–40] More recently, these bumps have been applied for fluxless flip chip assembly on green tape ceramic substrates[41–45] and on flexible circuits.[46–48] The self-alignment effect of this bump metallurgy for optoelectronic devices has also been described.[49,50] One essential aspect of the AuSn sandwich structure is the rapid interdiffusion of Au and Sn during which a series of AuSn intermetallic compounds are formed. Therefore, nonreflowed bumps can be used for only a limited time. At temperatures of 55 to 125°C a Sn layer of a few micrometers' thickness is consumed within only a few hours. Figure 15.6 shows metallographic cross sections of AuSn bumps with different types of intermetallics formed after aging at room temperature for 3 months and after accelerated aging at elevated temperatures. The growth of the compounds and the consumption of the Sn-rich phases is evident. Additionally, the formation of Kirkendall voids at the interface between the Au and the AuSn phase can be observed. A more detailed description of AuSn metallurgy is given in Sec. 15.3.3.

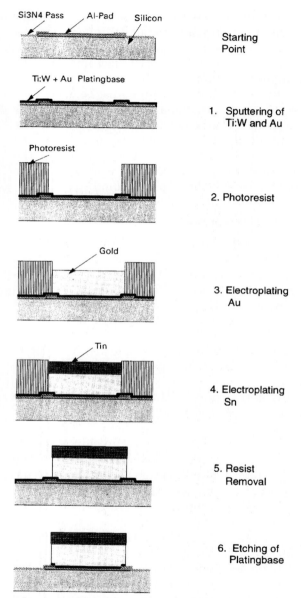

Figure 15.5 Flowchart for electroplating gold-tin solder bumps.

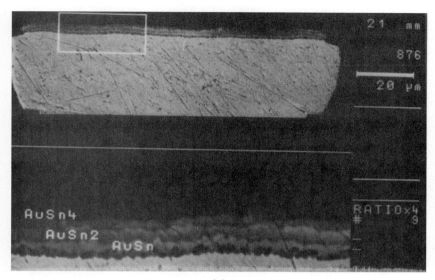

(a)

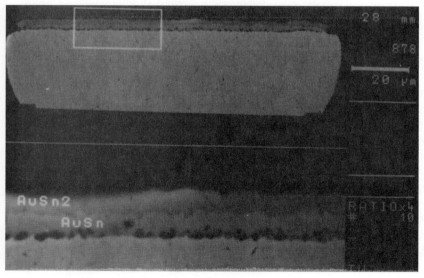

(b)

Figure 15.6 Metallographic cross sections of differently aged AuSn bumps showing the formation and growth of AuSn intermetallic compounds: (a) 6 weeks at room temperature; (b) 250 h at 75°C.

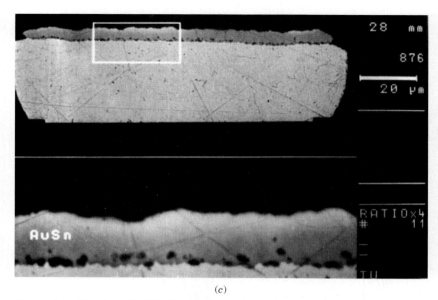

(c)

Figure 15.6 (*Continued*) Metallographic cross sections of differently aged AuSn bumps showing the formation and growth of AuSn intermetallic compounds: (c) 250 h at 100°C.

By a controlled reflow step in an inert atmosphere, the intermetallic compounds are transformed into the eutectic 80/20 AuSn alloy, which allows a fluxless solder process and an extended shelf life of the bumps.

The reflowed AuSn bumps consist of a hard core of Au located on the bottom of the bump, a ζ-phase layer, and a cap of eutectic 80/20 AuSn alloy. The amount of eutectic alloy is defined by the Sn-plating thickness during bumping. The chosen type of FC assembly process requires a defined amount of solder. For the formation of the eutectic solder composition, 1 μm of tin dissolves 1.5 μm of gold. For FC assembly, typically a tin thickness of 5 to 10 μm is chosen on a Au pedestal of 25 to 50 μm, whereas in TAB technology 3 to 4 μm Sn is typically used. Figure 15.7 shows cross sections of reflowed solder bumps with different thicknesses of the electroplated tin cap.

In order to avoid an excessive growth of the ζ phase during the reflow step, which causes a consumption of the eutectic AuSn for soldering, control of the reflow process and understanding of the metallurgical processes taking place at this time are essential. Figure 15.8 shows intermediate steps during the reflow process of this special bump metallurgy. The metallurgical reactions taking place during the reflow process at 300°C can be described as follows. During the heating of the Au-AuSn-AuSn$_2$-AuSn$_4$-Sn sandwich structure to a temperature of 300°C, the first tin-rich eutectic with 90 wt% Sn and a liq-

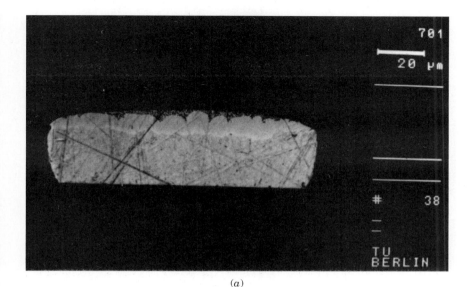

(a)

(b)

Figure 15.7 Cross sections of reflowed AuSn solder bumps with different thicknesses of the initial tin layer [(a) 3-μm Sn, (b) 5-μm S, (c) 10-μm Sn] over 25-μm gold. The structure of all reflowed bumps consists of a hard Au core, ζ phase, and the eutectic 80/20 AuSn solder cap.

uidus temperature of 217°C is formed. The compounds of this eutectic consist of Sn and $AuSn_4$ (Fig. 15.8a).

If tin is totally consumed in $AuSn_4$ by solid-state diffusion before reaching the melting temperature of 217°C, no Sn-rich eutectic will

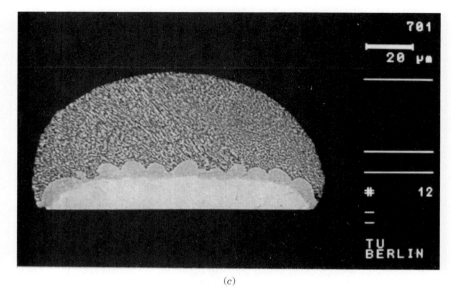

(c)

Figure 15.7 (*Continued*)

be formed. A special aspect of the eutectic formation is that the liquid eutectic alloy can wet the sidewalls of the bumps, as shown in Fig. 8a. By further solid–liquid diffusion, this first eutectic is fully consumed in $AuSn_4$. Furthermore, the $AuSn_2$ phase consumes $AuSn_4$ by solid–liquid diffusion. During all these processes, at the interface adjacent to Au, the phase AuSn is growing continuously by solid-state diffusion. At this interface Kirkendall voids can be observed (Fig. 15.8a). This effect is described in Sec. 15.3. After the full consumption of $AuSn_2$ only the compounds AuSn and Au remain. The initial formation of ζ and the kinetics of the phase growth during the reflow process are not yet clarified. Therefore, it is not clear at which step of the reflow process the first formation of ζ occurs. The investigations performed so far indicate a formation in the late state of the process by reaction between AuSn and Au (Fig. 8b). As AuSn and ζ are the components of the eutectic 80/20 alloy, the growth of ζ-phase adjacent to the AuSn interface implements the formation of the liquid eutectic alloy (Fig. 15.8c and 15.8d). The liquidus temperature of 280°C of this alloy is below the reflow temperature of 300°C. The growth of ζ phase during the reflow process by liquid–solid diffusion is investigated by Kallmayer et al.[50]

Figure 15.9 shows SEM images of AuSn bumps before and after reflow. The potential of this process for fine-pitch applications is illus-

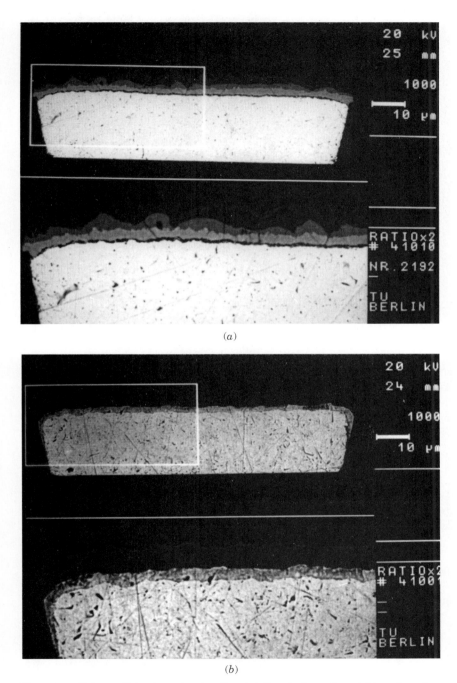

(a)

(b)

Figure 15.8 Intermediate steps showing the metallurgical reactions taking place during the reflow process of AuSn solder bumps: (a) step 1; (b) step 2.

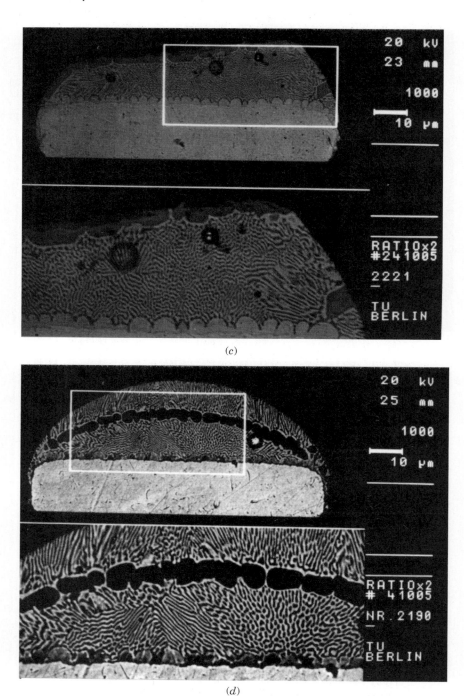

Figure 15.8 (*Continued*) Intermediate steps showing the metallurgical reactions taking place during the reflow process of AuSn solder bumps: (*c*) step 3; (*d*) step 4.

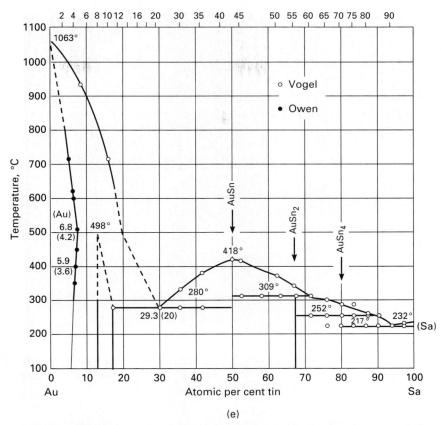

(e)

Figure 15.8 (*Continued*) Intermediate steps showing the metallurgical reactions taking place during the reflow process of AuSn solder bumps: (*e*) AuSn-phase diagram according to Hansen.[92]

trated in Fig. 15.10 by the bumps of a 10×10-μm pad size with 30 μm bump height for a 20-μm-pitch test chip.

15.2.3   Electroless nickel-gold bumps

Electroless plating has the highest potential for cost reduction of the bumping process in flip chip applications. Because the process provides a selective autocatalytic metal deposition directly on the aluminum pads of the wafers, no costly equipment for sputtering, photoresist imaging, and electroplating is required. As a maskless process, it has an additional advantage of being very flexible. In contrast to electroplating, which requires an extra cup for each wafer, electroless plating can be done in a batch, as it requires contact between only the electroless plating solution and the wafers.

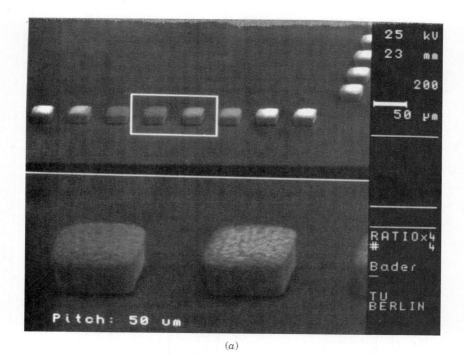

(a)

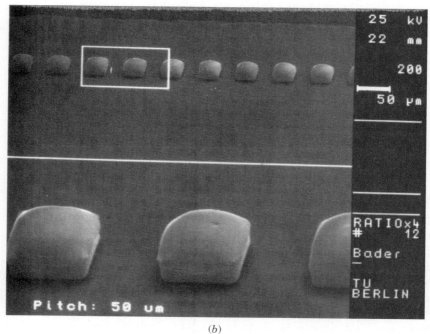

(b)

Figure 15.9 SEM images of AuSn solder bumps before (a) and after (b) the reflow process.

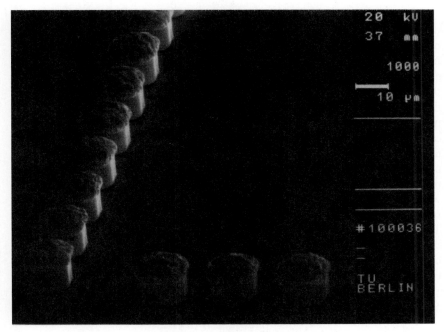

Figure 15.10   SEM of a AuSn bump with 10×10-μm pad size and 30-μm bump height.

For electroless plating, preconditioning of the Al pad is essential. Because of the high affinity of aluminum for oxygen, these surfaces exposed to air or to water-containing solutions are always covered with a thin layer of Al oxides and hydroxides. In order to remove this, a thin seed layer of a more noble metal must be deposited. Because of the amphoteric property of aluminum on dissolution, this procedure can be performed in either highly acidic or highly alkaline solution. Figure 15.11 shows the dissolution behavior of Al in different pH regions.[51]

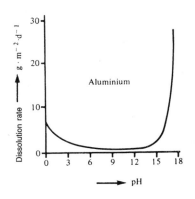

Figure 15.11 Dissolution rate of aluminum as a function of the pH value of a water-based solution according to Gellings.[51]

The electroless plating of aluminum with nickel is a standard process applied for corrosion protection and for tribological requirements of large workpieces. The use of these processes for wafer bumping of Al pads with a few micrometers' thickness is not possible because the dissolution rate of aluminum during the zincate pretreatment is too high. Therefore, the pretreatment must be modified in order to reduce and to control the dissolution of the Al pad metallization. Two types of pretreatment have been presented so far, based on either palladium[52–54] or zinc as seed metal.[34,55–65] In general the palladium process is more critical. In order to avoid the deposition of Pd seeds on the whole wafer, which would result in a nonselective nickel plating, the palladium solution must be stabilized[53] and will work only in a certain pH region. The zincate process is less critical with regard to the selectivity. The process flow is shown in Fig. 15.12. The process consists of cleaning of the Al pad and deposition of a seed layer of zincate followed by the electroless nickel deposition. Finally, a thin layer of gold is deposited, as a protective coating against oxidation and to improve the solder wettability.

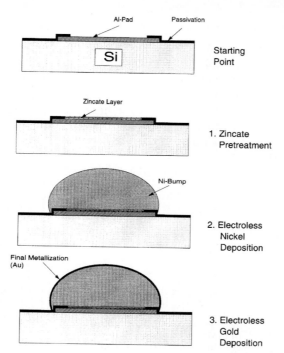

Figure 15.12 Process flow for zincate-based electroless deposition of NiAu bumps.

Precleaning.   In a first step the aluminum pad must be cleaned by removing the native oxide layers and other residues from the processing of the wafer. This pretreatment can be done by either wet or dry etchings. To reduce the cost of dry plasma etching, TU-Berlin and IZM-Berlin developed a wet chemical pretreatment of the Al pad.[60]

The precleaning has a central function in the whole process. Residues on the Al pad will result in an unhomogeneous deposition of coarse zinc grains in the subsequent zincate solution. According to Ostmann et al.,[59,60] this effect reduces the adhesion of the nickel bump and increases the contact resistance.

The zincate treatment.   A zinc film is deposited by an immersion process which can be chemically characterized as follows:

- Anodic dissolution of aluminum:

$$Al^0 + 3OH^- = Al(OH)_3 + 3e^- \qquad E^0 = 2.31$$

- Cathodic deposition of zinc:

$$Zn(OH)_4^{-2} = Zn^{2+} + 4OH^-$$

$$Zn^{2+} + 2e^- = Zn^0 \qquad E^0 = 0.763$$

The most important components of alkaline zincate solutions are sodium hydroxide and zinc oxide. The adhesion and morphology of the zinc film can be influenced by additives of copper, iron, or nickel ions and complexing agents such as tartrates or cyanides. The zincate treatment also has an important effect on the morphology and uniformity of the electroless nickel deposits.

Figure 15.13 shows the SEM images of different pretreatments. The impact of grain size and grain distribution on the adhesion of the nickel bump is evident in the corresponding bar charts of shear test values. The double zincate pretreatment has the finest grain size and results in shear test values of >200 cN (Fig. 15.13*f*). For the double treatment, the first zincate film is etched away, followed by a second dipping in the zincate solution. Another important aspect of zincate pretreatment is the dissolution rate of aluminum in the alkaline solution.

According to Ostmann et al.,[60] the exposure of 10 to 30 s results in a dissolution of 200 nm. This shows that for most wafers having an Al thickness of 1 to 2 μm the dissolution is controllable and not a critical parameter. The dissolution as a function of time is shown in Fig. 15.14.

Electroless nickel plating.   Typically two types of autocatalytic nickel-plating systems are supplied by commercial vendors, based on either nickel phosphorous or nickel boron. The difference consists in the reducing agent, which is based on hypophosphorous acid (H_3PO_2) and

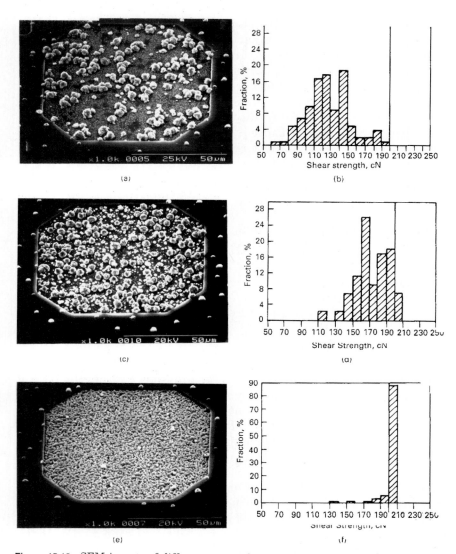

Figure 15.13 SEM images of different types of zincate pretreatment with the corresponding data from shear test. (*a*) Simple zincate pretreatment; (*b*) shear force distribution of the simple zincate pretreatment; (*c*) double zincate pretreatment; (*d*) shear force distribution for the double zincate pretreatment; (*e*) optimized double zincate pretreatment; (*f*) shear force distribution for the optimized double zincate pretreatment.

dimethylboramin (DMBA). During the reaction of the reducing agent, pure phosphorous and boron are produced to a certain amount and are codeposited in the nickel layer. Therefore the metal layers obtained from these plating solutions do not consist of pure nickel; they are NiP and NiB alloys.

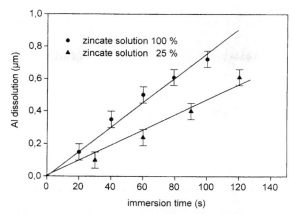

Figure 15.14 Dissolution of Al pad in the zincate pretreatment according to Ostmann et al.[60]

Because of the higher plating rates ($\leq$25 μm/h) electroless NiP was chosen for the process developed by TU-Berlin. The plating solution has a pH value of 4.5 and operates at 90°C. Figure 15.15 shows the bumps obtained with this process.

Final metal layer. As a final protecting and wettable metal layer, a gold flash (0.1 to 0.2 μm) can be deposited from an immersion plating bath. A minimal thickness of the nickel layer of 7 μm is required in order to seal the Al bond pad.

Possible applications. The electroless NiAu bumps can be applied for FC adhesives processes, where the bump metal provides sealing of the Al bond pad. Soldering is possible, but it requires the solder deposition on the substrate.

For the electroless solder deposition on nickel, no autocatalytic processes based on tin or on eutectic SnPb solder are available. Aschenbrenner et al.[61] presented a process solution with electroless solder. First a socket of 7 to 10 μm of electroless nickel is deposited, followed by an electroless copper layer and electroless solder as final layer. Because the achievable solder thickness is only a few micrometers, the process can be applied only for TAB-ILB soldering. For flip chip applications, a higher amount of solder is required. Therefore, different approaches using dipping methods in liquid solder—with ultrasonic energy[53,54] using wave soldering[65] or fine-pitch stencil printing of solder paste[66]—have been reported.

IZM Berlin developed the *meniscus solder deposition* process, which allows the deposition of different solder alloys such as Sn, 3.5 SnAg, 63/37 SnPb, and 80/20 AuSn on the electroless nickel bumps. Figure 15.16 shows SEM images of these solder bumps.

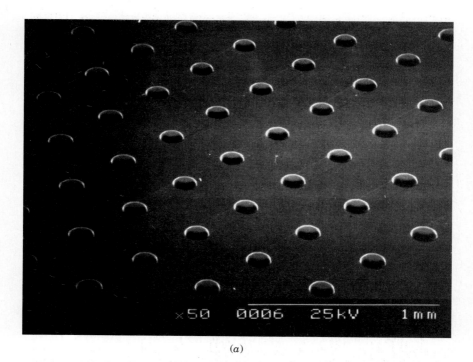

(*a*)

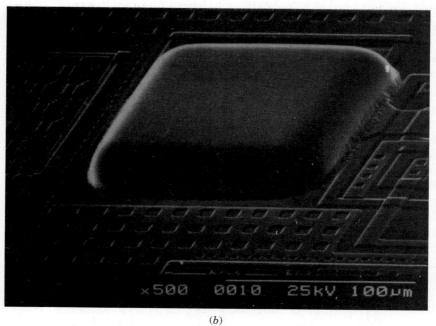

(*b*)

Figure 15.15   SEM views of electroless deposited NiAu bumps.

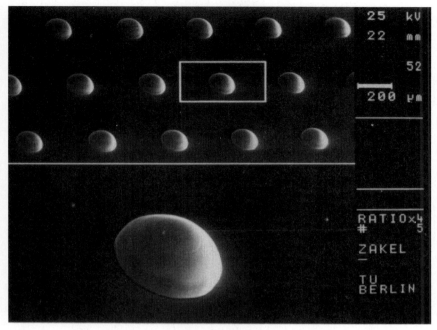

Figure 15.16 Electroless nickel bumps with a solder-cap obtained by the *meniscus solder deposition process.*

Another approach consists of stud bumping using solder wires of 98/2 PbSn.[63,67] This process is presented in Sec. 15.2.5. Recently manufacturers of solder materials, such as like Heraeus (Germany) have been delivering different types of solder ball material in diameters of 100 μm and smaller, and solder ball placement methods are also considered as a possible solution.[68,69]

Additional process requirements. For electroless plating of active circuits, the inner circuity is of essential importance. Ostmann et al.[60] showed that especially the ground pads of a circuit cause potential differences and thus will not be coated with metal. In order to avoid this effect, a photoresist coating of the wafer backside must be performed.

An additional process requirement is that the scribelines for wafer sawing be covered with a passivation; otherwise they will be coated with nickel. A possibility to avoid this is by covering the scribelines with a photoresist. This additional step, which requires a mask and photoresist deposition, photoimaging, and development, adds cost to the process. The extra cost can be avoided if this aspect is taken into account in the layout of the passivation mask for wafer manufacturing.

Minimal pitch. As a result of the maskless electroless deposition, the bumps have a mushroom shape. Therefore the pitch limitation of the

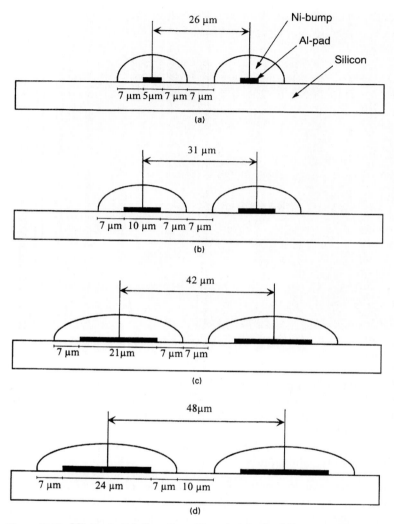

Figure 15.17 Minimal pitch for a 7-μm bump height for different pad designs.

process is determined by the required bump height. The lateral growth of the bumps is in the same range as the vertical growth. If we take into account that a minimal thickness of 7 μm is required for a protective sealing of the Al pad, a minimal pitch of 21 to 26 μm can be achieved with an adapted pad design shown in Fig. 15.17a and 15.17b. Figure 15.17c and 15.17d shows approaches with a more conservative pad design. The spacing between the Al pads is the same size as the Al pads themselves. With a minimal distance of 7 to 10 μm between the bumps, a minimal pitch of 42 to 48 μm can be determined.

Bump hardness. The high hardness of the nickel bumps with 10 wt% phosphorous causes some process limitations. The hardness is in the range of 500 to 550 HV. Therefore, the bumps are not suited for thermocompression processes. The deposited metal has an amorphous structure. The formation of crystalline nickel with Ni_3P precipitation is observed at temperatures above 300°C.[69,70] This causes an increase of hardness and an increase of the intrinsic stress of the deposit due to the different densities of the formed phases.

The influence of the phase transformation on nickel bumps has not been investigated thus far. For this reason <300°C is the recommended temperature for performing the soldering processes.

15.2.4 Processing of gold stud bumps

The process of Au stud bumping is performed with a modified wire bonder, which has the highest flexibility because it can be performed on single chips, whole wafers, and substrates. An important cost aspect for bumping is that this software programmable process permits the bumping of only tested good chips. It allows a series of bond pad metallurgies[71] and can be used for the aluminum pads of silicon devices as well as for gold pads of gallium arsenide devices.[72-74]

Kondoh et al.[75] developed a flip chip assembly method using gold balls on the chip site and an indium alloy on the substrate. The use of Au stud bumps for soldering using Sn or eutectic SnPb solder on the substrate has been presented.[47] Bonkohara et al.[77] and Larsen and Brock[78] introduced gold ball bumps for TAB inner lead bonding. Goodman and Metroka used gold ball bumps combined with a conductive adhesive.[76] Aschenbrenner et al. used gold ball bumps for a flip chip technology with a nonconductive adhesive film.[80,81] The use of Au stud bumps for a 70-μm-pitch application in combination with thermocompression of GaAs devices was demonstrated by Eldring et al.[82-84] Figure 15.18 is a schematic flowchart of the stud bumping procedure.

The process is a single-point bumping technology derived from ball bonding of gold wires. After formation of the ball and the first bond, normally the next step for conventional wire bonding is loop formation followed by the second bond. Instead of this, the ball bumping procedure will break off the wire above the ball (Fig. 15.18). For a controlled breaking of the wire above the heel, a palladium-doped wire with 1 to 2 wt% is applied. This is due to the recrystallization of the wire material in the heel area, which improves the defined breaking of the wire in this region during the tearoff step.[85]

Figure 15.19 shows a cross section of a stud bump with microhardness indentations. The special mechanical properties of this bump type due to the hardness gradient in the range of 50 to 80 HV is evident in this figure. Note especially the low-hardness region in the upper tail

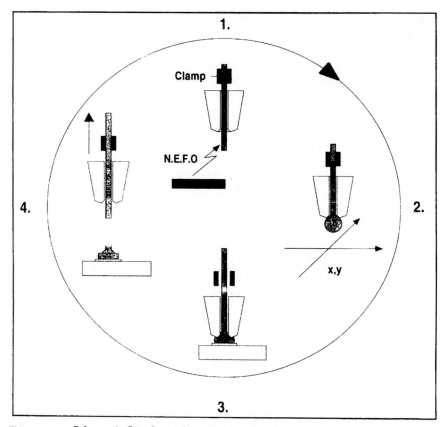

Figure 15.18   Schematic flowchart of Au stud bumping.

area of the stud bumps; this improves the bonding properties for thermocompression bonding, as its plastic deformation during the bond process provides an improved adhesion.[83–85] A comparison of different wire types—with and without palladium doping—is given by Eldring et al.[86] This shows that nondoped Au wires are not suited for stud bumping due to the undefined breaking behavior in the heel of the ball bump.

A special approach for stud bumping using standard Au wires was developed by Panasonic. In this process, the capillary performs a special movement after the ball bonding, during which the second bond is set on top of the ball. This allows a defined breaking of the wire.[87] Elenius[88] developed a process for known good die (KGD) for flip chip technology based on stud bumps. The chip is first ball-bonded on an intermediate substrate, thus allowing a full dynamic testing and burn-in of the IC. After this, the wires are broken above the balls with a special procedure, leaving a chip with stud bumps.

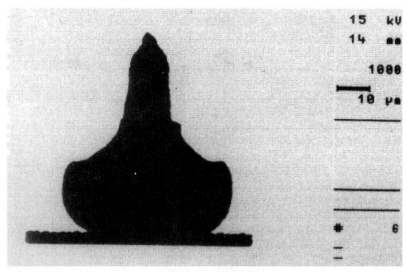

Figure 15.19 Cross section of a ball bump with hardness measurements. The hardness gradient of the ball bump is in the range of 80 HV (bottom) to 50 HV (top).

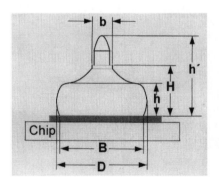

Figure 15.20 Scheme of stud bump geometry and sizes.

The geometry and dimensions of the stud bump are shown in Fig. 15.20. The bump sizes which can be obtained with this process are summarized in Table 15.2. A planarization of the stud bumps can be achieved by an additional coining step. Figure 15.21 shows a comparison of bumps both with and without coining. A change in bump height is also possible with the use of stacked bumps. Figure 15.22 shows some examples of possible stacked structures. With a high-accuracy bonding machine, five and even more bumps can be stacked with high precision. The use of this bump type on chip and substrate for FC thermocompression has been reported.[89]

TABLE 15.2   Bump Geometry Characterization
for 18-μm and 25-μm Wire Diameters

Wire diameter, μm	18	25
Diameter D, μm	40–60	70–85
Diameter b (hole of capillary), μm	25	33–37
Height h, μm	12–25	25–40
Height H, μm	22–35	40–55
Height h', μm	40–50	65–85

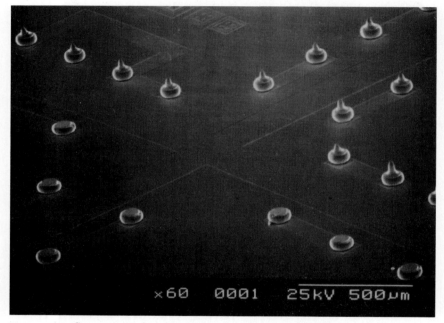

Figure 15.21   Comparison of stud bumps with and without coining.

Process advantages.   In summary, the gold stud-bumping process can easily be performed with modified wire-bonding equipment available from a series of manufacturers. Fully automatic bonders can achieve cycles of 100 ms per bond. The cost of the wire-bonding equipment and the required infrastructure are significantly lower than that of the equipment and cleanroom infrastructure required for electroplated gold bumps. Therefore this process should be taken into account for small- and medium-volume applications.

Compared to electroplated bumps, the stud bumps have no diffusion barrier between the Al pad and the Au bump. Therefore, the reliability aspects of Al–Au interdiffusion known for wire bonding must

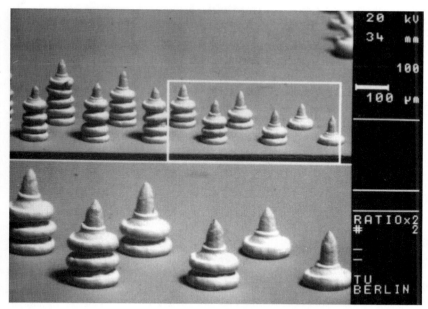

Figure 15.22 Demonstration of different possibilities for producing stacked stud bumps.

be taken into consideration. This is discussed in Sec. 15.3. Moreover, the Au bump does not provide hermetical sealing of the bond pad. For soldering, these bumps require fluxless processes, in order to avoid a contamination of the Al pad by flux residues, which can lead to corrosion during the life cycle of a product. In nonhermetical packages, additional underfilling and glob-top encapsulation is recommended, in order to protect the bond pad from corrosion.

15.2.5 Processing of solder stud bumps

Recently solder wires have been used to produce stud bumps for flip chip applications. The equipment required for ball bumping of solder wires consists of a modified wire bonder. The process steps are analogous to those of Au stud bumping (shown in Fig. 15.18).

To avoid oxide formation on the solder ball surface and to obtain defined geometry and surface characteristics, a shielding atmosphere is required during the flameoff process. This is done with special nozzle configurations around the bond capillary. Figure 15.23 shows the schematic arrangement of gas nozzles according to Jung et al.[67] A typical gas mixture applied for solder wire bumping consists of 90% argon and 10% hydrogen. Concerning the required Al-pad condition, two approaches have been presented so far. The first approach uses 97/3 SnAg solder wire, which is bonded directly onto the Al pad. No

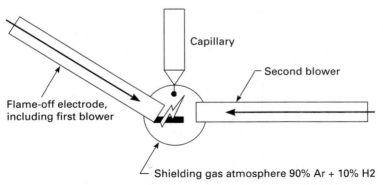

Figure 15.23 Scheme of a modified flameoff unit for solder ball bumping.[67]

preconditioning of the bond pad is required here according to Ogashiwa et al.[90]

However, these bumps cannot be used as solder bumps, because a reflow of the tin alloy would cause a total dissolution of the Al pad. The phase diagram between Al and Sn has a low melting eutectic, as shown in Fig. 15.24, which will cause a total consumption of the Al pad during a possible reflow process.

A second approach to solder stud bumping requires the preparation of the bond pad with a wettable metal. This can be done using the electroless nickel-gold process described in Sec. 15.2.3. The use of 98/2 PbSn wires in combination with electroless plating has been presented by several authors.[63,67,86,91]

Subsequent to the bumping process, a reflow in hydrogen or in an activated atmosphere can be performed. The melting of the solder ball results in a hemispheric bump shape with a defined bump height. Figure 15.25 shows bumps of 98/2 Pb/Sn over electroless deposited NiAu bumps before and after reflow.

Process advantages. In summary, stud bumping using solder wires offers the advantages of a flexible, software-programmable process. The bump metallurgy of 98/2 PbSn is resistant to fatigue and has properties comparable to those of C4 bumps. Therefore these bumps can be used not only for soldering processes in the high-temperature range (>300°C) for MCM-C and MCM-D substrates but also for FC soldering at lower temperatures (<300°C) on MCM-L substrates using eutectic solder on the substrate pads. Flux is required for all soldering operations.

On the other hand, this process requires an additional pad conditioning with electroless nickel gold. Solder wires are still significantly more expensive than gold wires, so the total cost of the process is higher than that of Au stud bumping.

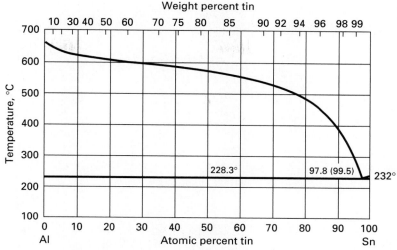

Figure 15.24 Phase diagram between aluminum and tin, according to Hansen.[92]

15.3 Metallurgical Considerations

This section presents the most important data on the applied metallurgies. The phase diagrams of all binary systems applied for the presented FC bumping and assembly process are well defined by Hansen[92] and more recently by Massalasky et al.[93] It must be pointed out that the degradation effects taking place at the interfaces between the binary solder materials and the substrate materials are due to the interaction of three, four, and sometimes even more numbers of different metals. Most phase diagrams of the corresponding ternary or quaternary systems have not yet been defined.

Important reliability issues are the growth of intermetallic compounds during the aging processes. The diffusion data from literature on the binary CuSn-AuSn, CuAu system and the recently defined ternary AuCuSn system are summarized in Table 15.3 (see also Fig. 15.26).

15.3.1 The CuAu system

The phase diagram of CuAu is shown in Fig. 15.27a. This system has been investigated frequently by various authors.[94–103] Tompkins and Pinnel[98] present a comparing overview of the diffusion coefficients. The interdiffusion coefficients which were determined in different investigations show considerable deviations, especially in the lower temperature range. These deviations are believed to be caused by different defect densities of the investigated materials, due to different methods of manufacturing. The transition from grain boundary diffusion to

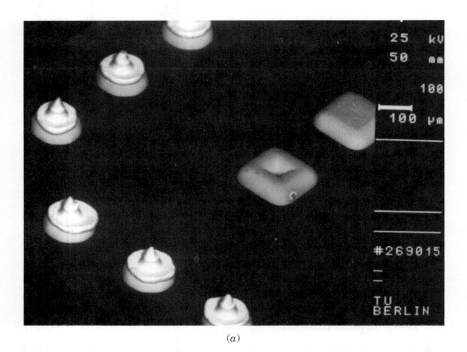

(a)

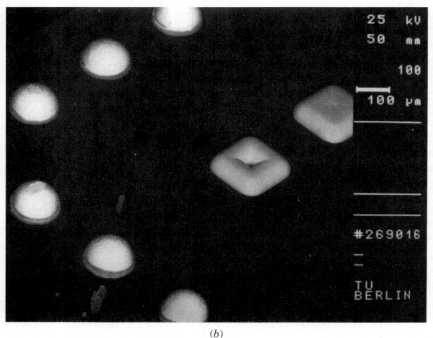

(b)

Figure 15.25 Stud bumps of 98/2 PbSn performed with a 30-μm wire diameter showing the bump shape as bonded (a) and after solder reflow (b).

TABLE 15.3 Data of the AuCuSn System from Different Literature Sources

Phase	$\ln D = \ln D_0 - E_a/(k_BT)$ $(D, cm^2/s; T, K)$	D_0, cm^2/s	E_a, eV	Temperature range, °C	Diffusing atom species	Ref.
		CuSn				
$Cu_6Sn_5(\eta)$	$\ln \bar{D} = -7.75-8{,}191/T$	4.29×10^{-4}	0.72	68–212	Cu, Sn	108
$Cu_3Sn~(\varepsilon)$	$\ln \bar{D} = -10.3-7{,}523/T$	3.39×10^{-5}	0.66	68–212	Cu, Sn	108
		AuSn				
$AuSn_4$	$\ln \bar{D} = -4.5-10{,}553/T$	5.09×10^{-3}	1.0	68–212	Au, Sn	108
$AuSn_2$	$\ln \bar{D} = -5.06-6{,}656/T$	6.34×10^{-3}	0.59	68–212	Au, Sn	108
AuSn	No data available	—	—	—	—	—
ζ	$\ln \bar{D} = -8.22-7{,}262/T$	2.7×10^{-3}	0.63	100–250	Au, Sn	40
Sn-~0%Au	$\ln D_{Au} = -5.14-6{,}032/T$	5.8×10^{-3}	0.52	125–232	Au($\parallel c$ axis)	114
Sn-~0%Au	$\ln D_{Au} = -1.83-9{,}977/T$	1.6×10^{-1}	0.86	125–232	Au($\perp c$ axis)	114
Au-~0%Sn	$\ln D_{Sn} = -3.24-19{,}141/T$	3.99×10^{-2}	1.65	689–1003	Sn	114
Au-~1.5%Sn	$\ln D_{Sn} = -4.57-19{,}548/T$	1.03×10^{-2}	1.69	749–855	Sn	114
		CuAu				
Cu-20 wt%Au	$\ln \bar{D} = -11.9-11{,}837/T$	6.86×10^{-6}	1.02	250–750	Cu, Au	97
Cu-20 wt%Au	—	—	0.48	25–250	Cu, Au	97
Cu-50 wt%Au	$\ln \bar{D} = -11-12{,}161/T$	17.0×10^{-6}	1.05	250–750	Cu, Au	97
Cu-50 wt%Au	—	—	0.56	25–250	Cu, Au	97
Cu-80 wt%Au	$\ln \bar{D} = -9.7-12{,}689/T$	60.85×10^{-6}	1.09	250–750	Cu, Au	97
Cu-80 wt%Au	—	—	0.51	25–250	Cu, Au	97
CuAu average	$\ln \bar{D} = -11.1-12{,}306/T$	15×10^{-6}	1.06	250–750	Cu, Au	97

TABLE 15.3 Data of the AuCuSn System from Different Literature Sources (*Continued*)

Phase	$\ln D = \ln D_0 - E_a/(k_B T)$ $(D,\ cm^2/s;\ T,\ K)$	$D_0,\ cm^2/s$	$E_a,\ eV$	Temperature range, °C	Diffusing atom species	Ref.
		CuSnAu				
$Cu_xAu_xSn_x$	$\ln \tilde{D} = -13.7 - 8742\,T^{-1}$	1.1×10^{-6}	0.76	100–200	Cu, Sn, Au	117
$\zeta_{1,2}\!: (AuCu)_{7-9}Sn$	$\ln \tilde{D} = -5.5 - 11322\,T^{-1}$	4.07×10^{-3}	1.0	100–200	Cu, Sn, Au	117
$\eta_1\!: Cu_5AuSn_5$	$\ln \tilde{D} = -14.4 - 6788\,T^{-1}$	5.34×10^{-7}	0.58	100–200	Cu, Sn, Au	117
$\eta_1 + \eta_2\!: Cu_5AuSn + Cu_4Au_2Sn_5$	$\ln \tilde{D} = -16.5 - 6004\,T^{-1}$	68×10^{-9}	0.54	100–200	Cu, Sn, Au	117
$\eta_2 + \eta_3\!: Cu_4Au_2Sn_{5+}\ Cu_3Au_3Sn_5$	$\ln \tilde{D} = -7.82 - 9232\,T^{-1}$	400×10^{-6}	0.8	100–200	Cu, Sn, Au	117
$\eta_3\!: Cu_3Au_3Sn_5$	$\ln \tilde{D} = -10.06 - 9446\,T^{-1}$	42.6×10^{-6}	0.74	100–200	Cu, Sn, Au	117

Key: E_a = activation energy; D_{AU} = partial diffusion coefficient of gold; D = diffusion coefficient; $\tilde{D}$ = interdiffusion coefficient.

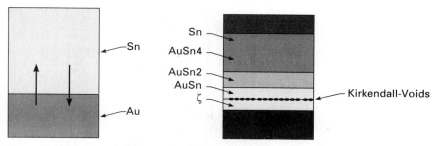

Figure 15.26 Intermetallic compounds and Kirkendall voids due to diffusion in an AuSn sandwich sample.

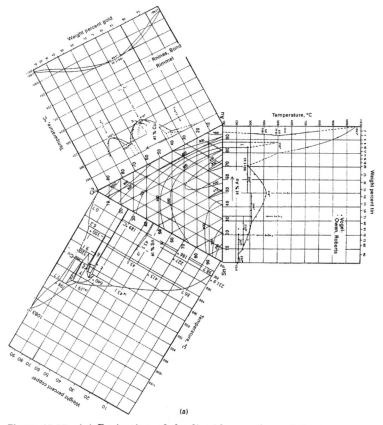

(a)

Figure 15.27 (a) Projection of the liquidus surface of the ternary AuCuSn system determined by differential scanning calorimetry (DSC) according to Ref. 117; binary systems according to Hansen.[92]

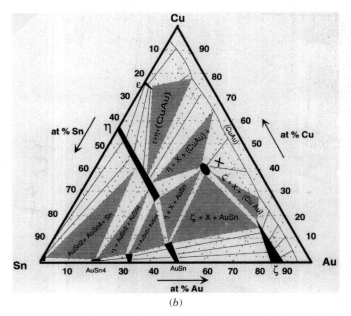

Figure 15.27 (*Continued*) (*b*) Isothermal cross section of the ternary AuCuSn system at 190°C according to Ref. 117.

bulk diffusion is taking place in the range from 200 to 250°C. The phases Cu_3Au, CuAu, and $CuAu_3$ are produced by diffusion within the concentration gradient. By the Kirkendall effect, voids are developing in the initial interface between Cu and Au.[95,96,98,100] The partial diffusion coefficient in copper-rich alloys is higher than in gold-rich alloys.

Because of the low diffusion coefficients in the temperature range below 200°C, no significant growth of the mentioned intermetallic compounds and no evidence of Kirkendall void formation was observed even after severe thermal ageing conditions of CuAu contacts for 2000 h at 200°C.[103]

15.3.2 The CuSn system

The phase diagram of CuSn is shown in Fig. 15.27*a*. Of all intermetallic compounds within the CuSn system,[92,93] only the phases Cu_6Sn_5 (η) and Cu_3Sn (ϵ) are stable below 350°C. The diffusion between Cu, Sn, and Sn/Pb alloys was investigated in numerous studies. Davis et al.[104,105] proved the formation of a multilayer structure from the phases Cu_6Sn_5 (η) and Cu_3Sn (ϵ) at the CuSn interface after tempering. These phases influence the solderability of the solder if an essential part of Sn has been consumed within these phases. The interdiffusion coefficients which were determined in different investigations show considerable deviations, especially in the lower temperature range. These deviations are attributed to the different defect densities of the investigated materials, due to different methods of manufacturing. The deviations of the interdiffusion coefficients determined by various authors are attributed to differences in Pb content of the solder,[104-110] to the method of solder deposition,[110] as well as to the method for determining the diffusion coefficient.[109,110]

Creydt[108] presented a comparison of the diffusion coefficients, and the activation energies determined by different authors show variations in the range of 0.65 to 0.82 eV. In a more recent investigation Kumar et al.[109,110] found a variation of 0.65 to 0.78 eV dependent on the lead content of the solder. The different diffusion coefficients and activation energies of both phases are evident in Table 15.3. No formation of Kirkendall voids was observed in these phases.

A schematic of the compound formation during diffusion in a CuSn sandwich sample is shown in Fig. 15.28.

15.3.3 The AuSn system

Creydt[108] investigated the diffusion in the AuSn system using AuSnCu layer systems. Because of the high Sn percentage in the test samples, only the Sn-rich phases $AuSn_4$ and $AuSn_2$ were formed. The diffusion coefficients are very high; within the investigated temperature range of 55 to 125°C the tin layer is consumed within only a few hours.[111,112]

Gregersen et al.[111] investigated the AuSn diffusion by RBS analysis. The original interface between Au and Sn was marked by implantation of argon ions. This method allowed the determination of the partial diffusion coefficients of gold and tin. Compared to tin, the partial diffusion coefficient of gold is higher by a factor of 3, thus supplying evidence of the Kirkendall effect within this system. The investigations are restricted to the Sn-rich part of the phase diagram in which the $AuSn_4$ and AuSn phases are found. Nakahara et al.[114] investigated the formation of voids due to the Kirkendall-effect. Tu and Rosenberg[116] demonstrated that by an interstitial diffusion mechanism some group Ib metals (copper, silver, gold) can diffuse rather quickly into group IIIb and IVb metals (indium, tin, lead). They also demonstrated that this diffusion involves a Kirkendall effect.[115]

Figure 15.26 shows the scheme of intermetallic compound formation and of Kirkendall void formation. In general the brittle intermetallic compounds $AuSn_4$, $AuSn_2$, and AuSn are critical for the mechanical properties of solder materials. A more critical effect for reliability is the formation of Kirkendall voids.

Carney et al.[40] investigated the Au-rich part of the phase diagram which includes Au-ζ and 80/20 AuSn, the growth of ζ below the liq-

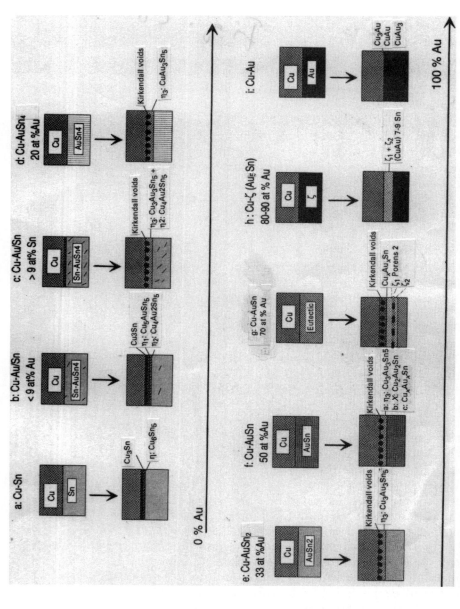

Figure 15.28 Scheme of phase formation and the interfaces of Kirkendall voids in sandwich samples between copper and AuSn alloys having different Au concentrations according to Ref. 117.

uidus temperature of the eutectic. The growth of the ζ phase at temperatures above 280°C by solid–liquid diffusion was investigated by Kallmayer et al.[50] No Kirkendall voids were observed in this part of the phase diagram.[36,37,40,50]

15.3.4   The AuCuSn system

A detailed investigation of the ternary system AuCuSn is presented in Ref. 117. It could be shown that the binary intermetallic compounds η (Cu_6Sn_5) and ζ (Au_5Sn) have a high solubility for the third element of the ternary system. The phase η forms stoichiometric variations: η_1, $Cu_5Au\,Sn_5$; η_2, $Cu_4Au_2\,Sn_5$; η_3, $Cu_3Au_3\,Sn_5$.[108] ζ also forms two ternary stoichiometric variations of the type $(CuAu)_{7-9}\,Sn$: ζ_1, $Au_{6-7}Cu_2\,Sn_{22}$; ζ_2, $Au_{24-25}Cu_5\,Sn_4$.[117,118] The other binary compounds of this system show only a small solubility for the third element. Figure 15.27a shows the projection of the liquidus surface of the AuCuSn system.

Figure 15.27b shows the isothermal cross section at 180°C with all the phases of the ternary system and the estimation of their solubility ranges. With the exception of one phase, the observed lattice parameters of the ternary alloys can be attributed to the binary compounds, which have a certain solubility for the third element forming mixed crystals with a ternary composition such as η and ζ. A new type of ternary intermetallic compound with independent lattice parameters and the stoichiometry $Cu_2Au_2\,Sn$ (phase X) was determined. The detailed data of phase X are described in Ref. 117.

The reliability-relevant aspects of this ternary system have been investigated using TAB ILB and OLB (inner and outer lead bond) samples.[118,120] It could be shown that a direct contact between copper and an AuSn alloy results in the formation of Kirkendall voids after the thermal aging if a critical Au concentration in the AuSn alloy of 10 to 90 wt% is reached. No Kirkendall voids are formed in the Sn-rich region below 10 wt% and in the Au-rich region above 90 wt%. Investigation of the degradation of adhesion in ILB and OLB samples show that the formed intermetallics do not cause a significant loss of adhesion. On the other hand, Kirkendall voids, which are formed as a secondary effect of diffusion, can cause catastrophic adhesion degradation and failures in the contacts. Figure 15.28 shows the compound types and the location of Kirkendall voids in the different concentration regions according to Ref. 117.

15.3.5   The AuSnPb system

For applications of solder joints between eutectic tin-lead alloy and gold surfaces, the liquidus temperature of the formed AuSnPb solder and the type of intermetallic compounds can be found in Fig. 15.29. It is well known that AuSn intermetallics in solder form acicular precip-

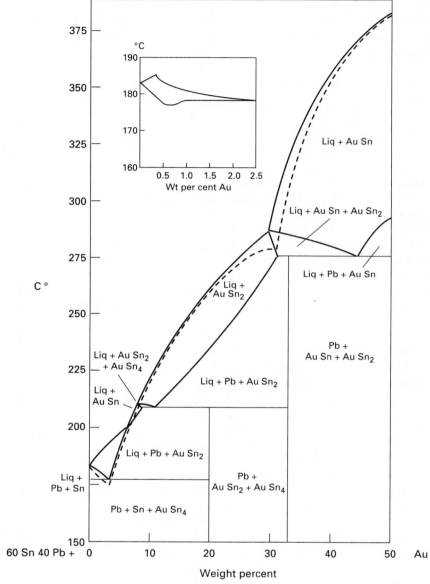

Figure 15.29   Pseudo-binary phase diagram between eutectic (63/37) SnPb + Au according to Reid and Goldie.[119]

itations and cause a significant reduction of the ductility of solder. According to Reid and Goldie,[119] the ductility is reduced to values near zero if a Au concentration of 8 to 10 wt% is reached in the solder.

Recent investigations in OLB contacts of TAB devices show that the presence of AuSn intermetallics in the fillet of the solder joints does not necessarily cause failure during thermal cycling and high-temperature storage. The failures found here with higher Au concentrations of >9 wt% are due to Kirkendall void formation and not to the effects of solder embrittlement.[4,20]

15.3.6   The AlAu system

Interdiffusion between the Al pad and the Au bump is important for mechanical stud bumps, which do not have a diffusion barrier. For electroplated Au bumps the bond process can induce cracks in the Ti:W diffusion leading to a local AlAu interdiffusion. This aspect can be observed for thermocompression bonding, which requires high bonding pressures and temperatures. The cracking behavior of the Ti:W barrier has been investigated in a series of publications on TC bonding of TAB devices.[121–124] The reliability aspects of AlAu in wire-bonded contacts are presented by Harman.[125] Further reading on Al–Au interdiffusion can be found in Refs 126–130.

15.4   Flip Chip Processes Using Electroplated Au, AuSn, and Electroless Ni/Au and Stud Au and 98/2 PbSn Bumps

This section presents some examples of the use of the five bump types: electroplated gold, electroplated AuSn, electroless NiP, stud Au, and stud 98/2 PbSn. The relevant data on the FC assembly processes are summarized in Table 15.1. For the FC assembly presented we must distinguish between processes requiring an individual bonding of each chip and aggregations of more than one chip in a reflow process. Figure 15.30 shows schematic characteristics of the two FC assembly processes.

The equipment for the first process requires high degrees of positioning accuracy and planarity between chip and substrate. It offers the possibility to perform a defined temperature force profile according to the applied FC joining method. For the collective solder reflow only a simple pick-and-place tool is required for positioning of the chip onto the substrate. For handling reasons, the chip must be fixed either by the use of flux or by tacking the chip onto the substrate with a minimal force and temperature. After this the chips are soldered in a reflow oven. Because of the self-alignment effect taking place during the solder reflow a high positioning accuracy can be achieved. Therefore, the first positioning step does not require a high degree of accuracy.

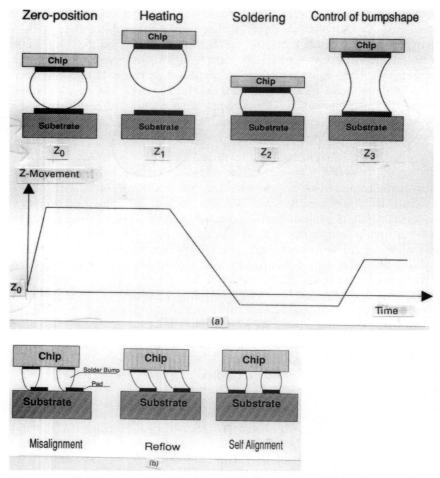

Figure 15.30 Single-chip bonding (a) and multiple-chip bonding (b) by collective reflow.

For testing solder joint quality during process evaluation, the use of test kits consisting of chips and corresponding substrates is recommended. For the evaluation of the FC process, electrical testing by using daisy-chain or four-point measurement structures can be performed. Mechanical testing of the whole chip is done by shear testing and vertical peel testing. These tests allow a general overview of the total adhesion of bump of a chip. A more detailed information on the adhesion of each bump with the possibility of a statistical evaluation is obtained by the shear test of single bumps. For this, the chip is selectively removed after the flip chip assembly. For a test chip with PSG passivation this is easily done using a 50% hydrofluoric acid solution.

Infrared microscopy is a simple nondestructive method which allows the evaluation of mechanical stress induced by the bonding process in the bond pad and in the semiconductor material.[131,132] This method requires a polished backside of the chip, which limits its applicability to specially prepared test chips.

15.4.1 Soldering processes using gold bumps

For this FC assembly, which is located in the tin-rich part of the phase diagram, Au is used as bump metallurgy and Sn—specifically 63/37 SnPb solder—as substrate metallization. Investigations were performed on both flexible circuits and rigid printed wiring boards. In contrast to the gold-rich 80/20 AuSn metallurgy presented in Sec. 15.4.2, the tin-rich metallurgy allows a reduction of the bonding temperature which avoids the softening of the polymer material. Some results of the use of this new type of metallurgy for FC bonding on polymeric substrates were presented by Tsukuda et al.[133] from IBM-Japan. Further investigations of the metallurgy and reliability were performed by IZM-Berlin.[47,134] The bump materials applied in these investigations are summarized in Table 15.4. They include both electroplated and mechanical stud bumps.

The material specifications of the rigid and flexible polymeric substrates are summarized in Table 15.5. The layout of the test chips and the corresponding test substrates includes a variation of chip size from 2.5, 5, 7.5, and 10 mm. The test kit includes chips with peripheral pads in 200 and 300-μm pitches and a test chip with an area bond pad configuration. For electrical measurements the chips have four-point measuring structures at the chip edges for quantitative measuring of the contact resistance and additional daisy chains in the middle

TABLE 15.4 Applied Bump Materials

Bump material	Type	Bump diameter, μm	Bump height, μm
Au	Electroplated	90	40
Au	Mechanical stud bump	80	50–60

TABLE 15.5 Applied Substrate Materials for FC Bonding of Electroplated and Mechanical Stud Au Bumps

Substrate	Material	Type	Cu thickness, μm	Final metallization, μm	Solder mask	Solder pad size, μm
1	Kapton E	Flexible	35	Sn 15	Yes	90×90
2	FR-4	Rigid	25	SnPb 25	Yes	100×100

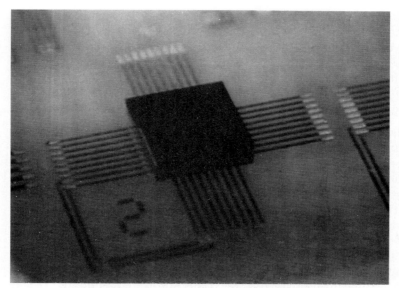

Figure 15.31 FC assembly of a 2.5-mm chip on a PWB substrate.

area of the chip for testing the electrical interconnections. Figure 15.31 shows a FC assembly of a 2.5-mm chip on the corresponding test substrate.

The stud bumps are produced by single-chip bumping using a modified semiautomatic thermosonic wire bonder from F&K Delvotec. Prior to the assembly, a slight coining is performed in order to obtain a uniform bump height. Figure 15.32 shows the SEM images of stud bumps.

The flip chip assembly is performed using the bonder FC 950 from Karl Suss with a tool temperature of 300°C and a bond force of 7 to 10 cN per pad. Figure 15.33 shows the result of the FC contact after selective removal of the chip on a flexible substrate of type 1 (Table 15.5). Because the applied interconnection metallurgy allows a reduction in soldering temperature, a reproducible contact without damaging the adhesive of the substrate material is possible. This is a significant improvement compared to the results obtained with the 80/20 AuSn metallurgy of the electroplated AuSn solder bumps (see Sec. 15.4.2).

Figure 15.34 shows the corresponding metallographic cross sections. The quantitative EDS analysis of the AuSn interface metallurgy shows that Sn is totally consumed and mainly the intermetallic compounds AuSn and $AuSn_2$ are formed. In the first reliability results of these contacts using an underfill from Hysol, type FP 4510, no failures occurred after ≤1000 thermal cycles from −55 to 125°C. These promising results indicate that this metallurgy can be considered for a fluxless FC assembly process.

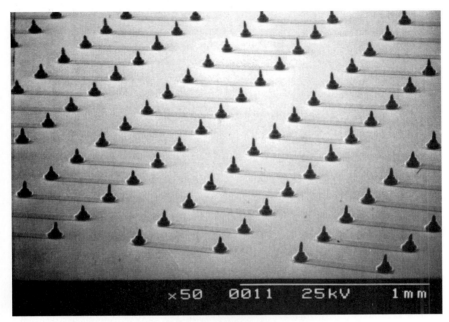

Figure 15.32 Area configuration of mechanical stud bumps.

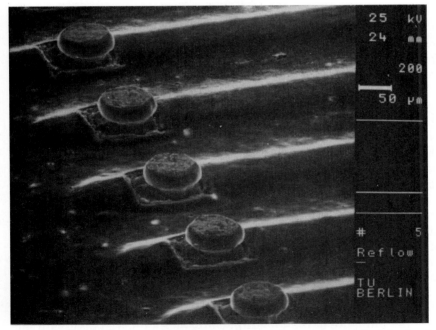

Figure 15.-33 SEM images of FC contacts using Au stud bumps and flexible circuits with adhesive after the selective etching of the chip (substrate type 1; see Table 15.5).

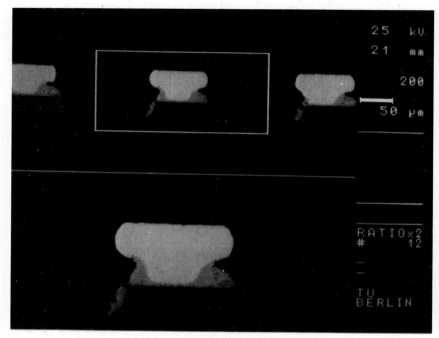

Figure 15.34   Cross sections of the contacts from Fig. 15.32.

Because of to the critical metallurgy discussed in Sec. 15.1, however, further detailed investigations are required in order to determine the reliability limitations of this metallurgy and to determine
failure models and life prediction.

Flip chip assembly on the FR-4 substrates is performed using electroplated Au bumps. Two assembly methods can be applied for this;
the first method uses a FC bonder with a heated tool and the second
one uses a reflow oven and nitrogen atmosphere. Figure 15.35 shows
the fluxless FC assembly of an area configuration with 572 bond
pads, which was successfully bonded with the FC 950 bonder at a tool
temperature of 300°C. For the second approach the chip is first positioned on the substrate using a placing machine. For handling reasons, a flux is necessary here. The chips are then reflowed in an oven
at 220°C in nitrogen atmosphere. Figure 15.36 shows a cross section
of the metallurgy obtained. At the interface between the gold bump
and the solder, the intermetallic compounds $AuSn$, $AuSn_2$, and $AuSn_4$
are formed in a sandwich-type structure.

For first estimation of the reliability of these contacts, different
chip sizes were FC-assembled for thermal cycling. In order to investigate the influence of the underfill, chips were prepared both with and
without underfill. The underfill material FP 4510 from Hysol was

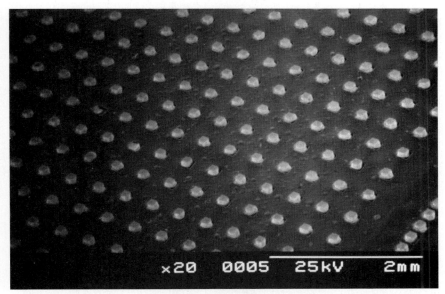

Figure 15.35 Area FC contacts with electroplated Au bumps on FR-4 substrates after the selective etching of the chip. The assembly is done without flux using the FC bonder.

Figure 15.36 Cross section of FC contact with electroplated bumps on FR-4 substrates with 25-μm 63/37 SnPb metallization. The assembly is done without flux using the FC bonder

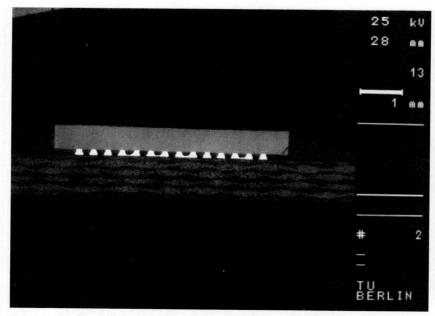

Figure 15.37 Cross sections of FC contacts on FR-4 material with the underfill FP 4510 from Hysol.

used. Figure 15.37 shows cross sections of the chips with the underfill and particle distribution of the filler material.

Thermal cycling was performed from −55 to 125°C with 10 min time at each temperature for up to 1000 cycles. During cycling a slight increase of contact resistance is observed, but no electrical failures were measured for the contacts even after 1000 cycles (Fig. 15.38). On the other hand, the chips without underfill show electrical failures already within the first 50 thermal cycles. These results indicate the importance of the underfill for the application of this new metallurgy.

15.4.2 Fluxless soldering using gold-tin solder bumps

The eutectic 80/20 AuSn alloy is suited for fluxless soldering. Because of the relatively high melting temperature of 280°C of the solder the applicable organic substrates are limited to materials with high T_g, such as the flexible substrate Espanex. This process requires gold or electroless NiAu as a final substrate pad metallization.[47] The measured shearforces and contact resistance values for the FC contacts on the applied thick-film pads are shown in Fig. 15.39. The mounting process is performed using a flip chip bonder with a tool temperature of 315°C and a bond pressure of 15 cN per bump. The bumps have an

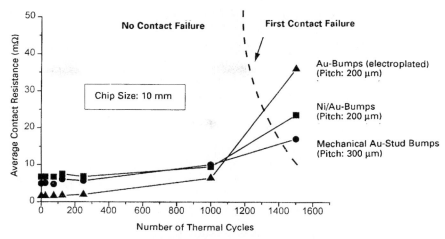

Figure 15.38 Electrical contact resistance for a 10-mm chip with Au bumps soldered to an FR-4 substrate (all chips were underfilled) during thermal cycling. No electrical failures were detected after 1000 cycles (-55 to $+125°C$).

octagonal pad size of 90 μm with a 25-μm Au socket and a 7- to 8-μm thick Sn cap. Prior to bonding, a reflow of the tin cap is performed according to the reflow process described in Sec. 15.2.2.

This bump metallurgy is also suited for fluxless soldering of alumina and green tape ceramic substrates. A fluxless FC-bonding process on thick-film patterns and via metallizations of green tape ceramic multilayers has been developed by IZM-Berlin.[41–45,135] The investigated thick-film metallizations are summarized in Table 15.6. The best mechanical and electrical results are achieved with AuSn bumps on PdAg thick-film metallizations. This system allows the highest adhesion and the lowest electrical contact resistance. The quality of these contacts is also evident in the low standard deviations shown in Fig. 15.40. The substrates consist of Du Pont's fired green tape of the 851A series. Figure 15.40 illustrates the applied test substrate layout. The reliability investigations of the FC contacts show that with a chip size of ≤5 mm the contacts can progress to 1000 cycles from -55 to 125°C. For higher chip sizes of 7.5 mm and 10 mm, an underfill must be applied. With the underfill Hysol FP 4511, 1000 cycles without electrical failures are obtained.

15.4.3   Flip chip soldering using electroless NiAu bumps

For this process, the solder must be applied on the substrate site. An evaluation of the reliability of these FC interconnections on FR-4-pointed wiring boards with electroplated eutectic SnPb solder pads (25 ±3 μm) is presented by Kloeser et al.[134] The bumps consist of 28-

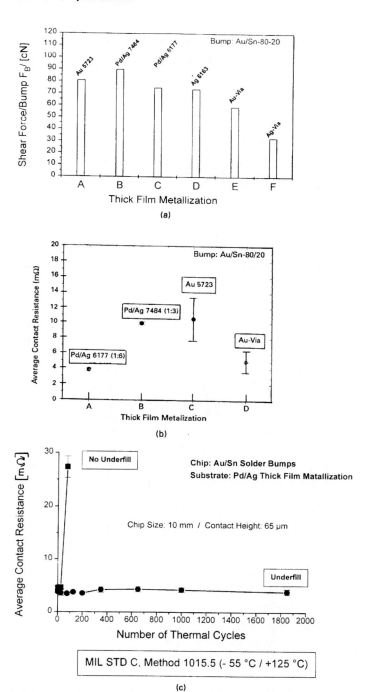

Figure 15.39 Shear force (*a*), electrical contact resistance (*b*), and example of the reliability performance (*c*) for fluxless FC joints on different thick-film metallizations.

TABLE 15.6   Substrate Metallizations for Fluxless FC Bonding and Specifications of the Applied Eutectic AuSn Solder Bumps.

No.	Pattern metallization	Vendor	Product name	Thickness, μm
A	Au	Du Pont	DP 5723	8–10
B	PdAg	Du Pont	DP 7484	9–11
C	PdAg	Du Pont	DP 6177T	6–8
D	Ag	Du Pont	DP 6163	10–12

	Via metallization	Vendor	Product name	Via diameter, μm
E	Au	Du Pont	5718D	200
F	Ag	Du Pont	6141D	200

Bump type	UBM	Bump metal
AuSn	200-nm Ti:W,	50-μm Au
Pad size: 90 μm, octagonal	200-nm Au	+
		7.5-μm, 8-μm Sn

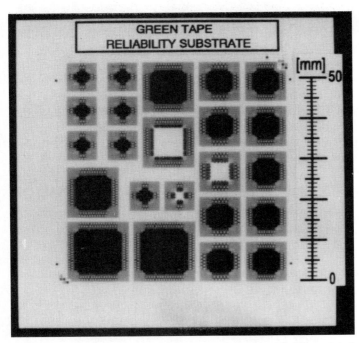

Figure 15.40 Green tape reliability test substrate with FC-mounted chips.

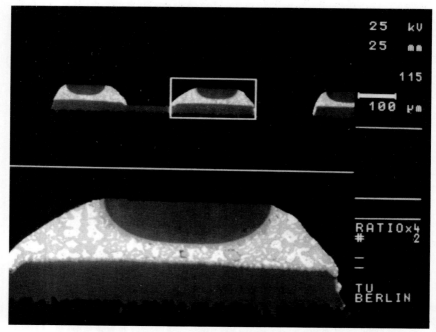

Figure 15.41   Cross section of a FC solder with electroless NiAu bumps on FR-4 boards.

μm electroless Ni with a 200-nm flash of Au. Flip chip bonding is performed by positioning the chip with a robotic pick-and-place machine by use of flux and a subsequent reflow process at 215°C in nitrogen atmosphere. For reliability testing, an underfill encapsulant is used. Figure 15.41 shows the cross section of this FC solder joint.

The underfill plays a key function in the long-term reliability of these contacts. With an adequate underfill, 1000 cycles without failure are reached for a 10-mm chip size. This is evident in Fig. 15.42 from the contact resistance as a function of the numbers of thermal cycles from −55 to 125°C.

15.4.4   Thermocompression bonding using gold bumps

The process requires gold bumps on the chip or the substrate and a correspondingly bondable, Au-plated surface. The use of electroplated Au bumps on aluminumoxide substrate for flip chip attach of GaAs devices has been reported.[135]

As a result of the hardness gradient of uncoined Au-stud bumps, having a low hardness in the upper region, adjacent to the TC bond interface, these bumps have properties for TC bonding. Because of the

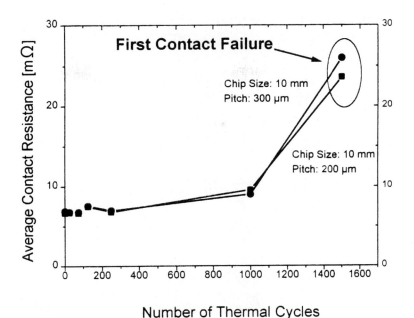

Figure 15.42 Electrical contact resistance for the electroless NiAu bumps soldered to FR-4 substrates.

required high bonding force of ≤100 cN for a 80-μm bump diameter and the bonding temperature of 300°C, the process is limited to rigid substrates such as aluminum or silicon. Additionally, the substrates must have a high planarity. For bonding a bonder with a high accuracy in the parallelism alignment is required. In order to avoid a predamaging of the Si or GaAs semiconductor material, the bonding force must be applied with a gradient. For brittle GaAs devices an optimal gradient of 5 to 10 cN/s was found. Figure 15.43 shows an example of a temperature–force profile. Figure 15.44 shows the change of shear force and of the bump geometry (height and diameter) as a function of the bonding force. Figure 15.45 shows an example of a bumped substrate with a minimal pitch of 70 μm and corresponding cross section of a FC attach to a GaAs FET (field-effect transistor).

15.4.5 Adhesive joining using gold and nickel-gold bumps

The bump metallurgies presented here allow the full range of possible adhesive joining processes. Figure 15.46 shows examples of anisotropic conductive adhesives, conductive adhesives, and nonconductive adhesives, in combination with electroplated gold, electroless nickel gold,

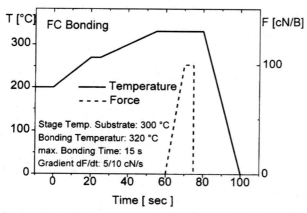

Figure 15.43 Example of flip chip bonding parameters for TC bonding of GaAs devices.

and stud gold bumps. The works performed by Aschenbrenner et al.[80,81] are focused on a new approach to flip chip interconnections using nonconductive thermosetting adhesives. The material applied is based on polymer materials used in anisotropic conductive adhesives.

The process advantage consists in the reduced material cost, because no conductive particles are needed. It is well known that flip chip technology increases the demand for a homogeneous particle distribution, size, and an increased conductivity in anisotropic conductive adhesives. This can be attributed to the small contact dimensions of the bumps compared to the contact pads of flexible circuits.

Nonconductive adhesive. The adhesive film investigated here consists of an insulating thermosetting/thermoplastic blend adhesive without conducting particles and fillers. The adhesive was fabricated in a dry-film format with film thicknesses of 25 and 50 μm. For repairing the interconnection, the whole flip chip joint is heated up to a temperature of 125°C in order to debond the IC. The detailed specifications and bonding parameters given by the manufacturer of this adhesive are summarized in Table 15.7.

Bonding process. The substrates used for flip chip assembly were 1-mm-thick PCBs (FR-4) and 25-μm flexible circuit (Espanex). The specifications of test chip and test substrate are summarized in Table 15.8. The conductor patterns consist of copper with a final metallization of electroless nickel gold. The PCB carries a solder mask as a finishing layer. The mask has no influence on the nonconductive adhesive process. The test vehicles carry interconnection tracks which allow monitoring of the electrical integrity by measuring the contact resistance (four-point probe) and transit resistance (daisy chain).

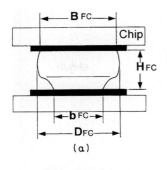

(a)

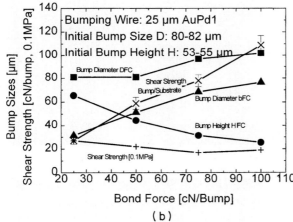

(b)

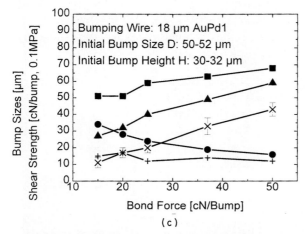

(c)

Figure 15.44 Bump size and shear strength as a function of flip chip bond force for initial bump diameters of 80 and 50 μm.

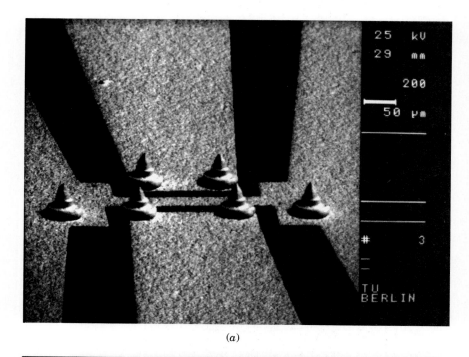

(a)

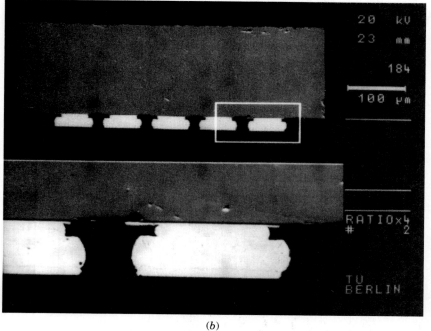

(b)

Figure 15.45 Flip chip bonding in minimal pitch of 70 μm using stud gold bumps: (a) SEM image of substrate bumping; (b) cross section of the flip chip attached GaAs-FET.

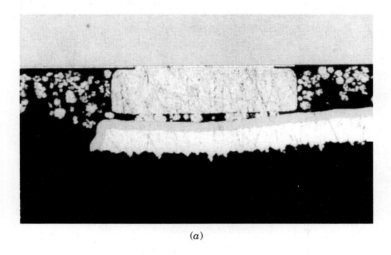

(a)

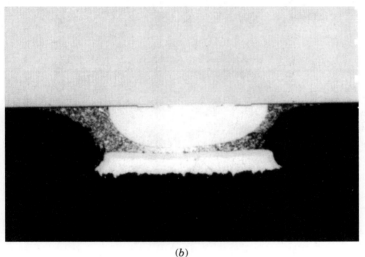

(b)

Figure 15.46 Cross sections of different adhesive flip chip processes with adapted bump metallization attached to printed-circuit boards (FR-4): (a) gold bump and anisotropically conductive adhesive; (b) electroless nickel bump and conductive adhesive.

The applied bump material for flip chip joining consists of stud gold bumps. The bump geometry, especially in the noncoined state, together with the low hardness in the upper tail region of the ball are very suitable for compensating substrate implanarities by the plastic deformation of the bumps. In this way, a nonplanarity of at least 10 μm over the chip size can be compensated. Figure 15.47 shows the geometry of the applied ball bumps. This is an important aspect especially

(c)

Figure 15.46 (*Continued*) (c) gold ball bump and nonconductive adhesive.

TABLE 15.7   Specifications of the Nonconductive Adhesive

Items	Specifications
Type	Thermosetting/thermoplastic film
Maximal operation temperature, °C	+ 125
Moisture absorption (85°C, 85%RH)	1.2%
Final bonding	20 s at 180°C; 20 kg/cm²

TABLE 15.8   Specification of Test Chip and Test Substrate

	Substrate 1	Substrate 2
Material	FR-4	FLEX (Espanex)
Size	2×2 in	2×2 in
Final metallization	CuNiAu	CuNiAu

IC	
Size	5.0×5.0 mm, 7.5×7.5 mm
I/Os	56, 84, 88, 136
Pitch	200 μm, 300 μm
Pad size	80 μm octagonal
Bumps	Au stud bumps

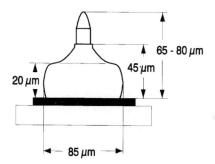

Figure 15.47 Ball bump geometry and dimensions.

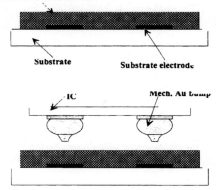

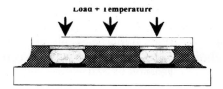

Figure 15.48 Schematic process flow for FC bonding with nonconductive adhesive.

for PWB, because increased planarity requirements can add considerable cost to this inexpensive polymeric material. For flexible circuits, electroplated Au bumps can also be applied.

The fundamental process flow for the nonconductive adhesive flip chip attachment is shown in Fig. 15.48. The process consists of three simple steps: first the nonconductive adhesive is applied to the substrate for the purpose of fixing the chip; then the gold ball bumps on the chip and the electrodes on the substrate are aligned; and finally, bonding the chip on the substrate with an appropriate load (≥ 20 kg/cm^2) and with a temperature of $\leq 180°C$ for 15 to 20 s is performed. Thus the IC is electrically connected to the substrate via compressed

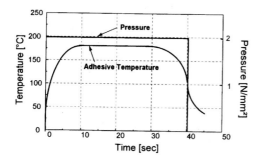

Figure 15.49 Bond parameters for FC assembly using a non-conductive adhesive.

and deformed gold ball bumps. The pressure on the bond site must be maintained until the chip is fixed by cooling the thermode. Figure 15.49 shows the schematic bond parameters for a bonding cycle. In the cross section of a gold ball bump attached to a PCB (Fig. 15.46c) an assembly force of 100 g per bump was applied. The diameter of the resulting contact area on the substrate is approximately 60 µm. To obtain a high reliability and interconnection stability in this bonding technology, a bonding load has to be selected by the thickness of the adhesive, by the bump height and by the parameters of the adhesives. The best results could be achieved with bond forces in the range of 80 to 100 cN per bump. Despite these high bond forces, no cracks occurred during bonding on either the chip or the substrate.

Reliability. The applied acceleration test method for thermal cycling involved 1000 cycles between −55 and +125°C, with a cycle duration of 30 min and constant humidity test at 85°C and at a relative humidity of 85% for up to 1000 h. After 1000 temperature cycles no open joints for the chips bonded to both substrates were observed. However, the degradation of the resistance of the 7.5-mm² chip increased slowly with the number of cycles, reaching a value twice as large as the initial condition (see Figs. 15.50a and 15.51a). The change in contact resistance can be attributed to the stress-induced contact spacing changes between the gold ball bumps and the electrodes of the substrate. In comparison with solder joints, where reliability problems arise from fatigue failures during thermal cyclings (CTE differences), adhesive joints tend to be more compliant and less susceptible to fatigue failure.

In contrast to temperature cycling, humidity testing caused higher contact resistance values (see Figs. 15.50b and 15.51b). The results for the flexible polyimide substrate (see Fig. 15.10b) exhibit a stability better than that of the printed wiring board. Possible causes for the high contact resistance are the change in volume when the adhesive swells in z-axis direction, due to moisture absorption or an oxidation effect.

An important point to obtain high reliability is to have compliance in the interconnection. This compliance is required to level out non-

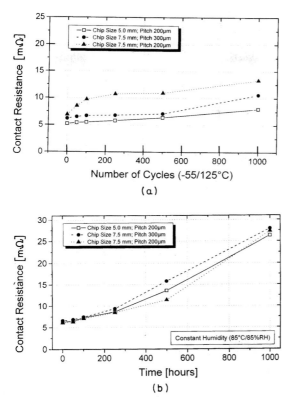

Figure 15.50 (*a*) Thermal cycle test results (−55 to 125°C) for nonconductive adhesive FC contacts on PCB (FR-4); (*b*) humidity test results (85°C, 85% RH) for nonconductive adhesive FC contacts on PCB (FR-4).

planarities of the substrate and thermal mismatch of the interconnection materials. Under operating conditions the thermal mismatch difference in z direction between the gold ball bump and the adhesive will expand the adhesive faster than the bumps. This leads to early failures when there is no compliance. Therefore the behavior of the flip chip interconnects drawing a temperature operating test was investigated. The specimen was heated up from 25 to 125°C, in increments of 25°C, and held at each temperature for measuring the contact resistance. Figure 15.52 shows the relationship between the contact resistance of the electrodes and the temperature. The resistance increases gradually for all devices from 5 mΩ to approximately 820 mΩ. No electrical contact failures were observed even above 125°C. The higher contact resistance disappeared when returned to room temperature.

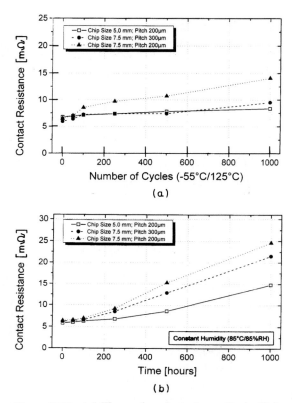

Figure 15.51 (*a*) Thermal cycle test results (−55 to 125°C) for nonconductive adhesive FC contacts on PCB; (*b*) humidity test results (85°C, 85% RH) for nonconductive adhesive FC contacts on PCB.

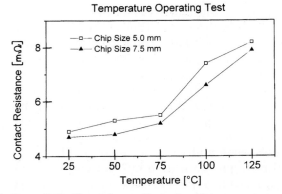

Figure 15.52 Dependence of transit resistance on temperature (substrate FR-4).

Adhesive Strength

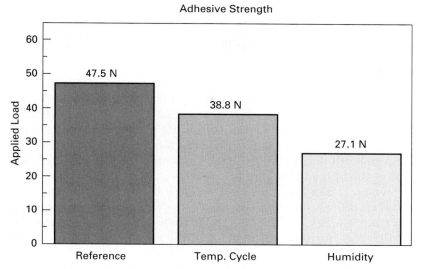

Figure 15.53 Peel strength of flip chip joints on FR-4 substrates before and after environmental tests.

The adhesive strength was evaluated by vertical tensile testing. The results in Fig. 15.53 show the influence of the environmental tests on the flip chip joint. For all samples the failure mode was an adhesion failure at the interface between the adhesive and the solder mask on the printed-circuit board. Cohesive failure was never observed in this test.

The results have shown that this bonding technique allows quality attachment of bare chips on low cost organic substrates.

15.5 Summary

It has been shown that the gold, gold-tin, and nickel-gold bump metallurgies cover the full range of possible flip chip interconnection techniques including soldering, thermocompression bonding, and adhesive joining. However, because these methods are quite new, the data available on the long-term reliability and degradation mechanisms thus far are limited to a relatively small number of publications. Because of the high flexibility and the potential for cost reduction, further effort will be done worldwide on these processes.

For soldering processes using the gold-tin metallurgy, the influence of Au–Sn interdiffusion and possible Kirkendall void formation on the long-term reliability of the FC contacts must be elaborated. The main advantages of this metallurgy consists in the possibility of a fluxless process and the self-alignment effect of the eutectic 80/20 AuSn alloy. The underfill plays a key role in the reliability performance of these FC interconnections; therefore, further work on adapted underfill materials is required.

For soldering processes using electroless nickel-gold bumps, the reliability of the nickel–aluminum interface and possible metallurgical reactions in the NiP matrix must be understood.

For the thermocompression process using gold bumps, the mechanical stress induced by the bond process and its impact on long-term reliability must be investigated. For the stud bumps, further work is required on the influence of Al–Au interdiffusion on the thermal fatigue behavior of the FC contacts.

Adhesive processes are quite new for FC interconnections. The material development on the adhesive site for the adapted joining processes plays a key role for the application of these processes. The degradation mechanisms taking place at the metal–polymer interface are of essential importance for the reliability of these contacts.

The reliability results presented in this chapter represent a summary of the data published so far. They show that the presented FC joining processes and metallurgies are promising alternatives to the standard FC soldering methods in the PbSn system.

References

1. Harman, G., "Wire Bonding towards 6σ Yield and Fine Pitch," *Proceedings of 42nd Electronic Components and Technology Conference* (ECTC) San Diego, p. 903, 1992.
2. Hamburgen, W. R., and J. S. Fitch, "Packaging a 150-W Bipolar ECL Microprocessor," *IEEE Transact. Components, Hybrids Manufact. Technol.,* **16**(1):28, Feb. 1993.
3. Atsumi, K., et al., "Inner Lead Bonding Technique for 500-Lead Dies Having a 90 μm Lead Pitch," *IEEE Transact. Components, Hybrids Manufact. Technol.,* **13**(1):222, 1990.
4. Zakel, E., J. Villain, and H. Reichl, "Reliability Improvement of OLB-Contacts with 75 μm Pitch by Optimization of the Gold Content," *Proceedings of 44th ECTC,* Washington, D.C., p. 883, 1994.
5. Matsuki, H., M. Yoshikawa, M. Minamizawa, T. Hamano, and K. Muratake, "High Density and High Performance A1N SPGA Package for VLSI," *Proceedings of the Technical Conference, International Electronics Packaging Conference, IEPS,* Atlanta, p. 308, Sept. 1994.
6. Miller, L. F., "Controlled Collapse Reflow Chip Joining," *IBM J. Research Develop.,* p. 239, May 1969.
7. Puttlitz, K. J., "Preparation, Structure, and Fracture Modes of Pb-Sn and Pb-In Terminated Flip Chips Attached to Gold-Capped Microsockets," *Proceedings of 40th ECTC,* Las Vegas, p. 360, 1990.
8. Ray, S. K., K. Beckham, and R. Master, "Flip-Chip Interconnection Technology for Advanced Thermal Conduction Modules," *Proceedings of 41st ECTC,* Atlanta, p. 772, 1991.
9. Hayden, T. F., and J. P. Partridge, "Practical Flip Chip Integration into Standard FR-4 Surface-mount Process," *International Flip Chip, TAB and Advanced Packaging Symposium,* San Jose (Calif.), p. 1, Feb. 1994.
10. Lau, J., T. Krulevitch, W. Schar, M. Heydinger, S. Erasmus, and J. Gleason, "Experimental and Analytical Studies of Encapsulated Flip Chip Solder Bumps

on Surface Laminar Circuit Boards," *Proceedings of NEPCON West Conference,* Anaheim, Calif., p. 184, 1993.

11. Totta, P., and K. Puttlitz, "Flip Chip Technology. Workshop on Flip Chip (C4), TAB and Ball Grid Arrays," *Proceedings of International Flip Chip, TAB and Advanced Packaging Symposium,* San Jose, Feb. 1994.

12. Giesler J., S. Machuga, G. O'Malley, and M. Williams, "Reliability of Flip Chip on Board Assemblies," *Proceeding of International Flip Chip, TAB and Advanced Packaging Symposium,* San Jose, p. 127, Feb. 1994.

13. Lowe, H.,"No-Clean Flip Chip Attach Process," *Proceedings of International Flip Chip, TAB and Advanced Packaging Symposium,* San Jose, p. 17, Feb. 1994.

14. Clementi, J., and G. Hill, "Flip Chip Encapsulation of Small C4's on Fine Pitches," *Proceedings of 1st International Flip Chip Symposium and 6th International TAB/Advanced Packaging Symposium, ITAP,* San Jose, p. 9, Feb. 1994.

15. O'Malley, G., J. Giesler, and S. Machuga, "The Importance of Material Selection for Flip Chip on Board Assemblies," *Proceedings of 44th ECTC,* Washington, D.C., p. 387, 1994.

16. Khadpe, S., "A Global Overview of Technology and Market Trends in TAB/Advanced Packaging," *Proceedings of 5th International TAB and Advanced Packaging Symposium, ITAB,* San Jose, pp. 1–8, 1993.

17. Khadpe, S., "An Overview of Technology and Market Trends in TAB and Flip Chip," *Proceedings of 1st International Flip Chip Technology and 6th International TAB/Advanced Packaging Symposium,* San Jose, p. 170, Feb. 1994.

18. Angelucci, T., "Wafer Bumping," in *Handbook of Tape Automated Bonding,* Lau, J., ed., Van Nostrand Reinhold, New York, 1992.

19. Meyer, D., A. S. Kohli, and H. Reis, "Metallurgy of Ti-W/Au/Cu System for TAB," *J. Vacuum Sci. Technol.,* **A3**(3):772–776, May/June 1985.

20. Fu, K. Y., and R. E. Plye, "On the Failure Mechanism of Titanium Nitride/Titanium Silicide Barrier Contacts under High Current Stress," *IEEE Transact. Electron. Devices,* **35**(12):2151–2159, Dec. 1988.

21. Engelmann, G., G. Azdasht, A. Eberle, O. Ehrmann, and R. Leutenbauer, "Alternative Bump Configurations and Low Stress Bonding," *Proceedings of 3rd International TAB Symposium ITAB,* San Jose, pp. 157–170, 1991.

22. Ehrmann, O., G. Engelmann, J. Simon, and H. Reichl, "A Bumping Technology for Reduced Pitch," *Proceedings of 2nd International TAB Symposium ITAB,* San Jose, p. 41, 1990.

23. Tovar, D., "Optimization of a Gold Bump Process by Statistically Designed Experimentation," *Proceedings of 3rd International TAB Symposium ITAB,* San Jose, pp. 1–15, 1991.

24. Nowicki, R. S., J. M. Harris, M. A. Nicholet, and I. V. Mitschell, "Studies on the TiW/Au Metallization on Aluminium," *Thin Solid Film,* **53**:195–205, 1978.

25. Nicholet, M. A., "Diffusion Barriers in Thin Films," *Thin Solid Film,* **52**:415–442, 1978.

26. Engelmann, G., O. Ehrmann, J. Simon, and H. Reichl, "Development of a Fine Pitch Bumping Process," *Proceedings of 1st Microsystems Technologies,* Reichl, H., ed., Springer-Verlag, Berlin, pp. 435–440, 1990.

27. Reichl, H., "Untersuchung der TAB-Technologie für die Kontaktierung von VLSI-Schaltkreisen," *Abschlußbericht zum F&E-Vorhaben, BMFT Projekt* 13AS0053, Feb. 1992.

28. Simon, J., "Plating of Gold for Semiconductor Applications," *INTERFINISH,* São Paolo (Brazil), pp. 183–192, 1992.

29. Wang, T. C., and D. Tovar, "Gold Bump Control Characterization Using Statistical Design of Experiment," *Proceedings of 4th International ITAB Symposium ITAB,* San Jose, pp. 249–259, 1992.

30. Liu, T. S., W. R. Rodrigues de Miranda, J. M. Montante, and P. R. Zipperlin, "A Review of Wafer Bumping for Tape Automated Bonding," *Solid State Technol.,* pp. 71–76, March 1980.

31. Chikawa, Y., K. Mori, S. Sasaki, T. Maeda, and M. Hayakawa, "Bumping Technology for Tape-Automated Bonding," *Proceedings of IMC,* Tokyo, pp. 227–232, 1984.

32. Engelmann, G., O. Ehrmann, R. Leutenbauer, H. Schmitz, and H. Reichl,

"Fabrication of Perfectly Three-Dimensional Microstructures by UV Depth Lithography," *Proceedings of Reprint from Laser-Assisted Fabrication of Thin Films and Microstructures,* SPIE-Society of Photo-Optical Instrumentation Engineers, vol. 2045, Washington, D.C., 1994.

33. Engelmann, G., O. Ehrmann, J. Simon, H. Reichl, "Fabrication of High Depth-to-Width Aspect Ratio Microstructures," *Micro-Electro-Mechanical Systems '92,* Travemünde, p. 93, Feb. 1992.

34. Zakel, E., J. Simon, and S. Schuler, "Neue Bumping-Verfahren für die TAB-Technologie," *Verbindungstechnik in der Elektronik,* vol. 2, p. 53, 1990.

35. Zakel, E., S. Schuler, and J. Simon, "The Application of an Eutectic Gold-Tin Cushion for TAB-Inner Lead Bonding Pressure," *Proceedings of 1st Microsystems Technologies Conference,* Reichl, H., ed., Springer-Verlag, Berlin, p. 400, 1990.

36. Zakel, E., J. Simon, and H. Reichl, "A Reliable Au-Sn Solder Bump Process for TAB Inner Lead Bonding with Reduced Bonding Pressure," *Proceedings of Japan IEMT Symposium,* Tokyo, p. 85, 1991.

37. Zakel, E., J. Simon, G. Azdasht, and H. Reichl, "Gold-Tin Bumps for TAB Inner Lead Bonding with Reduced Bonding Pressure," *Soldering & Surface Mount Technol.,* (12):27, Oct. 1992.

38. Steckhan, H. H., "TAB Inner Lead Bonding with Gold/Tin Solder Bump Technology," VLSI and GaAs-Packaging Workshop, Westford (Mass.), 1990.

39. Field, D. A., "A Tin Based TAB Assembly Process," *Proceedings of 11th IEMT Symposium,* San Francisco, pp. 31–35, 1991.

40. Carney, F., G. Carney, J. Heckman, M. Nagarkar, J. Schaper, M. Seddon, and E. Woolsey, "Development and Characterization of Tin Capped Gold Bumps in TAB," *Proceedings of SUR/FIN '94,* Indianapolis, 1994.

41. Zakel, E., J. Kloeser, J., Simon, and F. Bechthold, "Development of Flip-Chip Bonding Techniques on Green Tape Ceramic Substrates," *Proceedings of 3rd Microsystems Technologies,* Reichl, H., ed., VDE Verlag Berlin, p. 89, 1992.

42. Kloeser, J., E. Zakel, G. Engelmann, H. Distler, and H. Reichl, "Investigations of Flip-Chip Interconnections on Green Tape Ceramic Substrates," *Proceedings of 5th International TAB/Advanced Packaging Symposium, ITAP,* San Jose, p. 131, 1993.

43. Kloeser, J., E. Zakel, G. Engelmann, H. Distler, and H. Reichl, "Untersuchungen einer Flip-Chip-Bond-Technologie auf green Tape Keramik-Multilayer Substraten," *SMT/ASIC/Hybrid Konferenz,* Nürnberg, p. 521, 1993.

44. Kloeser, J, E. Zakel, F. Bechthold, W. Distler, and R. Reichl, "Development of Flip-Chip Attach Technologies on Green Tape Ceramic Substrates," *Microsystem Technologies '94,* Reichl, H., and A. Heuberger, eds., VDE-Verlag gmbh, Berlin, pp. 521–534, 1994.

45. Zakel, E., J. Kloeser, H. Distler, and H. Reichl, "Flip-Chip Bonding on Green Tape Ceramic Substrates," *Proceedings of the 9th European Hybrid Microelectronics Conference,* Nizza, p. 339, 1993.

46. Baggermann, A. F. J., and M. J. Batenburg, "Reliable Au-Sn Flip Chip Bonding on Flexible Prints," *Proceedings of 44th ECTC,* Washington, D.C., p. 900, 1994.

47. Zakel, E., J. Gwiasda, J. Kloeser, J. Eldring, G. Engelmann, and H. Reichl, "Fluxless Flip Chip Assembly on Rigid and Flexible Polymer Substrates Using the Au-Sn Metallurgy," *Proceedings of IEMT Symposium,* San Diego, p. 177, 1994.

48. Gupta, D., "Application of Thermal and Kinetic Modelling to the Reflow and FC-Bonding of Bumps Using Au-Sn-Au-Metallurgy," *Proceedings of 5th International Symposium on Flip Chip, TAB and Ball Grid Arrays,* San Jose, Feb. 1995.

49. Oppermann, H., E. Zakel, G. Engelmann, and H. Reichl, "Investigation of Self-Alignment During the Flip-Chip Assembly Using Eutectic Gold-Tin Metallurgy," *Microsystem Technologies '94,* Reichl, H., and A. Heuberger, eds., VDE-Verlag gmbh, Berlin, pp. 509–519, 1994.

50. Kallmayer, Ch., H. Oppermann, E. Zakel, G. Engelmann, and H. Reichl, "Experimental Results on the Self-Alignment Process using Au/Sn Metallurgy and on the Growth of the ζ-Phase during the Reflow," *Proceedings of 5th International Symposium on Flip Chip/TAB/Ball Grid Arrays,* San Jose, 1995.

51. Gellings, P. J., *Introduction to Corrosion Prevention and Control for Engineers,* Nijgh-Wolters-Noordhoff Universitaire Uitgevers B.V., Rotterdam, 1976.

52. Yamakawa, K., M. Inaba, and N. Iwase, "Maskless Bumping by Electroless

Plating for High Pin Count, Thin and Low Cost Circuits," *ISHM '89 Proceedings,* Baltimore, p. 620, 1989.

53. Uchida, M., K. Nozawa, and Y. Karasawa, "Interconnective Technology with Metallization for LCD," *Proceedings of 1991 Japan International Electronic Manufacturing Technology Symposium,* Tokyo, p. 97, 1991.

54. Inaba, M., K. Yamakawa, and N. Iwase, "Solder Bump Formation Using Electroless Plating and Ultrasonic Soldering," 5th International Electronic Manufacturing Technology Symposium IEEE/CHMT, 1988.

55. Simon, J., and H. Reichl, "Single Chip Bumping for TAB," *Proceedings of 2nd International TAB-Symposium ITAB '90,* San Jose, p. 49, 1990.

56. Simon, J., E. Zakel, and H. Reichl, "Electroless Deposition of Bumps for TAB-Technology," *Proceedings of 40th ECTC,* Las Vegas, p. 412, 1990.

57. Simon, J., E. Zakel, and H. Reichl, "Electroless Deposition of Bumps for TAB-Technology," *Metal Finishing,* p. 23, Oct. 1990.

58. Zakel, E., J. Simon, and H. Reichl, "Die Verwendung des Zinkatverfahrens zur stromlosen selektiven Erzeugung von Nickelhöckern für die TAB-Kontaktierung von integrierten Schaltungen," *DGO-Tagung,* Trier (Germany), p. 141, 1990.

59. Ostmann, A., J. Simon, and H. Reichl, "The Pretreatment of Aluminum Bondpads for Electroless Nickel Bumping," *Proceedings of IEEE Multi-Chip Module Conference MCMC-93,* Santa Cruz (Calif.), p. 74, March 1993.

60. Ostmann, A., R. Aschenbrenner, U. Beutler, and H. Reichl, "Alternative Metallization by Low Cost Electroless Plating for MCM Applications," *Workshop on MCM and VLSI Packaging Techniques,* Windsor, 1994.

61. Aschenbrenner, R., A. Ostmann, U. Beutler, J. Simon, and H. Reichl, "Electroless Nickel/Copper Plating as a New Bump Metallization," *3rd International Conference on MCM,* Denver, April 1994.

62. van der Putten, A. M. T., and J. W. G. de Bakker, "Geometrical Effects in the Electroless Metallization of Fine Metal Patterns," *J. Electrochem. Soc.,* 140(8):2221, August 1993.

63. Liu, J., "Development of a Cost-effective and Flexible Bumping Method for Flip-Chip Interconnections," *Hybrid Circuits,* (29), Sept. 1992.

64. Aintila, A., E. Järvinen, and S. Lalu, "Towards Low Cost High Density Bumping," *Proceedings IEMT,* Kanazawa (Japan), p. 33, 1993.

65. Aintila, A., A. Björklöf, E. Järvinen, and S. Lalu, "Electroless Ni/Au Bumps for Flipchip-on-Flex and TAB Applications," *Proceedings of 16th IEEE International Electronics Manufacturing Technology (IEMT) Symposium,* vol. 1, La Jolla (Calif.), p. 160, Sept. 1994.

66. Oberson, U., L. Belanger, and A. J. Brouillette, "Low Cost Bumping Process for Flip Chip," *Proceedings of ISHM,* Canada, 1994.

67. Jung E., J. Eldring, J. Kloeser, E. Zakel, and H. Reichl, "Flip Chip Soldering on PWBs Using Vapor Phase Reflow," *Proceedings of 5th International Symposium on Flip Chip/TAB/Ball Grid Arrays,* San Jose, 1995.

68. Treutler, C. P., "Dünnschicht-Multilayer-Technik und Flip-Chip Montage für optoelektrische Mikrosysteme," *Deutsche ISHM Konferenz,* Munich, 1993.

69. Azdasht, G., Germany, Patent P4200492.6-34.

70. Mallory, G. O., "The Relationship between Stress and Adhesion of Electroless Nickel-Phosphorus Deposits on Zincated Aluminum," *Plating and Surface Finishing,* pp. 86–95, June 1985.

71. Oyama, K., "Overview of Placement and Bonding Technology for 1005 Components and 0,3 mm-Pitch LSIs: An Equipment Perspective," *SMT Recent Japanese Developments,* IEEE Press, New York, p. 77, 1992.

72. Montgomery, C., "A Low Cost High Performance Interconnection Technique for GaAs Devices," *Abstracts of IEEE VLSI and GaAs Chip Packaging Workshop,* Scottsdale (Ariz.), 1991.

73. Montgomery, C., "Flip Chip Assemblies Using Conventionally Wire Bonding Apparatus and Commercially Available Dies," *Proceedings of ISHM,* Dallas, p. 451, 1993.

74. Goodman, C. E., and M. Metroka, "A Novel Multichip Module Assembly Approach Using Gold Ball Flip-Chip Bonding," *IEEE Transact. Components, Hybrids, Manufact. Technol.,* 15(4):457, 1992.

75. Kondoh, Y., T. Togasaki, and M. Saito, "A Flip Chip Interconnection Technique Using Indium Alloy Bumps for a Newly Designed CCD Module," *Proceedings of IMC,* Yokohama, p. 120, 1992.
76. Goodman, C. E., and M. Metroka, "A Novel Multichip Module Assembly Approach Using Gold Ball Flip-Chip Bonding," *IEEE Transact. Components, Hybrids, Manufact. Technol.,* **15**(4):457, 1992.
77. Bonkohara, et al., "Utilization of Inner Lead Bonding Using Ball Bump Technology," *Proceedings of ITAB,* San Jose, p. 86, 1992.
78. Larsen, E. N., and M. J. Brock, "Development of a Single Point Gold Bump Process for TAB Applications," *Proceedings of ICEMM,* p. 391, 1993.
79. Eldring, J., E. Jung, R. Aschenbrenner, E. Zakel, and H. Reichl, *Abstracts of SMT/ES&S-ASIC/Hybrid '95,* Nürnberg, 1995.
80. Aschenbrenner, R., J. Eldring, H. Reichl, J. Gwiasda, and E. Zakel, "Flip Chip Attachment Using Non-Conductive Adhesives and Gold Ball Bumps," *Proceedings of the Technical Conference, International Electronics Packaging Conference, IEPS,* Atlanta, p. 794, Sept. 1994.
81. Aschenbrenner, R., J. Gwiasda, J. Eldring, E. Zakel, and H. Reichl, "Gold Ball Bumps for Adhesive Flip Chip Assembly," *Adhesives in Electronics '94,* VDI/VDE Tagung, Berlin, Nov. 1994.
82. Eldring, J., E. Zakel, and H. Reichl, "Flip Chip Attach of Silicon and GaAs Fine Pitch Devices as Well as Inner Lead TAB Attach Using Ball Bump Technology," *Proceedings of the Technical Conference, International Electronics Packaging Conference, IEPS,* Vol. 2, San Diego, p. 304, Sept. 1993.
83. Eldring, J., E. Zakel, and H. Reichl, "Flip Chip Attach of Silicon and GaAs Fine Pitch Devices as Well as Inner Lead TAB Attach Using Ball-bump Technology," *Hybrid Circuits,* (34):20, May 1994.
84. Eldring, J., E. Zakel, G. Procoph, and H. Reichl, "Flip Chip Attachment Using Mechanical Bumps," Proceedings of 1st International Flip Chip Symposium and 6th International TAB/Advanced Packaging Symposium, ITAP, San Jose, p. 74, Feb. 1994.
85. Goldenberg, T., et al., "Ball Bumping-Procedure," U.S. Patent 4,717,066, 1988.
86. Eldring, J., E. Jung, R. Aschenbrenner, E. Zakel, and H. Reichl, "Mechanisches Bumping für die Flip Chip Technologie," *Proceedings of SMT/ES&S-ASIC/Hybrid '95,* 1995.
87. Leers, U., *SMD Trends und Technologien,* Brochure of PANASONIC Factory Automation Europe, 1993.
88. Elenius, P., "Au Bumped Known Good Die," *Proceedings of 1st International Flip Chip Symposium and 6th International TAB/Advanced Packaging Symposium, ITAP,* San Jose, p. 94, Feb. 1994.
89. Suwa, M., H. Takahashi, C. Kamada, and M. Nishiuma, "Development of a New Flip-Chip Bonding Process Using Multi-Stacked μ-Au Bumps," *Proceedings of 44th ECTC,* vol. 1, Washington, D.C., p. 906, May 1994.
90. Ogashiwa, C. R., et al., "Direct Solder Bump Formation Technique on Al Pad and Its High Reliability," *Jpn. J. Appl. Phys.,* **31**:761, 1992.
91. Scharr, T. A., R. K. Sharma, R. T. Lee, and W. H. Lytle, "Wire bumping Technology for Gold and Solder Wire Bumping on IC Devices," *Proceedings of 9th ISHM,* Nice, p. 351, 1993.
92. Hansen, M., *Constitution of Binary Alloys,* Genium Publishing, New York, 1985.
93. Massalasky, T. B., H. Okamoto, P. R. Subramanian, and L. Kacprzak, *Binary Alloy Phase Diagrams,* vols. I–III, 2nd ed., ASM International (USA), p. 11, 1990.
94. Butrymwicz, D. B., J. R. Mannig, and M. E. Read, "Diffusion in Copper and Copper Alloys," *J. Phys. Ref. Data,* **3**(2):527–602, 1974.
95. Pinnel, M. R., "Diffusion-Related Behaviour of Gold in Thin Film Systems," *Gold Bulletin,* (Johannesburg), **12**(2):62–71, April 1979.
96. Pinnel, M. R., and J. E. Bennett, "On the Formation of the Ordered Phases CuAu and Cu_3Au at a Copper/Gold Planar Interface," *Metal Transact. A,* **10A**:741–747, June 1976.
97. Pinnel, M. R., and J. E. Bennett, "Mass Diffusion in Polycrystalline Copper/Electroplated Gold Planar Couples," *Metallurg. Transact.,* **3**:1989–1997, July 1972.

98. Tompkins, H. G., and M. R. Pinnel, "Low Temperature Diffusion of Copper through Gold," *J. Appl. Phys.*, **47**:3804–3812, 1976.
99. Hall, P. M., J. M. Morabito, and N. T. Panousis, "Interdiffusion in the Cu-Au Thin Film System at 25°C to 250°C," *Thin Solid Films*, **41**:341–361, 1979.
100. Feinstein, L. G., and J. B. Bindell, "The Failure of Aged Cu-Au Thin Films by Kirkendall Porosity," *Thin Solid Films*, **62**:37–47, 1979.
101. Hall, P. M., N. T. Panousis, and P. R. Menzel, "Strength of Gold-Plated Copper Leads on Thin Film Circuits under Accelerated Ageing," *IEEE Transact. Parts, Hybrids, Packag.*, **11**(3):202–205, 1975.
102. Panousis, N. T., "Thermocompression Bondability of Bare Copper Leads," *IEEE Transact. CHMT*, **1**(4):372–377, 1978.
103. Zakel, E., and H. Reichl, "Investigations of Failure Mechanisms of TAB-Bonded Chips during Thermal Ageing," *IEEE Transact. CHMT*, **13**(4):856, 1990.
104. Davis, P. E., M. E. Warwick, and S. T. Mucket, "Intermetallic Compound Growth and Solderability of Reflowed Tin and Tin-Lead Coatings," *Plating Surface Finishing*, **70**:49–53, 1983.
105. Davis, P. E., M. E. Warwick, and P. J. Kay, "Intermetallic Compound Growth and Solderability," *Plating Surface Finishing*, **69**:72–76, 1982.
106. Unsworth, D. A., and C. A. Mackay, "A Preliminary Report on the Growth and Compound Layer on Various Metal Bases Plated with Tin and its Alloys," *Transact. Inst. Metal Finishing*, **51**:85–90, 1973.
107. Kay, P. J., and C. A. Mackay, "The Growth of Intermetallic Compounds on Common Basis Materials Coated with Tin and Tin-Lead Alloys," *Transact. Inst. Metal Finishing*, **54**:68–72, 1976.
108. Creydt, M., *Diffusion in galvanisch aufgebrachten Schichten und Weichloten bei Temperaturen zwischen 23°C und 212°C,* dissertation E.T.H. Zürich, 1971.
109. Kumar, K., and A. Moscaritolo, "Optical and AUGER Microanalysis of Solder Adhesion Failures in Printed Circuit Boards," *J. Electrochem. Soc.*, pp. 379–383, 1981.
110. Kumar, K., A. Moscaritolo, and M. Brownawell, "Intermetallic Growth Dependence on Solder Composition in the System Cu-(Sn-Pb-Solder)," *J. Electrochem. Soc.*, 2165, 1982.
111. Gregersen, D., L. Buene, T. Finstad, O. Lonsjo, and T. Olsen, "A Diffusion Marker in Au/Sn Thin Films," *Thin Solid Films*, **78**:95, 1981.
112. Buene, L., "Interdiffusion and Phase Formation at Room Temperature in Evaporated Gold," *Thin Solid Films*, **47**:159, 1977.
113. Buene, L., T. Finstad, K. Rimstad, and T. Olsen, "Alloying Behaviour of Au-In and Au-Sn Films on Semiconductors," *Thin Solid Films*, **34**:149–152, 1976.
114. Nakahara, S., R. J. McCoy, L. Buele, and J. M. Vandenberg, "Room Temperature Interdiffusion Studies of Au/Sn Thin Film Couples," *Thin Solid Films*, **84**:185–196, 1981.
115. Nakahara, S., and R. J. McCoy, "Kirkendall Void Formation in Thin-Film Diffusion Couples," *Appl. Phys. Lett.*, **37**:42, 1980.
116. Tu., K. N., and R. Rosenberg, "Room Temperature Interaction in Bimetallic Thin Film Couples," *Jpn. J. Appl. Phys.*, suppl. 2, pt. 1, p. 633, 1974.
117. Zakel, E., *Untersuchung von Cu-Sn-Au- und Cu-Sn-Metallisierungssystemen für die TAB Technologie,* thesis, Technical University, Berlin, 1994.
118. Zakel, E., and H. Reichl, "Au-Sn Bonding Metallurgy and Its Influence on the Kirkendall Effect in the Ternary Cu-Au-Sn System," *Proceedings of 42nd ECTC,* San Diego, p. 360, 1992.
119. Reid, F. H., and W. Goldie, *Gold Plating Technology, Electrochemical Publications Ayr,* San Diego, pp. 225–246, 1974.
120. Zakel, E., G. Azdasht, and H. Reichl, "Degradation of TAB Outer Lead Contacts Due to the Au-Concentration in Eutectic Tin/Lead Solder," *IEMT Symposium,* Japan, 1993.
121. Zakel, E., R. Leutenbauer, and H. Reichl, "Reliability of Thermally Aged Au and Sn Plated Leads for TAB Inner Lead Bonding," *Proceedings of 41st ECTC,* Atlanta, pp. 866–876, 1991.
122. Baggermann, A. J. F., and F. J. H. Kessels, "Hardness Reduction of Au Bumps for TAB Interconnections," *Proceedings of European IEMT Symposium,* Mainz (Germany), pp. 229–236, 1992.

123. Baggermann, A. J. F., "TAB Inner Lead Gang Bonding on Ni-Au Bumps," *Proceedings of 44th ECTC,* Washington, D.C., p. 938, 1994.
124. Beutler, U., E. Zakel, and H. Reichl, "Long Term Reliability of Inner Lead TAB Contacts—A Comparison between Gang and Single Point Bonding," *Proceedings of the International TAB and Flip Chip Symposium,* San Jose, 1994.
125. Harman, G. G., *Reliability and Yield Problems of Wire Bonding in Microelectronics, ISHM Monograph,* Reston, Va., 1989.
126. Majni, G., C. Nobili, G. Ottavani, M. Costato, and E. Galli, "Gold-Aluminium Thin Film Interactions and Compound Formation," *J. Appl. Phys.* **52:**4047, 1981.
127. Takei, W. J., and M. H. Francobe, "Measurement of Diffusion Induced Strains at Metal-Bond Interfaces," *Solid State Electron.,* **11:**205, 1967.
128. Kato, H., "Formation of Intermetallic Compounds in Al Thin Films Vapor-Deposited on Au-Plate," *Jpn. J. Appl. Phys.,* **26:**1786, 1987.
129. Philofsky, E., "Intermetallic Formation in Gold-Aluminum System," *Solid State Electron.,* **13:**1391, 1970.
130. Philofsky, E., "Design Limits When Using Gold-Aluminum Bonds," *Proceedings of 9th IRPS Conference,* p. 11, 1971.
131. Alpern, P., and R. Tilgner, "IR Microscopic Observation of Plastic Coated TAB Inner Lead Bonds Degrading during Thermal Cycling," *IEEE Transact. Components, Hybrids, Manufact. Technol.,* **15**(1):114, Feb. 1992.
132. Weiss, S., E. Zakel, and H. Reichl, "A Reliable Thermosonic Wire Bond of GaAs-Devices Analyzed by Infrared Microscopy," *Proceedings of 44th ECTC,* Washington, D.C., p. 929, 1994.
133. Tsukuda, Y., N. Watanuki, S. Okamoto, and Y. Syonji, "The Design of Flip Chip Joint by Other Metal Bump-Flip Chip Technology," *Proceedings of 8th IMC,* Boston, 1994,
134. Kloeser, J., E. Zakel, A. Ostmann, J. Eldring, and H. Reichl, "Cost Effective Flip Chip Interconnection on FR-4-Boards," *Proceedings of IMC,* Boston, 1994.
135. Richter, H., A. Baumgärtner, G. Baumann, and D. Ferling, "Flip Chip Attach of GaAs-Devices and Application to Millimeter Wave Transmission Systems," *Microsystem Technologies '94,* Reichl, H., and A. Heuberger, eds., VDE-Verlag, Berlin, pp. 535–543, 1994.

Assurance Technologies for Known Good Die

Larry Gilg

16.1 Introduction

16.1.1 Advanced packaging

Advances in reducing size and increasing functionality of electronic systems have been due primarily to the shrinking geometries and increasing performance of integrated-circuit (IC) technologies. Recently, development efforts aimed at reducing size and increasing functionality have focused on the first level of the electronic package. The result has been the development of multichip modules (MCMs), a technology in which bare IC chips are mounted on a single high-density substrate that serves to "package" the chips, as well as interconnect them.

Traditionally, the first level of electronics packaging, the IC package, has provided a number of benefits for the system assembler. The IC package provides protection from the environment, especially humidity and contamination. Package sizes are relatively large-scale, allowing ease of manufacturing and processing. The package provides a mechanism for thermal management, that is, a means to conduct heat generated by the IC away through the metal connections to the chip or to the backside (via thermal die attach). Electrical connection to the package leads is facilitated, and the package can be mounted temporarily in a socket for ease of testing or debugging.

However, the benefits of the first level of electronics package come at a cost. The electrical performance of any package is degraded by the additional capacitance, inductance, and lead length added to the IC bond pads. In addition to the performance limitations of the IC package itself, the larger footprint of the package implies that the next level of interconnect will also be suboptimal because of size and fanout of the IC interconnections. Figure 16.1 shows the performance

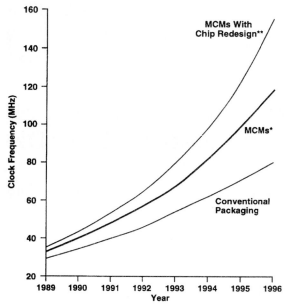

Figure 16.1 Performance gain of traditionally packaged ICs versus MCMs (* assumes a 4-ns gain; ** assumes a 6-ns gain). (*Source: nChip.*)

advantage that is realized by packaging a system using MCM versus conventional packaging. The 60-Mhz conventionally packaged system of 1994 could realize approximately a 25 percent speed increase (to ≈75 Mhz) by packaging the system as a multichip module (MCM). Even greater performance improvements could be realized if the IC were redesigned to take advantage of the benefits of the MCM interconnect.

The size, weight, and typical board space of several IC packages are shown in Fig. 16.2. As the figure makes apparent, eliminating the first level of package, and directly attaching the chip to the board holds a number of attractions, namely, higher I/O densities, lower profile, shorter lead length, and lower weight. So, given that the technologies for assembling bare ICs directly to boards have been developed recently,[1] what are the obstacles to utilizing bare die for advanced packaging?

The issue of fully conditioning a bare die to achieve the quality and reliability levels of traditionally packaged ICs is a major obstacle to the widespread acceptance of the advanced packaging concepts that eliminate the IC package to realize dramatic performance improvements.

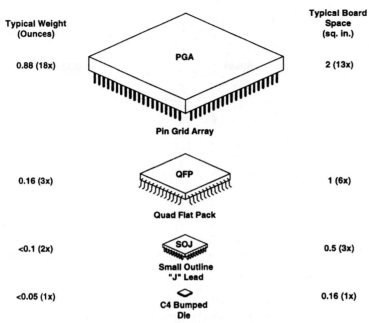

Typical Weight
(Ounces)

Typical Board
Space
(sq. in.)

0.88 (18x)

PGA

2 (13x)

Pin Grid Array

0.16 (3x)

QFP

1 (6x)

Quad Flat Pack

<0.1 (2x)

SOJ

0.5 (3x)

**Small Outline
"J" Lead**

<0.05 (1x)

0.16 (1x)

**C4 Bumped
Die**

Figure 16.2 Various IC packaging technologies: C4 size and weight comparisons. (*Source: Motorola/IBM.*)

16.1.2   KGD definition

Test and reliability screening of bare (or minimally packaged) die is not a new requirement for IC makers; however, it has yet to be satisfactorily addressed. The IC industry currently does an excellent job of routinely testing and screening traditionally packaged chips, although this can at times be a costly process. The resulting high-quality, high-reliability devices allow board-level products to be assembled with a high probability of success. However, the probability of successfully assembling multiple bare die on an MCM is decidedly poorer than that of the packaged ICs on a PC board because of the difficulties associated with conditioning the die appropriately.

The following probability formula gives the expected yield of an assembled module:

$$Y_b = 100(P_c)^n$$

Y_b = predicted board yield
P_c = probability that the IC functions correctly
n = number of ICs

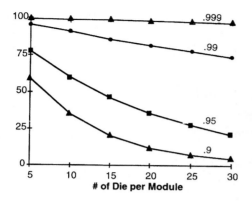

Figure 16.3 Plot of $Y_b = 100(P_c)^n$ for several probabilities of known good die. This plot makes the simplifying assumption that all die in a module have the same probability of being known good.

Figure 16.3 plots Y_b against n for several values of P_c. Packaged ICs can approach 99.999 percent probability (considered to be known good) of performing correctly for a specified time in the final application.[2] This probability index means that less than 10 parts out of one million will fail to perform their function correctly throughout a minimum guaranteed lifetime. The ability to fully test at speed and over temperature and to eliminate weak components with burn-in is not generally available or cost-effective for bare die. This significantly lowers the probability that a device will perform as specified over its expected lifetime.

Figure 16.3 shows the effect of lower KGD probability on the assembled MCM. Even with a 90 percent probability of KGD—which is typical of wafer probe results for mature products[3]—the resulting yield of the assembled module is unacceptable for systems with more than a few chips.

In addition, the MCM final test coverage required to identify the low-yielding MCMs must be extremely high to avoid escapes.

Figure 16.4 shows the MCM defect level[2,2a] or percentage of defective modules in a lot after final test as a function of the fault coverage of the test. These are modules that escaped detection because of the less-than-perfect final test. Notice that MCM defect level is a function of both final test fault coverage and the quality level of the incoming die. Figure 16.5 shows a comparison of several incoming die yields with MCM final test fault coverage.[3]

The incoming die quality levels that are less than 99 percent cause unacceptably high MCM failure rates, with attendant rework costs. But even worse, the poor yield indicates that faulty MCMs are not detected at final test, and thus are likely to cause system-level failures later in the life cycle, perhaps in the field. So, for most applications, KGD can be defined as a > 99 percent probability that the die in the lot are free of defects, both actual and latent. This level of KGD is

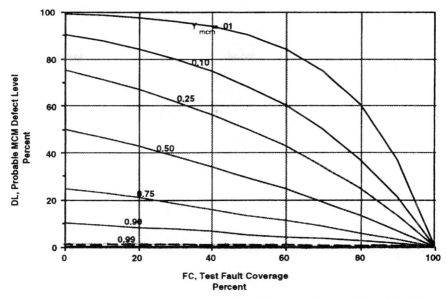

Figure 16.4 Defect level of outgoing MCMs as a function of MCM yield and final test fault coverage.

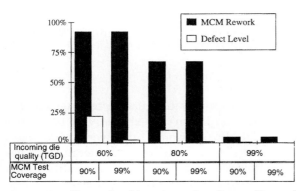

Figure 16.5 Example of how incoming die quality and MCM final test coverage affect rework and defect levels. This example is for a MCM incorporating five chips with identical yield.

considered the absolute minimum probability that can be tolerated in an MCM assembly process. The SIA roadmap of 1992 indicates that this KGD level (10,000 ppm) is achievable in the mid-1990's time-frame (see Fig. 16.6).

Thus, the KGD problem can be summarized as follows:

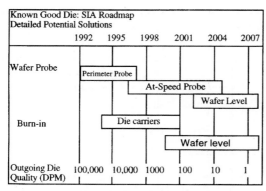

Figure 16.6 KGD technology milestones as envisioned in the SIA roadmap (1992).

- Unless incoming die quality is at least 99 percent, the percentage of die that are actually defect-free, the yield of all except the simplest MCMs will be unacceptable.

- Low-yielding MCMs require an exceedingly high fault coverage at final test to detect faulty modules. High fault coverage is very costly to achieve at MCM final test.

In addition to these test issues, burn-in of bare die is also a fundamental issue for KGD. The following sections will highlight the issues of test and burn-in as they relate to KGD.

16.1.3 IC test

The complexity of VLSI circuits prohibit exhaustive testing. For example, it takes 2^{64} clock cycles (minimum) to exhaustively test a 64-bit adder. If the clock rate is 50 MHz, it would take over 11,500 years to test all possible combinations! Obviously, test engineers have developed techniques which allow effective testing, with quality levels approaching 100 percent, which take much less time, but no one claims to be able to exhaustively test VLSI circuits in a production environment. So, as a matter of fact, all ICs, whether packaged or bare, have a finite probability of containing defects. This probability depends on the effectiveness of the testing done to the IC during manufacturing and assembly.

Fault coverage, a measure of test effectiveness, is usually quoted for a particular logic fault model, the single stuck-at fault. This simple fault model assumes that any line in a circuit may have a fault which causes it to remain permanently either at logic 1 or at logic 0. Fault coverage then is a measure of how many lines in the circuit are checked for stuck-at faults with a particular test program.[4] Single stuck-at fault coverage grades can sometimes be misleading in today's VLSI CMOS circuits. For example, the single stuck-at model does not directly take into account defects which occur only at the rated speed of the device. Hence, there is a finite probability that a device which passes a test program that is graded at 100 percent fault coverage for single stuck-at faults will not work in a particular application, at speed, over temperature or at power supply extrema.

Failure modes. The mechanisms of semiconductor failure can be classified into three main areas.[5]

1. Electrical stress (in-circuit) failures;
2. Intrinsic failure mechanisms;
3. Extrinsic failure mechanisms.

The first category of failure mechanism is termed event related, and is directly related to equipment design in which the IC is assembled, or to handling damage due to electrostatic discharge (ESD). This failure mechanism is very important to MCM assemblers, as they have more control over this than the other mechanisms, but the other two are more germane to the discussion of KGD, as they are under the control of the IC manufacturer.

The second category of failure mechanism is inherent to the die itself and—design flaws notwithstanding—is usually the result of defects introduced during fabrication of the wafers. These may be crystal defects, results of contamination, flaws in the deposition layers, or damage from mishandling during processing. These are the key mechanisms which determine the quality and reliability levels of bare die as shipped to customers.

The third category, the extrinsic failure mechanisms for IC failure, refer to the defects introduced during the "backend" IC processing. Traditional packaging (bonding), test, burn-in, mark, pack, and ship all contribute failure mechanisms, usually to more traditional packaged parts. Failure mechanisms which are unique to KGD may be introduced at this point, as the KGD backend processes usually are somewhat different than for traditional packaged devices.

Operational tests. Operational tests are those which detect defects in the IC. There are a number of tests which may be applied to an IC in a manufacturing environment. For the purposes of this chapter, these tests are characterized as follows:

1. *Functional test.* A functional test is performed to verify that a circuit performs its intended function. This is sometimes termed *truth-table verification* for digital circuits. The functional test performed at wafer probe test is usually done at reduced clock speed, due

to the moderate performance nature of traditional needle-type epoxy ring probe cards, and is used primarily to determine whether the circuit behavior is correct. The final test program, especially for the packaged device, usually incorporates a full at-speed functional test to ensure device behavior over the full rated speed range.

2. *DC parametric test.*   DC parametric tests are those in which steady-state voltages and currents are applied to certain I/Os of the device and corresponding voltages and currents are measured at other I/Os. These tests are used to detect open and short circuits, input/output levels, noise margins, static and dynamic supply current levels, leakages, etc. These test may be performed at both wafer probe test and final test.

3. *AC parametric test.*   AC parametric test are those used to measure the frequency-dependent characteristics of the circuit. These characteristics are propagation delay, setup/hold times, duty cycles, clock period, signal timing, etc. These tests are usually not done at wafer probe test due to the poor electrical environment offered by a standard probe card. High-performance probe cards are becoming available in the industry which offer the capability of high-speed test, so AC parametric testing (as well as at-speed functional testing) is becoming an option for wafer probe test, especially for bare die which do not need burn-in-type stress tests. These bare die could then be fully tested and may be adequate meet the quality and reliability specifications for certain applications at lower cost than fully conditioned KGD.

4. $I_{DD,Q}$ *testing.*   This is a technique for detecting certain faults in CMOS circuits by monitoring the quiescent current of the device between clock edges. Once all transistors in a CMOS circuit have switched, no appreciable current should be flowing in the internal logic circuits. Any value of current detected above a threshold indicates a problem. There is some evidence that $I_{DD,Q}$ testing identifies certain reliability problems that burn-in would normally eliminate.[5a]

5. *Built-in self-test (BIST).*   Of course, a number of ICs now incorporate built in test, where test structures are built into each die to enhance its testability. These structures can be accessed from the I/O pins of the IC and provide a variety of levels, coverages, and effectivity values.

6. *Visual inspection.*   Although not an electrical test, visual inspection is being used to increase the probability of KGD. Visual inspection criteria for bare die is an issue that is not currently resolved, especially for current VLSI memory products which incorporate redundancy and laser repair. Differentiating between strictly cosmetic effects and defective die is not straightforward.

Screening.   Screening ICs is the process of applying an accelerating stress to the device and then using operational tests to detect failures.

Individual defective die can be detected by 100 percent screening of a batch. Process or design related problems can be detected by sample screening. This is an important distinction for KGD. If the reliability of individual devices must be known, then a 100 percent screen will be required. Burn-in stress screening is probably the most common screen for detection of reliability defects in individual devices. This is usually a 100 percent screen used in production of leading edge IC devices. Careful attention to the design of stresses for screening tests is necessary to ensure that defective devices are stressed to failure but that the useful life of the remaining devices is not adversely affected. Defects may be introduced through an improper screen.

Tests to detect faults.   Table 16.1 (summarized from Ref. 5) lists some failure modes which are typical of VLSI devices. The table also lists the appropriate operational tests to detect the faults which typify the failure mode and suggests a screen which will accelerate the defect. Of course, preventing the failure mode from occurring through improved process control is always the best choice, but this is not always economically or operationally feasible.

16.1.4   Burn-in

Burn-in is the most common production technique for improving the reliability of IC devices as delivered to the customer. Dynamic burn-in is the screening technique of choice for leading-edge microprocessors and memory devices.

Burn-in is a reliability screening tool used in the microelectronics industry to reduce the risk of early failures of IC devices (see Fig. 16.7[6] and Ref. 6a). The entire population of devices is aged by the application of accelerated stress to make the weak devices fail.[7] In addition to temperature-related stress, dynamic burn-in typically is done with maximum power supply voltages applied. Random patterns are applied to the inputs of the device to cause internal nodes to toggle. The goal is to stress all nodes during burn-in, which may have a duration of a few hours up to a few days, depending on the level of maturity of the device and the level of reliability required.

There is no consensus today among IC manufacturers as to whether burn-in is a requirement for VLSI devices. There is a claim that current early failure rates are low (<5000 ppm).[8] Certainly, for mature products, burn-in is not typically done for commercial or industrial use, although burn-in may be used for mature products to identify "rogue" lots, that is, wafer lots which have a much higher than average failure rate due to processing anomalies (see Fig. 16.8).

Systems for high-reliability use, especially nonrepairable systems, still require substantial IC burn-in to achieve acceptable field failure rates. Studies published by IBM indicate that burn-in effectiveness,

TABLE 16.1   Summary of Failure Modes in IC Devices[5]

Life-cycle period	Failure mode	Fault	Detection	Lifetime region (see Fig. 16.7)	Screen
Design	Parasitic elements	Latchup	Functional test	Event-related	Possible destructive test of sample
Fabrication	Ionic contamination	Degradation	AC, DC parametric test	Early life, wearout	Burn-in
	Gate oxide breakdown	Short and open circuits	Functional test	Early life	Burn-in with overvoltage
	Corrosion	Open and short circuits	Visual inspection, functional test	Wearout	Temperature-humidity bias tests
	Surface charge spreading	Degradation	AC, DC parametric test	Wearout, early life	Burn-in
	Piping	Parametric shifts	AC, DC parametric test	Wearout	Burn-in
	Dislocations	Threshold shift	AC, DC parametric test	Wearout	
	Slow trapping	Threshold shift	AC, DC parametric test	Wearout	
Assembly	Microcracks	Open circuits	Functional test	Early life	Burn-in
	Electrostatic discharge	Short and open circuits	Functional test	Event-related	None
		Degradation	AC, DC parametric test	Event-related	None
	Storage	Degradation	Functional test	Wearout	None
Use	Hot electrons	Degradation	AC, DC parametric test	Wearout	Low-temperature life test
	Corrosion	Open and short circuits	Visual inspection, functional test	Wearout	Temperature-humidity bias test
	Electromigration	Open and short circuits	Functional test	Early life/ event-related, wearout	Burn-in
	Contact migration	Short-circuited junctions and open-circuit contacts	Functional test	Wearout	Burn-in
	Mechanical stress relief migration	Short and open circuits	Functional test	Wearout	High-temperature bake
	Radiation	"Soft" errors, threshold shift, activation of parasitic elements	Functional test, AC/DC parametric tests	Event-related	
	Electrical overstress	Short and open circuit	Functional test	Event-related	None

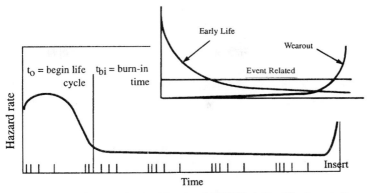

Figure 16.7 Traditional "bathtub" curve for IC lifetimes. The insert shows the three distinct models of device lifetime. Burn-in is designed to screen the early life failures from the population. Time t_0 is the time the device is manufactured. The failure rate shows an initial increase, then a steep decline during the early life, then is constant during the long normal operating period, again exhibiting a sharp increase at the end of life. This bimodal distribution is caused by a population of weak devices which are a result of anomalies in the fabrication and manufacturing processes. This weak population can be accelerated to failure through the application of appropriate stresses for a relative short period as part of the manufacturing process. The resulting population of normal devices appears to have the failures in time beginning at t_{bi}.

defined as the ratio of hazard rate without burn-in to hazard rate with burn-in equivalent to t_{bi} hours, reduced the reliability failures of IC devices by close to 2 to 3 times through 3000 fielded hours.[9]

Jensen and Petersen[10] differentiate between devices that are congenitally weak; that is, the weakness was manufactured in (freak population), and devices which have unwittingly been subject to some stress subsequent to manufacture but prior to being put into service (infant mortality). This distinction has not been important to traditional packaged ICs, as devices receive burn-in after most postfabrication operations have been completed, and burn-in stress can be applied to accelerate both the freak population and the infant mortalities to fail. However, as burn-in strategies aimed at die level burn-in, either singulated bare die or die still in wafer form become available, more post-burn-in processing and handling will occur. The question of how this will affect infant mortalities experienced in the field is an open question, and will need to be analyzed as part of any KGD strategy.

Another consideration for burn-in of KGD is the thermal environment that the device sees. At present, burn-in of bare die requires the use of a temporary carrier to provide the function of the IC package, namely, environmental protection, thermal management, and electrical access to chip power and I/O pads. However, it is not a given that

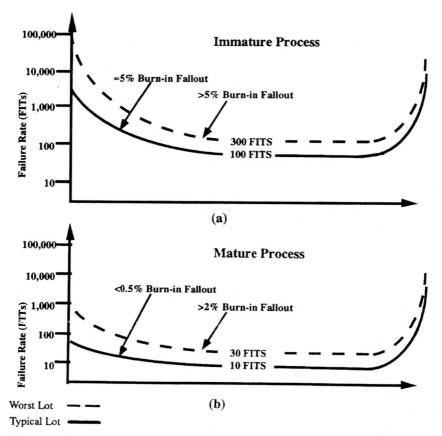

Figure 16.8 Curves for typical and worst-case (rogue) wafer lots for (*a*) immature IC processes and (*b*) mature IC fabrication processes. (*See David Maliniak, "Known-Good Die Poised to Take Off," Electronic Design, Nov. 21, 1994.*)

a temporary carrier will provide the same characteristics as a permanent package. The thermal resistance of the path from the die junctions to the ambient ($R_{\Theta ja}$) may be quite different, necessitating a different burn-in regime (time, temperature, and atmosphere) which could require a separate backend process for KGD production.

16.2   Methods of Assuring KGD (Improving the Odds)

16.2.1   Process control

The move to 6σ processing has improved semiconductor yields and reliability over the past few years.[11] As the information in Table 16.1 in-

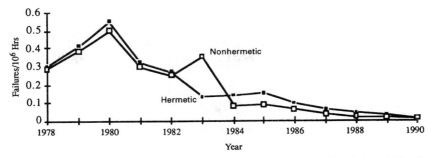

Figure 16.9 Historic IC failure-rate data. The total devices tested is close to 400 million over 12 years. These data were collected from an equipment maker by the Reliability Analysis Center, Rome Air Force Base.

dicates, some of the most prevalent failure modes (contamination, corrosion, gate oxide breakdown, etc.) for VLSI devices are a result of the wafer fabrication process and can be minimized through effective process control.

The data in Fig. 16.9 are indicative of sound manufacturing and process control improvements over the past few years. Vertically integrated companies that control the IC fabrication process as well as assembling modules can use improvement processes to raise KGD probabilities. By correlating failures at the MCM or system level with testing done at the wafer level, defects can be migrated back to the front end and detected at wafer test.[12] Early identification of defective devices is the most cost-effective approach to improving the probability of known good ICs.

16.2.2 Statistical sampling

IC suppliers may provide statistical probabilities of known good die by packaging a sample of die in a wafer lot, performing exhaustive test and burn-in on the packaged sample, and applying the statistics of the sample to certify the entire lot. This technique is effective only in detecting whether the batch (wafer lot) has been fabricated such that the statistics vary significantly from the "normal" population. To achieve KGD to an acceptable level (i.e., >99 percent) requires that test and burn-in statistics be obtained on a significant sample of devices. Figure 16.10 shows the probability of accepting a lot with a lot tolerance percent defective (LTPD) of 0.10, or a probability of .90 that any lot accepted will have at least 99 percent true good die (TGD). The various sample size and lot accept criteria are plotted. For an accept on 0, reject on 1 defect criteria, a sample size of 230 devices is needed to obtain the .90 probability of accepting a lot that has at least 99 percent TGD. This assumes that defects are evenly distributed throughout a

Operating Characteristic Curves for 1% LTPD

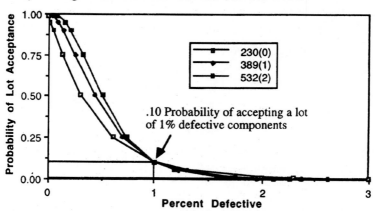

Figure 16.10 This plot shows the probability that a lot of ICs have a certain percentage of defective units based on packaging and fully testing a sample of devices from the lot. The legend shows the number of parts sampled and the pass criteria, that is, 230 parts sampled with accept on 0 rejects, 389 parts samples with accept on 1 or less rejects, and 532 parts sampled with accept on 2 or less rejects. This particular set of curves is based on the criteria of less than 10 percent probability of accepting a lot of more that 1.0 percent defective units.

large population and randomly sampled in each lot. It also assumes that a test is available to detect all possible defects.

This is not generally the case. A test program that detects all possible defects on a VLSI device is not possible within the constraints of time, cost and complexity. This is especially true of wafer test, where the probe card provides a very difficult electrical environment that is hostile to performance type testing (at-speed, over temperature, full functionality, etc.).

If the probability of obtaining a fatal defect is the same on one chip as on another, then the probability of finding a chip with no defects is modeled by the Poisson yield equation[13]

$$TGD = e^{-\lambda}$$

where TGD is the percentage of true good die for a given λ, the average number of fatal defects per chip. However, it is rarely possible to detect all fatal defects using finite test resources available for production test, especially at the wafer level. So, the KGD probability can be thought of as an estimate of TGD, the percentage of devices without defects out of the total number of die in the lot after a given level of testing, assuming the devices failing the test have been removed from the lot. As an example, for an estimate of 99 percent TGD the average number of fatal defects per chip remaining in the lot will be .01, or

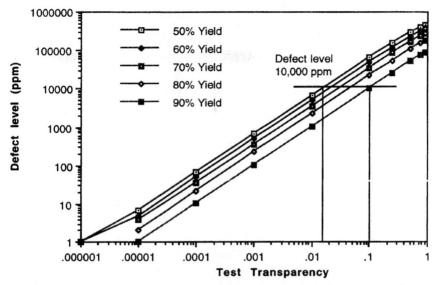

Figure 16.11 The plots of defect levels against test transparency for selected yields. Test transparency is defined as (1−fault coverage). The plot of different yields versus test transparency gives the expected defect levels (undetected defects). The line through the yield curves shows the test transparency necessary to achieve 10,000-ppm outgoing die quality for various yields, where yield is the number of devices passing the test divided by the total number of devices tested.

10,050 ppm defective. To predict that the die remaining after wafer test are no more than 10,050 ppm defective, the yield of the lot after wafer test and the test transparency (probability that the wafer test is unable to identify a chip that is faulty) need to be taken into account. Figure 16.11 is a plot of defect level versus test transparency for selected yields. The defect level is defined as

$$DL = 1 - Y^{TT}$$

where TT is test transparency (or 1−fault coverage) and Y is the yield for the test at wafer probe. So, given a wafer test that yields 80 percent, that is, the test identifies 20 percent of the lot as being defective; how many defects remain in the lot? As an example, if there are 1000 devices entering the test, 800 will pass the test (80 percent yield). With a test transparency of .10:

$$DL = 1 - .8^{.10} = 0.022$$

So, multiply the apparent good devices by the defect level: 800 · 0.022 ≈ 18 defective devices which the test did not catch. So an estimate of the percentage of TGD will be

$$\frac{800 - 18}{800} = 97.75 \text{ percent}$$

To increase the probability that the lot contains 99 percent TGD for the previous example, an improvement in the quality of the test will have to be made, given the same quality level of the lot. With an improved test transparency, even though a lot may yield 80 percent, the estimate of TGD can achieve the 99 percent level. So, for an 80 percent yielding lot,

$$DL = (1 - Y^{TT})$$

and, for a 99 percent estimate of TGD (or a defect level of .01)

$$.01 = (1 - .8^{TT})$$

$$TT = .045$$

so test coverage will have to improve to >95 percent.

Another way to look at defect levels is, for example, that if 10 percent of the lot is defective, and if the test transparency is .10, the test will fail to detect 10 percent of the defective devices. So $(.10)^2 = .01$ or 1 percent defect level of 99 percent probability of KGD. Said another way, this means that the test coverage (1—test transparency) at wafer test should be greater than 90 percent to achieve a KGD probability of 99 percent for lots yielding at least 90 percent. This is typically achieved only for very mature products.[2] Now, however, it may be expected that the final test program used to test the sample of packaged devices will also be less than perfect. If the final test of the packaged ICs in the sample also has fault coverage less than 100 percent, then the defective die in the lot after wafer test may not be detected in the testing of the sample. That is, the LTPD curve will give an optimistic estimate of the quality of a lot using the results of a less than 100 percent final test.

Of course, there is no reason to believe that defects are evenly distributed over a wafer lot or even over a single wafer. In fact, defects on a wafer are not uniformly distributed, but are known to exhibit clustering.[14] Using this fact to relate die which pass wafer test with the number of nearest neighbors which also pass wafer test can lead to close to an order of magnitude improvement in defect levels for the same test transparency. For example, if clusters of die on a wafer show 90 percent yield while the overall average yield is 50 percent, the expected defect level (for a test transparency of .10) of the 90 percent cluster is about 10,000 ppm while the average defect level (50 percent yield) is about 70,000 ppm (see Fig. 16.12).

Each die has eight neighbors. By "binning" each die into a separate category based on the results of the test on nearest neighbors, higher

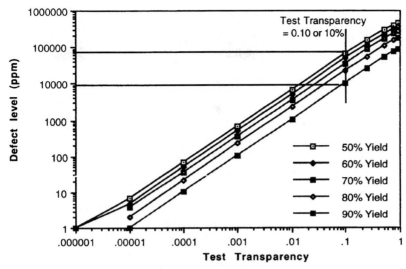

Figure 16.12 The plots of defect levels against test transparency for selected yields. Shows the defect levels resulting from a test transparency of 0.10 for various yields.

probabilities of being known good will be associated with die whose neighbors all pass the tests. These die are then good candidates to be selected for bare-die applications.

In summary, statistical methods require fault coverage at probe testing that is much more complete than typically done for packaged parts, with attendant costs and type I testing errors. In addition, a fairly large sample of devices must be packaged and tested to demonstrate that the probability of KGD is acceptable. Using the fact that defects cluster to bin the devices seems to hold promise of improving the probability of KGD using fairly standard wafer test routines, but the capability of binning at wafer test is not generally available in wafer handlers today. More importantly, the issue of burn-in is not addressed with any of these methods.

16.2.3 Test every die

The most effective method of assuring KGD is to test every device to the quality and reliability levels demanded by the particular application. The various quality levels of bare die can be classified as follows:

Level 1. Standard DC wafer probe. These die receive no additional screening above the normal probe used in commercial packages, usually DC parametric and low-speed functional test only. This level usually requires that the die supplier be willing to share yield data with customers so they will have an understanding of the quality levels being produced at the moment. A 100 percent visual inspection could be performed to be compatible with die used in military standard packages.

Level 2. Standard DC wafer probe with lot acceptance testing performed. A sample of die from the lot is packaged and screened to provide statistical assurance that the lot is typical of the population. The lot acceptance does not provide confidence that individual devices will meet the quality and reliability targets, but the statistics on the lot will predict how the aggregate of die will perform with respect to the population.

Level 3. AC/DC probe at temperature. This test regime may be similar to the final test used to screen the packaged equivalent product. This test can provide the same quality levels as the equivalent packaged device, and it may provide some level of early life failure screening, although typically not as extensive as die level burn-in.

Level 4. Full test and burn-in. This option usually will require that the die be assembled into a temporary die carrier and conditioned identically to the equivalent packaged device. This level is the most commonly agreed upon definition of KGD.

16.2.4 Test and burn-in carriers

As noted above, the solutions which promise the highest probabilities of producing KGD is to assemble the bare die into temporary "carriers" which serve as a temporary single-chip package that allows traditional final test and burn-in infrastructure to be used to achieve quality levels and reliability levels comparable to traditionally packaged ICs. In general, with these methods the die is mounted in a carrier that has the same form and function as a single-chip package. Temporary electrical connection is made to the bond pads and the device is qualified through test and burn-in processes identical to the traditional packaged part. ATE test equipment, component handlers, burn-in boards, burn-in ovens, and loaders can be used. Once the die is qualified, electrical connections to the bond pads are released and the die is taken from the carrier. The result is a fully tested, qualified IC device with specifications comparable to those of an equivalent packaged part. These technologies can be divided into categories based on the type of die-to-carrier connection: permanent, semipermanent, or temporary.[15]

Permanent carrier approaches. These approaches take advantage of the fact that some minimal packaging approaches enhance the capability to do test and burn in. Ruggedized packages, wider pitch, and

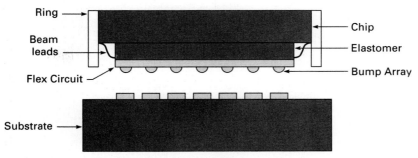

Figure 16.13 Example of a chip-scale package consisting of a flex circuit with a rerouted bump array, bonded to the I/O pads on the chip through beam leads. The rerouted bump array can be held in contact to pads on a substrate for test and burn-in by applying force on the backside of the package.

gold contacts can help solve many of the problems with handling and contacting bare die.

Tape-automated bonding (TAB) is an established technique for testing integrated circuits. The IC is bonded to a lead frame which fans out the electrical connections to peripheral pads on tape, which may be readily contacted. TAB carriers are available which allow the device to be socketed for test and burn-in.

Another minimal packaging approach, the chip-scale package, is, like TAB, designed to be permanently assembled with the die into the next-level interconnect. The chip-scale package is designed to take up little more real estate on the board than the chip itself. In the chip-scale package approach, the die bond pads are routed to a standard footprint, usually an area array, of bumps, which provide mechanical and electrical connections to a substrate (see Fig. 16.13). Thermal management can be provided through the backside of the die. The chip scale package, with its gold-plated ball grid array, can more readily be tested at speed and burn-in by pressure contacting to an area array substrate. A major advantage to the chip scale package, from a KGD point of view, is that the chip to substrate connection can remain fixed through die shrinks, upgrades, etc.

Semipermanent carriers. Semipermanent carrier approaches consist of making a nonstandard (or rerouted) metallurgical connection (wire-bond, C4, etc.) to a reusable standard package type; sending the assembly through test and burn-in; then breaking the connection and removing, inspecting, and shipping the die. Several semipermanent carrier technologies are currently available for preparing KGD, including temporary TAB connection, temporary wire-bond (see Fig. 16.14), and reduced radius removal (R3) based on the IBM C4 flip chip technology (see Figs. 16.15 and 16.16).

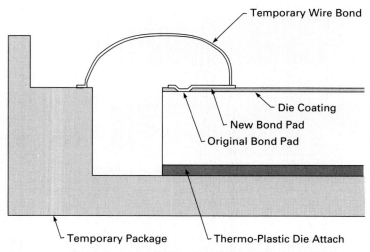

Figure 16.14 Semipermanent KGD carrier. The wafer is coated with dielectric material and the bond pads enlarged. Wires are temporarily bonded to the enlarged pad and the die attached to the temporary carrier. After test and burn-in, the wire bonds are "clipped" off, and the die is removed from the temporary package.

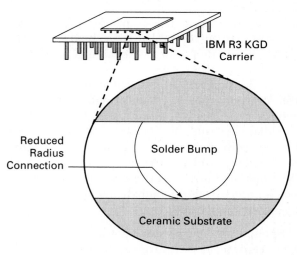

Figure 16.15 Semipermanent KGD carrier. The IBM R3 carrier uses a C4 process to join a bumped die to reduced radius pads on a ceramic substrate. After test and burn-in, the chip is sheared off the substrate with a lateral force. The bumps remain intact on the die.

Figure 16.16 SEM photo showing the solder bumps of chips which have been removed from the IBM R3 carrier. The bumps on the left side of the photograph have been reflowed subsequent to removal. The bumps on the right have not been reflowed. The row of bumps on the right edge are not contacting the blanket aluminum so they are charging in the SEM and appear to glow.

These technologies rely heavily on established processes and tools to condition the die. In some cases, semipermanent carrier technology is targeted exclusively toward specific final assembly methods, such as TAB or R3 for solder-bumped die. Concerns with the cost-effectiveness and final assembly limitations of some semipermanent carrier methods have resulted in heightened interest in temporary carrier solutions that are applicable to all die, regardless of final assembly methods.

Temporary carriers.   Temporary carrier approaches contain microprobe sets built into carriers that are made to "look" like standard packages. The die is held in alignment to the probe set with force to ensure reliable electrical contact to the IC bond pads (see Fig. 16.17). The main technical challenges with the probes are compliance to nonplanar pads on the die, penetration of the native aluminum oxide present on the IC pads, and maintaining adequate contact without causing damage, which could preclude the next assembly operation.

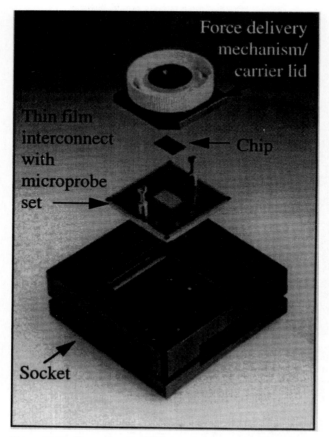

Figure 16.17   Example of a KGD temporary carrier. The chip is assembled to a probe set on the thin-film interconnect. The force delivery mechanism lid latches to the carrier. The entire assembly is then inserted into the IC socket. (*Courtesy of Texas Instruments, Inc.*)

Pad penetration.   A temporary carrier approach requires some form of "scrub" or penetration through the native oxide on the Al pad of the IC, typically on the order of 50 to 80 Å, before sealing itself off. Traditional needle and blade type probe cards achieve scrub by forcing the tip of the probe to move laterally across the surface of the Al pad with a vertical force "overdriving" the probe (see Fig. 16.18).

This form of scrub may be used effectively for peripheral bond pads on an IC. Many KGD approaches currently being developed use some form of z-axis scrub in which the microprobe is designed to be used with a piercing vertical on-axis force (see Fig. 16.19).

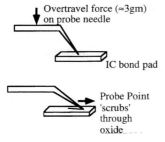

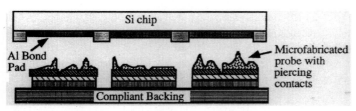

Figure 16.18 Overtravel force on probe needle causes horizontal motion that "scrubs" through native oxide on aluminum IC bond pad.

Figure 16.19 Microfabricated probe set makes reliable contact to the aluminum IC bond pad by piercing through the oxide layer on the surface of the pad.

Compliance. Scrubbing actions on the bond pad require that some form of compliant member be available to equalize the force on each contact. For traditional epoxy ring needle probe cards, the spring constant of the needle provides compliance. For temporary carriers, especially those that rely on z-axis scrub to make reliable, low resistance contact, this problem is nontrivial. Since force is applied on-axis through the probe structure, the compliance must be built into the substrate on which the probe set is mounted—unless some form of compliant or deformable probe contact capable of z-axis scrub is available. This compliant substrate must be capable of providing two conflicting functions—the ability to transmit probe forces independently in the z direction while maintaining x- and y-direction positional accuracy. In addition, the compliant members must be capable of maintaining these properties during burn-in temperature excursions typically to 150°C. Many temporary carrier approaches use a form of membrane as the probe set substrate. Thin films, laminates, organic, and inorganic membranes have all been proposed as KGD carrier interconnect substrates. These membranes may be backed by an elastomer to provide the compliance required for the probe set.

Pad damage. Another major problem to be addressed by both semi-permanent and temporary carrier methods is bond pad damage after the removal of the die from the carrier. A primary concern is determining how many retests a given approach is capable of before next-level assembly is precluded. The generally accepted industry standard—MIL-STD-883, method 2010, sec. 3.1.1.1—states the criteria for rejection as 25 percent of the passivation underlying the bond pad being exposed. This serves as a reference to determine acceptable pad damage. A probe should be capable of ≥ 2 touches on the pad before unacceptable damage occurs. At no time should the probe touch down on the top or sides of the top passivation on the die.

Alignment.   Bare-die handling and alignment is the next critical issue. The die must be accurately aligned to the probe. If mechanical alignment is used, standard practice for wafer saw tolerances must be improved. If not, vision systems will be required. If mechanical die alignment is required, i.e., using the sides of a die for alignment to a probe set, the following issues must be taken into account:

- *Saw cut dimensions and placement accuracy.*   Saw cut width must be maintained within tolerance (≈ 0.1 mil) over the life of the blade. The distance of the cut to a designated die feature must also be maintained within tolerance.

- *Saw alignment accuracy.*   The placement of the saw cut center-line should be within the maximum tolerance of intended location.

- *Sawn edge camber.*   Total deviation (or bevel) of sawn edge from a straight vertical cut should not be more that a small percentage (≈ 5 percent) of the wafer thickness.

As IC pad pitches and sizes become smaller, improved saw drift control and placement accuracy may be required for die that must be mechanically aligned onto a probe set in a temporary KGD carrier.

16.2.5   Wafer-level burn-in

If an IC which is targeted for MCM applications needs burn-in to meet reliability targets today, the industry accepted solution is to burn-in at the die level using one of the carrier technologies described above. There is a consensus that while die-level carrier technologies that take advantage of existing test and burn-in infrastructure are a requirement today, the long-term cost-effective solution to KGD is to perform test and burn-in at the wafer level.

In fact, wafer level test and burn-in offers the promise of reducing costs for all die, regardless of whether the final package format is single-chip or multiple-chip modules. While wafer level test probe cards, capable of high-performance testing of a die at a time on a wafer are currently becoming available, there is no technology generally avail-

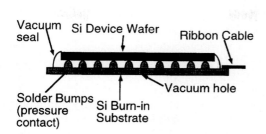

Figure 16.20 Wafer-level burn-in probe card. (*Source: David B. Tuckerman, et al., "A Cost-effective Wafer-level Burn-in Technology," Proceedings of the 1994 International Conference on Multichip Modules, April 1994.*) A silicon burn-in substrate makes contact with the device wafer through solder bumps deposited on the substrate. The silicon-to-silicon probe technology reduces one of the major risks of wafer level probing, that of CTE mismatches during temprature excursions.

able to provide dynamic burn-in to multiple die on a wafer. The issues of thermal management, mechanical alignment through temperature and over time, routability, interface to ATE and burn-in electronics, and reliable contact to die I/O contacts are extremely challenging (see Fig. 16.20).

16.3   Cost of Producing KGD

As with most technology development, as solutions begin to appear cost issues become a leading factor. One objective of both die suppliers and users is to reduce the cost of KGD to a level comparable to package part costs for test and burn-in. The major cost driver for KGD technologies is the number of reuses that the carrier supports.[16] To make KGD carriers cost-effective, a technology should support at least 100 reuses, unless there is a breakthrough in the today's cost of the carriers.

There are also cost factors which result from the particular die being conditioned, regardless of carrier technologies selected. The die complexity, a function of size and number of I/Os, is a critical cost driver for carrier technologies which achieve a high (>100) number of reuses.

Although cost issues need to be addressed at the system level to determine effective tradeoffs, a first-order measure of cost-effectiveness is parity with a conventional packaged devices. This is more likely to be achieved with large, complex die, especially those shipped in ceramic packages. Other KGD carrier independent factors, such as test and burn-in strategies (e.g., a decreased burn-in time), are more likely to have a consequential effect on the relative cost of KGD. As the carrier technologies currently being developed take advantage of the industry infrastructure for test and burn-in, selection of a specific KGD carrier technology should be made based on ease of insertion into present process/product line and reliability of the technology.

16.4   KGD Standards

16.4.1   EIA Standard for KGD*

This standard was created by a joint MCC/SEMATECH task group in 1992 to facilitate the procurement and use of high-reliability semiconductor microcircuits or discrete devices provided in bare-die form, commonly known as "Known-Good Die" (KGD). The standard was submitted to JEDEC for adoption as a standard. TAB and flip chip addenda were added by the task group in 1994 and sent to JEDEC for standardization.

The standard provides requirements and guidance to KGD suppliers in regard to the high levels of as-delivered performance, quality, and long-term reliability expected of this type of device. It also reflects the special needs of KGD product customers in terms of design and application data. The standard is applicable to KGD products used in both commercial and military applications.

The KGD standard also reflects an understanding on the part of die users that quality and reliability cannot always be assured in the same fashion as for conventionally packaged microcircuits. KGD customers take on a significant responsibility for the proper application and long-term environmental protection of this type of product. The extent to which KGD suppliers shall warrant die product is highly dependent on customer capability and adherence to strict quality controls. Cooperation between suppliers and users is essential.

Scope.   This standard provides guidelines and requirements for known-good semiconductor die (KGD) used in other than conventionally packaged microcircuit or discrete formats. The die described are intended to be high-quality, reliable bare die in die form only, for use in a variety of user-defined applications (multichip modules, hybrid circuits, memory cards, etc.). While the standard allows negotiation between supplier and user to establish specific requirements for performance, quality and reliability, it is important to recognize, in the case of military and aerospace applications, the minimum requirements described in relevant military specifications.

---

*Portions of this section were taken from the *EIA Standard for Known Good Die*, version 2.2, as modified on May 18, 1993.

The standard is limited to KGD consisting of a single microcircuit or discrete device connected using conventional wire bonding or high-density interconnect. KGD are intended to be equivalent to or better than their corresponding packaged parts in terms of electrical and reliability performance (unless specifically noted by the supplier). Per requirements mutually agreed on by the KGD supplier and user, the KGD supplier shall implement and demonstrate testing and screening required to assure this performance. This standard deals only with KGD supplied in individual die form, for which the user accepts responsibility for providing and assuring final environmental protection (e.g., hermetic sealing).

16.4.2 Die Information Exchange (DIE) Format*

Although obtaining devices which meet quality and reliability specifications is challenging, the information pertaining to the die may be more difficult to obtain. There are a host of data which are required for a user to assemble MCMs or COB using bare die that haven't been necessary for using packaged ICs. Under sponsorship of an ARPA contract, an industry group called the DIE Industry Group, which includes MCM assemblers, die suppliers EDA vendors and foundries, has developed an information exchange format for bare die, called the Die Information Exchange (DIE) Format The DIE Format is a human and computer-sensible interchange format for information about unpackaged integrated circuits (ICs). The format can convey information about die devices (components) which are used by MCM designers and foundries. The information is intended for direct use in EDA and other computer-based software tools. Human readability has been maintained to aid in the verification and use of the information independent of the available tools. It is expected that the information in this format will be created and distributed by either the die manufacturer or a die broker.

The DIE Format conveys the physical and functional characteristics of an unpackaged die—that is, those characteristics needed for place and route, thermal analysis, electrical signal analysis, power distribution design, physical bonding, behavior, test, and timing analysis. Existing formats are referenced and used where beneficial to avoid creating another standard for the same information.

Use of the DIE format is not intended to be a replacement for a

---

*The information contained in this section was excerpted from the Die Information Exchange (DIE) *FormatReference Manual,* revision 1.0.

data sheet nor to represent all the information needed to understand a die. The focus is on conveying that information which can be directly processed and is either time-consuming to collect or error-prone to enter into the tools. In limited cases, information important to the end user but not computer sensible has been included in the format to facilitate understanding and use of the data. Information that is time-consuming to manually collect, difficult to enter, or not generally available has a priority for inclusion. Information about bare die generally falls into this category.

An important aspect of the format are the *levels of compliance*. In some cases, it may be difficult for an information provider to gather or generally release all the information needed for a given device. Therefore, the format categorizes information into one of three levels of compliance. Each higher level provides more detail about the device with level 0 being the basic, rudimentary data. It consists of the common, indisputable information required by any die user. It represents the minimum set of information which must be conveyed in a block about a die. Although level 0 information may be expanded in future revisions of the specification, it is intended to be the most static level of information defined.

Compliance level 1 contains information which is either (1) derivable from level 0 information, (2) not needed except by MCM users pushing the edge of technology, or (3) not as well accepted by the general industry as level 0 information is. Level 1 may contain sensitive information which is not generally distributed to the community at large. In a few cases, level 1 contains refinements of level 0 settings (e.g., adding a tolerance to a value supplied in a level 0 setting). It is expected that level 1 type die information will be under tighter distribution rules or used as a competitive edge by some die suppliers to better support the user base.

Compliance level 2 contains information either highly proprietary in nature or so unique that it will probably only be delivered under special request directly from the producer to the consumer. Examples may be measurement data for a given processing lot of die.

Each compliance level is a superset of the one before it. Therefore, they build on one another. To be level 1–compliant, a block must contain all level 1 and 0 information.

16.5 Summary

The known good die (KGD) problem is an information problem. Die suppliers have traditionally packaged ICs with lead frame, molding compound, and stenciling while minimizing the information content of test and inspection and keeping the results proprietary. The costs of collecting the data and analyzing the information has been hidden to the customer, to some degree, by the cost of the package. Also, users have traditionally been somewhat willing to pay for extra screening that high-reliability components require. These die are usually packaged in higher-cost hermetical packages, though, so the cost of the higher quality information was, again, not as apparent to the customer.

Now, however, IC customers value the lead frame and molding compound of traditional ICs less. They believe they can do a better job, or gain competitive advantage by "packaging" the IC themselves. However, neither supplier nor customer have developed an appreciation of the fundamental problem associated with bare-die sales, namely, that the information is much more difficult (costly) to obtain (and the users are not willing to pay the premium required). Or, stated another way, the test and inspection data that exist are not optimally "informing" the supplier or user of the condition of the die. The question of how to best gather (and provide) the information that provides knowledge that a die is "known good" is behind the search for methods to provide bare die with the same quality and reliability levels as packaged ICs.

The solutions which are emerging today in the industry take advantage of the historically excellent record of the IC industry to provide high-quality packaged ICs. These emerging solutions use "temporary" die carriers to perform the functions of a traditional IC package, namely, ease of handling and electrical access, ruggedness, protection from contamination and adverse ambient environments, and an effective means for providing thermal control of the die junctions in operation. These methods used to achieve the high quality and reliability of today's packaged ICs rely on adding redundancy to the test plan in the form of additional tests, screens, and inspections. The problem with implementing these redundant tests and screens at the bare-die level is the additional cost of the temporary die carrier, the cost and time to load and unload the die, and the probable cost of having unique processing parameters [e.g., different burn-in temperatures for bare die due to a quite different thermal path $(R_{\Theta_{ja}})$ for the KGD carriers]. The net result is that KGD carriers are seen as an interim solution to the KGD problem. A longer-term cost-effective solution lies in conditioning die at the wafer level. However, technical challenges such as thermal management of kilowattage of power generated by powering all die on a wafer, CTE mismatch between probe contact mechanism and the wafer, reliable contact to aluminum bond pads through temperature excursions, etc. must be met before wafer-level KGD will become a reality.

Figure 16.21 presents a plot of actual and projected MCM sales from 1992 to 2000.

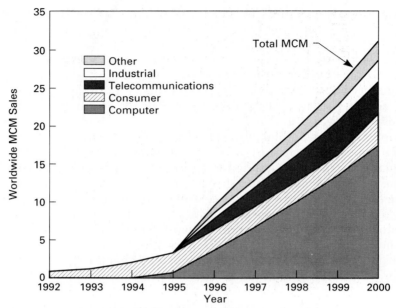

Figure 16.21 Worldwide MCM forecast (fully populated value). (*Source: Toshiba/EMR.*)

References

1. Crowley, R., E. J. Vardaman, and I. Yee, *Worldwide Multichip Module Market Analysis,* TechSearch International, Inc., July 1993.
2. Hagge, J. K., and R. J. Wagner, "High Yield Assembly of Multichip Modules through Known-Good IC's and Effective Test Strategies," *Proceedings of the IEEE,* pp. 1965–1994, Dec. 1992.
2a. Williams, T. W., and N. C. Brown, "Defect Level as a Function of Fault Coverage," *IEEE Transact. on Computers,* pp. 987, 988, Dec. 1981.
3. Landzberg, Abraham H., ed., *Microelectronics Manufacturing Diagnostics Handbook,* Van Nostrand-Reinhold, New York, 1993.
4. Hnatek, E. R., *Integrated Circuit Quality and Reliability,* Marcel Dekker, New York, 1987.
5. Amerasekera, E. A., and D. S. Campbell, *Failure Mechanisms in Semiconductor Devices,* Wiley, New York, 1987.
5a. McEuen, S. D., "Reliability Benefits of I_{DDQ}," *J. Electronic Testing* **3**: 327–335, 1992
6. Kuo, W., and Y. Kuo, "Facing the Headaches of Early Failures: A State-of-the-Art Review of Burn-in Decisions," *Proc. IEEE.,* **71**(11):1257–1266, Nov. 1983.
6a. Denison, W. K., "Predicting the Reliability of State of the Art Microcircuits," RADC-TR-89-177, 1989.
7. Woods, M. H., "MOS VLSI Reliability and Yield Trends," *Proc. IEEE,* **74**(12):1715–1729, Dec. 1986.
8. Flaherty, J. M., "A Burnin' Issue—IC Complexity," *Test and Measurement World,* **13**(11):61–64, Oct. 1993.
9. Huston, H. H., M. Wood, and V. M. DePalma, *Burn-in Effectiveness–Theory and Measurement, 29th Annual Proceedings, 1991 RPS,* pp. 271–276, April 1991.

10. Jensen, F., and N. E. Petersen, *Burn-in: An Engineering Approach to the Design and Analysis of Burn-in Procedures,* Wiley, New York, 1982.
11. Reliability Analysis Center newsletter.
12. Eastman, J., W. Creighton, A. Laidler, and T. Leung, "Defect Migration of Multichip Modules Using Structural Test," *Proceedings of ISHM-IEPS International MCM Conference,* pp. 230–235, Spring 1994.
13. Ferris-Prabhu, A. V., *Introduction to Semiconductor Device Yield Modeling,* Artech House, 1992.
14. Singh, A. D., "On Optimizing VLSI Testing for Product Quality Using Die-Yield Prediction," *IEEE Transact. CAD ICs Systems,* **12**(5), May 1993.
15. Vasquez, B., and S. Lindsey, "The Promise of Known Good Die Technologies," *Proceedings of ISHM-IEPS International MCM Conference,* pp. 1–6, Spring 1994.
16. Murphy, C. F., "Known Good Die Selection Tradeoffs: A Cost Model," *Proceedings of ISHM-IEPS International MCM Conference,* pp. 261–265, Spring 1994.

Chapter 1

Burn-in and Test Substrate for Flip Chip ICs

Glenn A. Rinne

17.1 Introduction

The useful lifespan of integrated circuits is limited by a number of factors, some easily controlled and others less so. Integrated-circuit and electronic equipment manufacturers have devised a number of strategies for preventing unreliable circuits from reaching the customer. One of the most common strategies is burn-in: the accelerated aging of circuits to detect latent defects. Accelerated aging of flip chip ICs has been problematic because of the need to make reliable electrical connections to large arrays of small, soft solder bumps. The burn-in and test substrate (BATS) developed at MCNC addresses this problem using temporary solder joints to connect the IC to a reusable test carrier.

17.2 Principles of Burn-in

The empirical measurement of electronic component reliability typically exhibits the classic "bathtub" curve. The initial high failure rate of components decreases rapidly as the weak components [so-called early-life failures (ELF)] are removed from the population (Fig. 17.1).

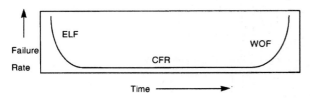

Figure 17.1 The "bathtub" curve.

The population then exhibits random failures at a nearly constant failure rate (CFR). Finally, as the components age still further, the failure rate increases as a result of wearout failures (WOF).

17.2.1   Failure distributions

For most production operations, the primary concern is the reduction or elimination of early-life failures. Empirically derived failure curves have been fitted to the Weibull distribution function[1]

$$f(t) = \frac{\beta}{\lambda}\left(\frac{t}{\lambda}\right)^{\beta-1} \exp\left[-\left(\frac{t}{\lambda}\right)^{\beta}\right] \tag{17.1}$$

where β = shape parameter
λ = characteristic lifetime
t = time

For example, the Weibull distribution for a population with a shape parameter of 1 and a characteristic lifetime of 1000 will exhibit the ELF curve shown in Fig. 17.2, where time is in seconds of operation.

This distribution is useful in predicting the anticipated failure rate at any given time. A more useful application of the distribution is the cumulative failure distribution obtained by integrating $f(t)$ over time (Ref. 1, p. 231):

$$F(t) = 1 - \exp\left[-\left(\frac{t}{\lambda}\right)^{\beta}\right] \tag{17.2}$$

This distribution allows the prediction of the fraction of the population that will have failed up to a given time (t).

In Fig. 17.3, the line $F(t) = 0.632$ is highlighted because it is the intercept with $t = \lambda$, where λ is the characteristic lifetime. Therefore λ

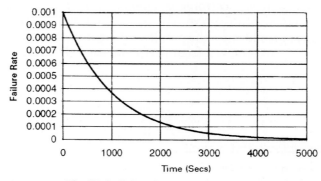

Figure 17.2   The Weibull distribution.

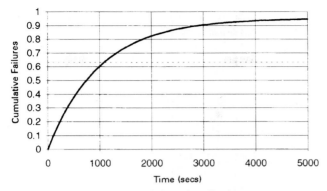

Figure 17.3 The cumulative failure distribution.

is defined as the time t at which 63.2 percent of the population has expired.

For cost-effective manufacturing, it is important that these ELF be identified as early as possible, certainly prior to shipment. However, it is impractical to operate every component for thousands of hours prior to shipment; therefore it is necessary to enable detection of latent ELFs in a cost-effective manner.

17.2.2 Reaction rates

In 1883, before the electron was discovered, the Swedish physicist and chemist Svante August Arrhenius postulated the existence of ions in solutions in order to explain electrolysis.[2] While investigating the dissolution of metal ions in 1887 Arrhenius determined that the rate of chemical reaction is dependent on temperature and an exponential factor called the *activation energy*. This relationship allowed Arrhenius to predict the rate of reaction at any temperature if the rate was known at two different temperatures. The acceleration of a reaction as a function of temperature is calculated using a relationship known as the *Arrhenius reaction-rate formula*

$$A_{i,j} = \exp\left[\frac{E_j}{K}\left(\frac{1}{T_i} - \frac{1}{T_0}\right)\right] \tag{17.3}$$

where $A_{i,j}$ = acceleration factor at temperature T_i for an activation energy E_j
T_0 = baseline temperature (usually ambient)
T_i = second temperature (usually elevated)
E_j = activation energy

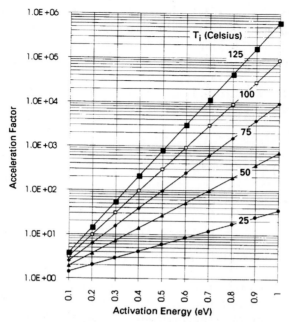

Figure 17.4 Acceleration factors as a function of activation energy.

For reactions characterized at $T_0 = 0°C$, the acceleration factors for activation energies up to 1 eV are shown in Fig. 17.4 for various elevated temperatures.

As an example, components with defect mechanisms having an activation energy of 0.6 eV will accumulate 1000 equivalent operating hours for every hour at elevated temperature when operated at an elevated temperature of 100°C. This, then, is the basis for burn-in.

In practice, the conditions of burn-in include environmental requirements other than temperature. Applied power, biases, electrical stimuli, and other requirements are often specified to accelerate defect mechanisms that are also affected by voltage or current. Therefore, electrical connections to the component are required during the elevated-temperature burn-in.

17.3 Flip Chip Burn-in Considerations

The need for electrical connections during burn-in is most problematic for bare die, particularly solder-bumped flip chips. Maintaining electrical connections to a large array of soft solder bumps during elevated-temperature burn-in can be extremely difficult. Since the mechanisms that cause these connections to fail are accelerated by increased temperature, the connections are often less reliable than the ICs being tested. The net result is that the data collected from burn-in is contaminated or even dominated by connection failures.

An effective burn-in methodology for solder bumped flip chip must consider the following criteria: (1) reliable electrical connection, (2) negligible solder-bump damage, and (3) cost. Reliable electrical connections to large arrays of soft solder bumps must take into account the following factors:

1. Alignment
2. Contact area
3. Normal force
4. Differential expansion
5. Coplanarity
6. Oxidation
7. Solid-state diffusion

17.3.1 Alignment

The alignment of a large number of contacts to an array of solder bumps is difficult to achieve and maintain in a burn-in environment because the tolerance requirement for contact placement is more stringent than for flip chip assembly placement. This is the result of the need to maintain sufficient contact area during test and burn-in, as shown in Fig. 17.5.

17.3.2 Contact area

The contact area is important for a number of reasons beyond the obvious issue of contact resistance. Flip chip solder bumps of the standard 125 μm diameter are usually rated at 0.1 to 0.5 A per bump. Small contact areas (Fig. 17.6) may result in current densities that can lead to localized elevated temperatures that, in turn, lead to solid-state diffusion or even melting of the solder and wetting to the contact.

Contact area is also affected by insufficient coplanarity and differential expansion. Coplanarity problems (Fig. 17.7) arise from a variety of causes including solder-bump volume variation, IC topology, contact height variation, and insufficient normal force.

17.3.3 Differential expansion

Differential expansion becomes a problem during elevated-temperature testing and burn-in. If the contact support material has a coefficient of thermal expansion (CTE) different from that of the IC or the

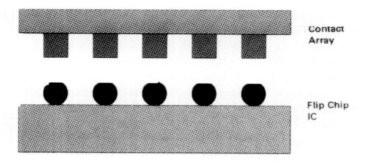

Contact
Array

Flip Chip
IC

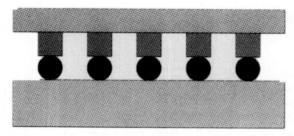

Figure 17.5   Array of contacts.

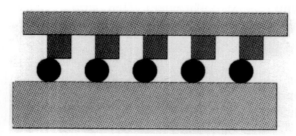

Figure 17.6   Contact area in misaligned condition (simplified).

Figure 17.7   Noncoplanarity.

Figure 17.8 Differential expansion.

IC is operated at a temperature different from that of a CTE-matched support, the positions of the contacts relative to the solder bumps can vary, leading to reduced contact area, as shown in Fig. 17.8.

17.3.4 Oxidation

Even with adequate contact area, maintaining satisfactory electrical connections to solder bumps is made challenging by several chemical and metallurgical processes. The most challenging of these is the natural tendency of tin to oxidize, forming an insulating layer, especially at elevated temperature. It is critically important, therefore, to prevent oxygen from reaching the contact area during the test and burn-in process. This is most often accomplished by performing all testing and burn-in in an inert atmosphere, such as pure nitrogen.

17.3.5 Solid-state diffusion

Another concern is the solid-state diffusion of metals from the contacts into the solder or from the solder into the contacts. The first case can affect the metallurgy of the solder bump, changing fatigue resistance. The second case can affect the metallurgy of the contact, limiting the useful life.

As an example, for a gold contact held against a solder bump at elevated temperature, diffusion of the gold into the tin of the solder can be calculated in the following manner. The rate of diffusion D through an infinite plane at temperature T is given as

$$D_T = D_0 \, e^{-(E_a/RT)} \tag{17.4}$$

where D_0 = diffusion constant for gold into tin
   E_a = activation energy for this solid-state diffusion reaction
   R = universal gas constant

Although this model ignores the spreading effects that would be present in a point contact system, this does provide a conservative first-order approximation. From the CRC tables, $D_0 = 0.16$ cm/s, $E_a = 0.769$ eV, and $R = 7.909 \times 10^{-22}$ kg m²/s².

At a typical burn-in temperature of 150°C (423°K), the gold diffusion front will progress at $D_T = 1.13 \times 10^{-6}$ μm/s. Over a typical burn-in duration of 160 h (5.76×10^5 s) the diffusion front will travel approximately 0.651 μm into the solder, potentially affecting the metallurgical properties of the surface of the solder bump. Since the contact is usually made at the top surface of the bumps as shown in Fig. 17.5, the affected surface is precisely the area that must later successfully wet to the substrate pad during assembly. As for the effect on the metallization of the contact, since the solid solubility limit for gold into tin at 150°C is approximately 0.14 atomic percent (at%),[3] the contact will lose 0.324 μm of gold from the surface. Whether this constitutes a concern for probe lifetime, of course, depends on the thickness of the gold on the contact.

17.3.6 Bump integrity

Since the temperatures during normal operation, testing, and burn-in are all above half the melting point of solder (room temperature = 298 K, testing temperature = 343 K, burn-in temperature = 423 K, melting point of eutectic = 456 K, and the melting point of 95/5 PbSn = 591 K), the plastic and creep properties of solder must be considered. Under the constant load of the contact normal force, the solder will undergo plastic deformation and creep deformation that will reduce the height of the bump and increase the diameter as shown in Fig. 17.9.

A simple geometric calculation will show that two solder balls separated by a space equal to their diameter will, when deformed to 50 percent of their height, touch each other. This, then, sets an upper limit on the allowable deformation during burn-in. Allowing a margin of safety, a more reasonable design limit would be 35 percent deformation during burn-in.

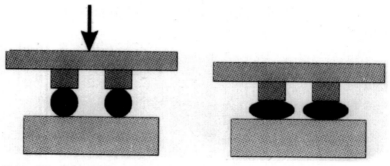

Figure 17.9 Solder deformation.

17.3.7 Plastic and creep deformation

To calculate the expected deformation during test and burn-in, the conditions of burn-in (temperature and applied contact normal force) and the mechanical properties of the solder (Poisson's ratio, shear modulus, and temperature coefficient of shear modulus) must be known. Using the empirically derived relationships for time-independent plastic flow, transient creep, and steady-state creep,[4] an estimate of bump deformation can be calculated. The contact normal force is the only variable affected by the burn-in contact scheme, so an analysis of deformation as a function of contact normal force can be useful in selecting an appropriate solution. For 97.5/2.5 PbSn solder bumps 125 μm in diameter, exposed to a 160-h burn-in at 150°C, the analysis results in the relationship shown in Fig. 17.10, which plots inelastic strain (deformation) ε_{in}, as a function of the applied normal force in grams.

Most contacting systems depend on high normal forces to maintain low contact resistance. For example, IBM's "COBRA probe" area array probe system uses a normal force of 25 g.[5] However, from Fig. 17.10 it can be seen that the highest acceptable normal force that will cause less than 35 percent deformation is less than 6 g. When noncoplanarity is taken into account, this maximum force will occur at the first contact/bump pair that touch as the contacts approach the die. All other contact/bump pairs will have lower normal forces. Low contact forces can result in unreliable connections, particularly during the rigors of burn-in. One solution is to use a compliant contact support material such as so-called membrane probes. The difficulty here is the competing requirements of flexibility and routing density. For fine-pitch (125-μm-diameter bumps on 250-μm centers) flip chip

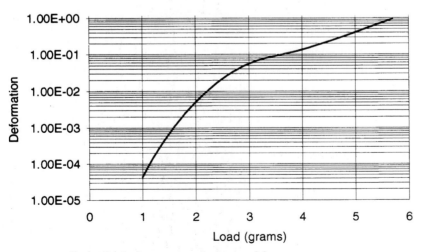

Figure 17.10 Inelastic strain versus contact normal force.

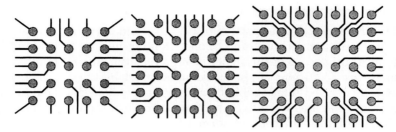

Figure 17.11   Routing of 5×5, 6×6, and 7×7 arrays.

arrays, "one between" routing (one conductor line between any two bumps) is about the best that current membrane technology can support on a single conductor layer. As the size of the flip chip array increases, the problem of routing the signals at the interior of the array out to the periphery where the tester connections are made becomes severe. This is illustrated in Fig. 17.11, where one-between routing rules prevent connection to the center bump of a 7×7 array.

The obvious solution of adding additional routing layers, however, reduces flexibility and compliance of the membrane, thus aggravating the deformation problem.

17.4   Attributes of the Ideal Flip Chip Burn-in and Test Solution

The net result of this discussion is that the ideal solution for flip chip burn-in and testing would have the following attributes:

1. Self-alignment with wide placement tolerance
2. Large contact area
3. Low contact resistance
4. Zero applied normal force
5. Large noncoplanarity tolerance
6. Gastight connections
7. Negligible solid-state diffusion
8. High routing density

17.5   The Burn-in and Test Substrate

One approach that addresses the flip chip test and burn-in described above is to use MCNC's *Burn-in and test substrate* (BATS) method,[6,7] shown in Fig. 17.12. This method, developed by the MCNC Electronic

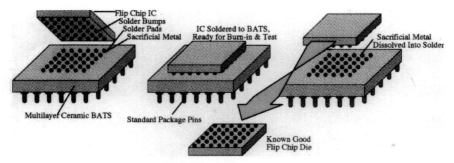

Figure 17.12   The burn-in and test substrate.

Technologies Division, provides a temporary metallurgical connection between the flip chip IC and a reusable carrier. The connection is made temporary by the inclusion of a solder-soluble sacrificial metal layer at the solder–substrate interface. This layer produces a strong, low-resistance solder joint that can withstand the rigors of burn-in and testing. On completion of the testing, the strength of the solder joint can be dramatically reduced by inducing the sacrificial metal to dissolve into the solder bump, whereupon the solder is left in contact with the nonwettable base metallurgy of the substrate. The greatly reduced strength of the joint facilitates the damage-free removal of ICs, even with very high I/O counts. MCNC has practiced this method with hundreds of test chips having up to 1679 solder bumps without damage.

The first-generation BATS is a multilayer ceramic substrate with a cofired thick-film metallurgy that is not wettable by PbSn solders. The top surface is patterned with an array of pads matching the pattern of the IC to be tested, usually a standard grid. Each top surface pad is routed to a pin or pad on the bottom surface that matches a JEDEC standard package pattern such as a pin grid array (PGA). This allows a single substrate to support any IC with a matching bump grid while conforming to industry-standard sockets and handling equipment. Prior to mounting an IC to the BATS, a thin layer of sacrificial metal is deposited over the nonwettable top surface pads. The nonwettable metal is selected to minimize solid-state diffusion. An IC is then placed onto the substrate with the solder bumps coarsely aligned with the pads. A traditional solder reflow is performed, self-alignment due to surface tension of the solder occurs, and the IC is metallurgically joined to the substrate as shown in Fig. 17.13. Testing, burn-in, and retest are then performed.

Because the connection is a true solder joint, all the requirements for a flip chip burn-in solution are met. The contact area is the same

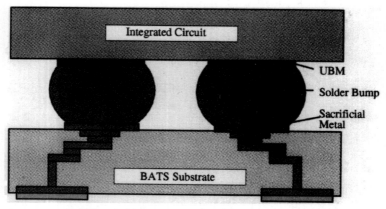

Figure 17.13   BATS with IC mounted.

as the pad area (equivalent to the bump diameter), the controlled collapse tolerates large noncoplanarities, and the connection has low resistance and is gastight. No normal forces are required to maintain the connection. Since the substrate is rigid, it can be a multilayer interconnection with tremendous routing density potential.

After the final test is complete, the BATS assembly again undergoes the reflow process, at which time the sacrificial metal dissolves into the solder bump, leaving the solder in contact with the nonwettable cofired metallurgy. This causes the solder to dewet from the substrate surface, dramatically reducing the strength of the solder joint. The IC can then be removed from the substrate as shown in Fig. 17.14. Since the sacrificial metal is very thin, the volumetric fraction

Figure 17.14   Removing the IC from the BATS.

of sacrificial metal in the solder bump is on the order of 10^{-4}. Depending on the sacrificial metal used, the contamination limits for lead-tin solder ranges from 0.02 percent to several percent.[8] The BATS design maintains a safety margin of at least 100 percent to avoid contamination issues. Using this temporary solder connection technique, it is possible to fabricate a reusable carrier for flip chip KGD. After the IC is removed from the BATS, the sacrificial metal is redeposited in a batch process and the test substrate is ready for reuse.

One unique advantage of this technique is that solder bumps can be transferred to bare die, provided a suitable solder-wettable terminal is predeposited on the I/Os. This eliminates the need for solder deposition processes at the wafer level. Using MCNC's electroplated solder-bump deposition process, solder bumps are formed on top of the sacrificial metal on the BATS substrates. When a bare die is placed on the substrate and reflowed, the solder forms a strong intermetallic bond with the UBM on the IC, as shown in Fig. 17.15. The remainder of the processing proceeds as described above.

Another benefit of this technique is derived from the electrical performance of flip chip connections. For very high-performance integrated circuits, it is difficult to accurately test the high-frequency characteristics because of the parasitic reactances of the test fixturing or temporary packaging. However, since the size of flip chip solder bumps is very well controlled, the electrical performance of flip chip connections is very consistent. Therefore, a calibration test chip can be fabricated containing the RF calibration structures necessary to characterize the BATS substrates. This test chip would have solder bumps of the same dimensions as those on the IC to be tested. By temporarily mounting the test chip to the BATS substrate, the high-frequency reflection and transmission characteristics of the BATS can be measured. After removal of the test chip, the effects of the BATS on electrical tests can be subtracted from the measurements using any of the various deembedding techniques.

The initial research and development of the BATS technique was conducted using vacuum-deposited thin-film metallurgy on silicon substrates, as shown in Fig. 17.16. A special bumped die test vehicle

Figure 17.15 BATS for solder-bump transfer.

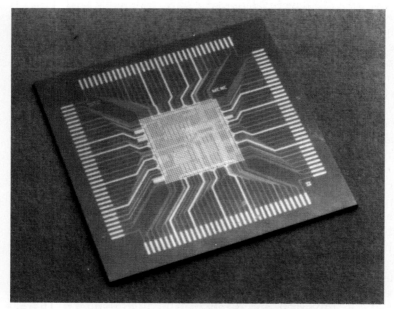

Figure 17.16 Thin film on silicon BATS demonstration vehicle.

(BDTV) was designed to provide parametric test data. The test die, shown mounted to silicon BATS in Fig. 17.17, has 1679 solder bumps and contains test structures such as daisy chains, strain gauges, four-wire bump resistance monitors, heaters, and temperature-sensing diodes. These structures facilitated the optimization of the BATS technique.

Recent development efforts with this technology have focused on decreasing the cost of the BATS process. Early cost modeling was based on a reusable ceramic single-chip carrier design. The cost model included the amortized cost of the carrier, chip placement, reflow, cleaning, chip removal, and substrate reconditioning for reuse. The estimated cost was normalized to cost per I/O pin to facilitate comparison with other technologies. The model predicts a cost of $0.01 per pin for ICs with 200 pins. Using the model as a guide, the significant cost factors have been targeted for cost reduction.

The next-generation of BATS will be a multichip burn-in board, as shown in Fig. 17.18. The BATS multichip burn-in board will be designed to match the outline and connectors of standard burn-in chambers, and the flip chip ICs will be bonded directly to the burn-in board. For ICs designed for testability using the IEEE 1149.1 Boundary Scan architecture, a test access port will be provided. In other cases, probe pads will allow access to individual die during testing. Once die are

Figure 17.17 1,679 I/O test chip mounted on a silicon BATS.

Figure 17.18 Low-cost BATS burn-in board and shipping tray.

bonded to the burn-in board, the burn-in board serves as the test socket, burn-in tray, and shipping and storage tray. When needed for flip chip assembly, the burn-in board is sent through a reflow furnace to cause the solder to dewet from the pads. The burn-in board can then be used for parts presentation to robotic pick-and-place equipment.

17.6 Summary

Testing and burn-in of solder-bumped flip chip ICs presents many challenges—mechanical, electrical, and chemical. Solutions to the burn-in problem must carefully address these challenges to be successful. The BATS technology seeks to address these by forming metallurgical connections to the solder bumps that can be readily disconnected when required. With the technical challenges overcome, the economic factors are the new focus for this technology.

References

1. Tummala, R. R., and B. T. Clark, "Multichip Packaging Technologies in IBM for Desktop to Mainframe Computers," *Proceedings of the 42nd Electronic Components Technology Conference,* **1**:225–359, May 1992.
2. *Compton's Encyclopedia,* online edition, downloaded from America OnLine, Nov. 11, 1994.
3. Matijasevic, G. S., C. C. Lee, and C. Y. Wang, "Au-Sn Alloy Phase Diagram and Properties Related to Its Use as a Bonding Medium," *Thin Solid Films,* p. 223, 1993.
4. Darveaux, R., and K. Banerji, "Constitutive Relations for Tin-Based-Solder Joints," *Proceedings of the IEEE 42nd Electronic Components Technology Conference,* pp. 538–551, 1992.
5. *IBM C4 Product Design Manual,* vol. 1, *Chip and Wafer Design,* Issue: A, pp. 9–8.
6. Adema, G. M., C. J. Berry, N. G. Koopman, G. Rinne, E. K. Yung, and I. Turlik, "Flip Chip Technology: A Method for Providing Known Good Die With High Density Interconnections," *Proceedings of the 3rd International Conference and Exhibition on Multichip Modules,* Denver, April 1994.
7. Koopman, N. G., G. A. Rinne, I. Turlik, and E. K. Yung, "Method for Testing, Burn-in And/Or Programming of Integrated Circuits," U.S. Patent 5,289,631, March 1, 1994.
8. Manko, H. H., *Solders and Soldering,* McGraw-Hill, New York, pp. 73–78, 1979.

Author Biographies

John H. Lau is the president of Electronics Packaging Services. His current research and development activities cover a broad range of electronics packaging and manufacturing technology.

He received the B.E. in Civil Engineering from National Taiwan University, the M.A.Sc. in Structural Engineering from the University of British Columbia, the M.S. in Engineering Mechanics from the University of Wisconsin, the M.S. in Management Science from Fairleigh Dickinson University, and the Ph.D. in Theoretical and Applied Mechanics from the University of Illinois.

Prior to founding Electronics Packaging Services in 1995, he worked for Hewlett-Packard Company, Sandia Research Laboratory, Bechtel Power Corporation, and Exxon Production and Research Company. He has more than 25 years of research and development experience in applying the principles of engineering and science to the electronic, petroleum, nuclear, and defense industries. He has authored and coauthored over 100 technical publications in these areas, is the author of twelve book chapters, and is the author and editor of eight electronics packaging books.

Nick Koopman received a B.S. in Metallurgical Engineering from Lafayette College (1960) and an M.S. and a Ph.D. in Metallurgy and Materials Science from the Massachusetts Institute of Technology (1963 and 1966). He was employed for 23 years with the IBM General Technology Division, East Fishkill, New York, specializing in computer microelectronics interconnection metallurgy development. As manager of Bonding and Interconnections, Nick had responsibility for developing the C4 flip chip technology for IBM.

Since 1990 Nick has been with MCNC Electronics Technologies Division in Research Triangle Park, North Carolina. The current assignment is Program Manager, Fluxless Soldering. He has published extensively in the area of interconnections with 40 papers and holds 48 inventions. He has received numerous informal awards, a Division award, 4 Invention Plateau awards, and Best Paper awards at Packaging ITL, the Electronic Components and Technology Conference, and at Surface Mount Technology Association. He has chaired an internal technical seminar series on Advanced Metallurgy/ Technology (AMET) for many years, and has taught interconnection technology courses at numerous IBM sites and over CNET. The basis for one course has been published in a book (*Microelectronics Packaging Handbook*). He has been a guest lecturer at Duke, North Carolina State, University of North Carolina Chapel Hill, and Georgia Tech. He is on the editorial board of the journal *Processing of Advanced Materials* and an associate editor of the *International Journal of Microelectronics Packaging*.

Sundeep Nangalia is an associate member of technical staff at MCNC's Electronics Technology Division, Research Triangle Park, North Carolina. He received a B.S. in Metallurgical Engineering from Andhra University, Visakhapatnam, India, and an M.S. in Material science from North Carolina State University in 1993.

He is currently working on a dry fluxless soldering process. He has worked extensively in flip chip related areas and was also a member of a task force on a novel known-good-die approach for flip chips. His research interests lie in the fields of assembly and electronic packaging. He has numerous award-winning publications and one patent.

Ravi Sharma received his Ph.D. in Materials Science at Marquette University in 1977. Prior to joining Motorola in 1984, he was with Bendix Corporation where he was responsible for automotive exhaust sensors development. As Manager of Process and Interconnect Technologies Group at Motorola he spearheaded the development of several flip chip interconnect systems based on Pb-free solders and other new composite materials. He has contributed extensively in both fundamental and applied research aspects of electronic ceramics, chemical sensors, electronic thin films and package interconnect technologies. Dr. Sharma is a corecipient of 1987 Edward C. Henry Award for the most significant contribution to electronic ceramic literature published in the *Journal of American Ceramic Society* during the preceding ten years. He is currently Manager of Chemical Sensor Package Development for Motorola, Toulouse, France.

Ravi Subrahmanyan received his Ph.D. in Materials Science at Cornell University, Ithaca, New York, in 1990, and is currently with the Advanced Interconnect Systems Laboratory, Motorola Inc., Tempe, Arizona. Since 1985, he has been actively involved in solder fatigue research, and he has several publications to his credit which describe the fundamental aspects of the phenomenon. Dr. Subrahmanyan is currently involved in materials research and development related to design, processing and reliability assessment of advanced electronic package interconnect systems for new products.

Tom C. Chung is Tandem Computers' representative at Microelectronics and Computer Technology Corporation (MCC) in Austin, Texas. Dr. Chung has more than 16 years of experience in the field of advanced computer packaging and interconnect technologies, including surface mount, tape-automated bonding, flip chip, ball grid array (BGA), advanced single-chip packaging, and multichip module technologies. He worked on semiconductor assembly, test, packaging, and interconnect-related technology developments for United Technologies Corporation and MCC for five years, respectively, before joining Tandem Computers Inc. in 1989 as a development engineering manager in the Advanced Computer Packaging and Interconnect Department. Since early 1993, Dr. Chung has taken the responsibility as Tandem's representative at MCC to not only manage technology-development-related projects and activities but also lead a group to develop flip chip and BGA-based single-chip and MCM technologies for high-performance computer workstations and multiprocessors. Dr. Chung has a B.S.E. degree from NCKU in Taiwan, an M.S.M.E. from Texas Tech University, an M.S. in Engineering Management, and a doctorate degree in engineering from Southern Methodist University, Dallas, Texas. He holds four U.S. patents and has published more than 25 technical papers. He is also a coauthor of three technical books, the *Handbook of Tape Automated Bonding, Chip on Board Technologies for Multichip Modules,* and *Ball Grid Array Technology,* which were published in 1992, 1994, and 1995, respectively.

Tom Dolbear a is Senior Product Development Engineer at Advanced Micro Devices in Austin, Texas. Prior to joining AMD he was a member of the technical staff in MCC's High Value Electronics Division. He joined MCC in July 1988. He obtained a B.S. in Mechanical Engineering from The University of Texas at Austin in 1987 and his M.S. in Mechanical Engineering from Stanford University in 1988. Mr.

Dolbear's area of concentration at MCC has been in the thermal management of electronic systems. While at MCC he developed liquid- and air-cooled heat exchangers for electronics cooling and integrated them into electronic systems. In addition, he also conducted numerical and experimental analyses to improve the thermal conductance of copper/polyimide interconnect as well as to understand in-process mechanical stresses during fabrication. He has developed compliant, ultralow resistance thermal interface materials and used finite element modeling to provide a greater understanding of MCC's laser bonding process and of flip chip bonding of ICs using adhesives. Mr. Dolbear was the task leader for the thermal task in the MCC Workstation Open Systems Project addressing the cooling design for 175-W MCMs for 1996 workstations. He was the project leader for MCC's Low Cost Adhesive Flip Chip Project, task leader for BGA thermal and reliability aspects of MCC's Parallel Processing Project. Mr. Dolbear holds six patents and has three others in application. He has published six papers in the area of electronic packaging. He is a member of ASME, Pi Tau Sigma, Tau Beta Pi, ASM, and IEPS.

Richard Douglas Nelson joined MCC in 1985 and is a senior member of the technical staff in MCC's High Value Electronics Division. He has pursued research in air and liquid cooling for electronics, thermal interfaces, nondestructive testing of TAB bonds, and modeling of thermal and thermo-mechanical aspects of electronic packaging. Specific technical interests cover both the analytic and experimental aspects of electronic system cooling, stress analysis, solder mechanics, adhesive bonding, die attach, acoustic emission, and noise control for electronic systems. While at MCC, Dr. Nelson has been named as inventor or coinventor of 12 granted U.S. Patents. Dr. Nelson received his Ph.D. from the University of California at Berkeley in 1968. His doctoral thesis concerned frictional anisotropy exhibited by single crystal metals in ultrahigh vacuum. From 1968 to 1977, Dr. Nelson taught mechanical engineering at the University of Alaska, Fairbanks, and conducted research on sea ice mechanics. From 1978 to 1983, he was a member of the technical staff at Hewlett-Packard Laboratories doing research in high-density magnetic disk and tape recording. He left Hewlett-Packard in 1983 to work at Seagate Technology, where his work focused on thermal deformations in small disk drives. Dr. Nelson began working in the MCC Packaging and Interconnect Program in 1985, shortly after its inception and was responsible for the establishment of its thermal and mechanical analysis capabilities.

Yinon Degani received his B.Sc. in Chemistry and Physics in 1982 and his Ph.D. in Chemistry in 1985 from the Hebrew University of Jerusalem. He joined AT&T Bell Laboratories in 1985 where he is currently a member of technical staff in the Materials and Technology Integration Laboratory. Yinon's research topics include: photoelectrochemistry of semiconductors, electrochemistry of enzymes, device fabrication for optical fiber communication, and chemistry of solder pastes and fluxes. Yinon's research is currently centered around solder-based electronic assembly technologies.

T. Dixon Dudderar is a Distinguished Member of Technical Staff in the Materials and Processing Research Laboratory of the AT&T Bell Laboratories, Murray Hill, New Jersey. He received his B.S. in Mechanical Engineering from Lehigh in 1957 and worked on the design of guidance systems and the Telstar satellite antenna at Bell Laboratories, Whippany, New Jersey, until 1961. After receiving his Ph.D. in Engineering Mechanics from Brown University in 1966, Dudderar rejoined Bell Laboratories where his areas of research activity have ranged from experimental studies of the mechanics of materials and fluids to fiber optics and holographic interferometry. Current interests include research on the mechanical properties of electronic materials and the development of advanced techniques for the assembly of high-performance and/or high-density electronics, with an emphasis on manufacturability and yield, environmental compatibility, long-term reliability and cost reduction. Dudderar is a Fellow and President of the Society for Experimental Mechanics, and a member of the National Academy of Engineering, the New York Academy of Sciences, the American Academy for Mechanics, Sigma Xi, the International Society for Hybrid Microelectronics, the International Society for Optical Engineering (SPIE), and the American Society for Metals.

Robert C. Frye received his B.S. in Electrical Engineering from M.I.T. in 1973. From 1973 to 1975 he was with the Central Research Laboratories of Texas Instruments, where he worked on charge-coupled devices for analog signal processing. In 1975 he returned to M.I.T., and received a Ph.D. in Electrical Engineering in 1980. Since then, he has been with AT&T Bell Laboratories. His past research activities there have included thin-film semiconductor devices and neural network implementation and applications. More recently, his work has focused on advanced electronic interconnection technology and multichip modules.

He is currently a Distinguished Member of Technical Staff in the Electronic Packaging Research Department. He is a member of IEEE, ISHM, IEPS, and the Materials Research Society.

King L. Tai joined Bell Labs in 1967. In 1976, he earned his Ph.D. from Stevens Institute of Technology. He is an AT&T Bell Lab Fellow. He is a supervisor of Microinterconnect & System Prototyping Research. His current research activities are focused on low-cost and high-performance packaging and interconnection technologies. His past interests include submicron lithography and mass transport in thin films.

Maureen Y. Lau is a member of the technical staff in the Electronics Packaging Research Department at AT&T Bell Labs, since 1988. She is responsible for new processes and fabrication methods for leading edge packaging technologies, to date, primarily MCM-D. Upon joining Bell Labs in 1982, she was involved in high-speed silicon NMOS and bipolar-CMOS integrated circuit research. Maureen Y. Lau received a B.S. degree from the State University of New York at Binghamton in 1982, and an M.S. degree from Stevens Institute of Technology, Hoboken, New Jersey, in 1987.

Byung Joon Han is a member of the technical staff of the Physical Science and Engineering Research Division of the AT&T Bell Laboratories, Murray Hill, New Jersey. After receiving his Ph.D. from Columbia University in 1988, he joined Bell Laboratories where his areas of research activity have ranged from experimental studies of the mechanical properties of polymers to photosensitive dielectrics and electronic package design. Current interests include research into the thermal properties of high-temperature polymeric materials, advanced printed circuit board design and structures, electronic package design and analysis, and the development of advanced techniques for the processing of high-performance and/or high-density microelectronics. Han is a member of the Sigma Xi, the international Society for Hybrid Microelectronics, the Materials Research Society, and the American Institute of Chemical Engineers.

Richard H. Estes received his degree in Chemistry from the University of Massachusetts in 1975. He was a chemistry instructor for six years prior to taking a position at Epoxy Technology, Inc. in February 1981. Mr. Estes has held the positions of Quality Control Manager, Technical Service Manager, and is currently the Vice

President of Technical Operations at Epoxy Technology. Areas of responsibility include technical services and quality control, as well as supervising R&D in the development of new materials and processes for applications in the semiconductor and hybrid microelectronics industries and the optoelectronics/fiber optics industries. Mr. Estes has authored several technical papers on the technology of adhesives, is a member of ISHM, SEMI, and IEEE, and holds patents in the field of flip chip technology.

Frank W. Kulesza, president of Epoxy Technology, Inc. is a graduate of Northeastern University, receiving a B.S. in Chemical Engineering in 1950. Prior to his founding of Epoxy Technology in 1966 he held positions at Borden Chemical Company and IBM Corporation. Mr. Kulesza has authored and presented a number of technical papers on epoxies and their applications, both here and abroad. Mr. Kulesza was primarily responsible for and is well known as the pioneer in the use of electrically conductive epoxies being employed for die attach in the microelectronic industry. He was presented the Technical Achievement Award by ISHM in 1989 for his contributions to the microelectronic industry in relation to epoxy die attach. More recently, he was granted a patent on PFC, a polymer flip chip processing technology—an efficient and cost-effective method of flip chip assembly without the use of solder. He is a corporate member of ISHM and SEMI and holds memberships in the American Chemical Society as well as the Society of Plastic Engineers.

Mark Breen is a member of the technical staff with AMD, focusing on technology transfer and implementation of dry etch processes. Mr. Breen was a member of the technical staff with MCC's High Value Electronics (HVE) Division from 1986 to early 1995. At MCC, he worked as a process engineer in the microfabrication group specializing in polymerprocessing, photopolymer processing, and thick-film photolithography. He worked on polymer coating applications for projects that included displays, chip-on-board packaging, holographic information storage, and high-density interconnect. For two years (1989 to 1990), Mr. Breen worked as a quality engineer in the fabrication group. He organized a course on designed experiments and continues to work with members of the technical staff on experimental design and analysis. Prior to joining MCC, Mr. Breen worked at MA/COM and Duracell (both in Burlington, Massachusetts). Mr. Breen received his B.S. in Chemical Engineering from the University of Notre Dame (South Bend, Indiana) in 1983. He published various articles about MCC's Low Cost Interconnect Project. He is a member of SPIE and IEPS.

Randy L. German is an etch engineer for Cypress Semiconductor in Round Rock, Texas. Mr. German was a member of the technical staff in the High Value Electronics Division of MCC from 1989 to 1994. His work included research in electroless metallization of integrated circuit devices for flip chip applications, direct metallization of dielectric materials, and anisotropic wet etching of copper. Mr. German was responsible for all wet etch processes and electrolytic and electroless metallization.

Mr. German received a B.S. in Chemistry from Oklahoma State University. He received an MCC Outstanding Achievement Award for his work in electroless metallization. He has one U.S. Patent and has coauthored papers concerning electroless metallization and compliant bumps for advanced packaging technologies.

Kathryn Keswick is a staff member working on the development and transfer of dry etch processes with Applied Materials, Inc. in Austin, Texas. Ms. Keswick was a member of the technical staff with MCC's High Value Electronics (HVE) Division from 1990 to 1994. She has worked at MCC since 1990. Her work included research in electroless metallization of dielectrics, dielectric materials for MCM-D applications, novel new wafer bumping processes, chemomechanical polishing, and nonhermetic encapsulants. She developed a dry etch process for patterning niobium on SiO_2 for fabrication of high-density niobium/SiO_2 substrates. Other processes she developed at MCC include: dry etch of silicon, inorganic dielectrics, polyimide etches, depassivation etch for rework, and failure analysis of ceramic packages. Before joining MCC, Ms. Keswick was employed with ARDEX, Inc., where she was a process engineer. Her work included developing processes for etching and laminating flexible thin film circuits. Ms. Keswick received a B.S. in Chemistry from the University of Texas at Austin. She is a coauthor of articles on novel wafer bump techniques and low-cost approaches to MCM-D fabrication. She was the lead author on "Compliant Bumps for Adhesive Flip Chip Assembly," which was nominated for the best paper at ECTC 1994.

Rick Nolan is an operations manager with Horizon Battery in San Marcos, Texas. Mr. Nolan was Senior Member of Technical Staff and Laboratory Manager with MCC's High Value Electronics (HVE) Division from 1990 to 1994. His work included research in tape-automated bonding (TAB), TAB rework, flip chip technology, and bare chip test and burn-in technology. Mr. Nolan was the Project Leader of HVE's Flip Chip Technology Development Project, researching methods for implementation of flip chip with commercially available integrated circuit (IC) chips. He was also the project leader for a NIST/ATP award program which has produced revolutionary improvements in flip chip on glass technology, and another NIST/ATP program which developed assembly and interconnection technology for thin IC films. He managed the development of many key MCC proprietary processes, such as laser TAB bonding and gold TAB outer-lead-bonding rework. Before joining MCC, Mr. Nolan was employed by Cincinnati Microwave, where he was Manufacturing Engineering Manager, responsible for $10 million per year, 5 million component placement per week surface mount assembly operations. While at Cincinnati Microwave, he led the manufacturing design team that introduced two new radar receiver products and twelve contract design/assembly products into production. Mr. Nolan received his B.S.E.E. from Virginia Tech in Blacksburg, Virginia. He has been the National Program Chairman of IEPS. He currently holds two U.S. patents and has published several articles relating to advanced packaging technology.

Diana Carter Duane is the Laboratory Operations Manager/Senior Member of the Technical Staff with MCC's High Value Electronics (HVE) Division. She has been with MCC since 1984. Ms. Duane's work at MCC includes the development of low-stress silicon nitride for device passivation, pinhole-free oxide for interlevel dielectric, near-hermetic device passivation, copper/epoxy high-density interconnect, and low-stress sputtered films. Team technical accomplishments include the development of gold- and solder-bump technology, development of copper/polyimide processes for MCM-D, low-cost HDI, development of thick- and thin-film die prep and substrate prep processes for single- and multichip packaging, and development of integrated die prep, bonding, and assembly processes for a variety of technologies including flip chip, TAB, C4, DCA, and BGA. Before joining MCC, Ms. Duane was employed by Motorola as a startup engineer for a CMOS fabrication facility. Ms. Duane received her B.S. in Electrical Engineering from Georgia Institute of Technology and published several articles relating to chip prep and packaging technologies.

Hiroaki Date is a member of the technical staff of the Electronics Packaging Laboratory, Personal System Laboratories, Fujitsu Laboratories Ltd. He received his B.S. in Industrial Chemistry from the Science University of Tokyo in 1987 specializing in organic chemistry.

Yuko Hozumi is a member of the technical staff of the Electronics Packaging Laboratory, Personal System Laboratories, Fujitsu

Laboratories Ltd. He received his B.S. in Materials Science from Himeji Institute of Technology in 1991 specializing in metallurgy.

Hideshi Tokuhira is a member of the technical staff of the Electronics Packaging Laboratory, Personal System Laboratories, Fujitsu Laboratories Ltd. He received his B.S. in Chemical Engineering from Himeji Institute of Technology in 1991 and his M.S. in Chemical Engineering from Himeji Institute of Technology in 1993 specializing in mechanical engineering.

Makoto Usui is a member of the technical staff of the Electronics Packaging Laboratory, Personal System Laboratories, Fujitsu Laboratories Ltd. He received his diploma from Kanagawa Technical High School in 1957 specializing in organic chemistry.

Eiji Horikoshi is a member of the technical staff of the Electronics Packaging Laboratory, Personal System Laboratories, Fujitsu Laboratories Ltd. He received his B.S. in Materials Science from Tohoku University in 1987 and his M.S. in Materials Science from Tohoku University in 1980 specializing in metallurgy.

Takehiko Sato is a member of the technical staff of the Electronics Packaging Laboratory, Personal System Laboratories, Fujitsu Laboratories Ltd. He received his B.S. in Materials Science from Tohoku University in 1969 specializing in metallurgy.

Itsuo Watanabe is a leader and research scientist with Shimodate Research Laboratory of Hitachi Chemical Co., Ltd. Dr. Watanabe has been employed by Hitachi Chemical since 1982. He has been responsible for developing anisotropic conductive adhesive films for flip chip technologies (chip on glass, chip on board). He was also a visiting scientist of the Massachusetts Institute of Technology (Department of Materials Science and Engineering) from 1987 to 1989. He has also worked on conducting polymers, organic optical recording materials, and polymeric materials for optical communication. He received his B.S. and M.S. in Chemistry from Utsunomiya University and his Ph.D. in Polymer Science from Kyoto University. His doctoral research was concerned with syntheses, thin-film formation, and electrical and optical characteristics of conducting polymers.

Naoyuki Shiozawa is a research scientist with Shimodate Research Laboratory of Hitachi Chemical Co., Ltd. Mr. Shiozawa has been employed by Hitachi Chemical since 1991. He has been responsible for developing anisotropic conductive adhesive materials for surface mount technologies. He has also worked on plasma display in the Display Development Section of Oki Electric Industry Co., Ltd. from 1987 to 1991. He received his B.S. and M.S. in Polymer Chemistry from Yamagata University.

Kenzo Takemura is a research scientist with Shimodate Research Laboratory of Hitachi Chemical Co., Ltd. Mr. Takemura has been employed by Hitachi Chemical since 1983. He has been responsible for developing anisotropic conductive adhesive films for flip chip technologies (chip on glass, chip on board). He has also worked on thin-film technologies for electroluminescent display and PWB. He received his B.S. in Physics from the Science University of Tokyo.

Tomohisa Ohta is a senior research scientist with Shimodate Research Laboratory of Hitachi Chemical Co., Ltd. Mr. Ohta has been employed by Hitachi Chemical since 1974. He has been responsible for developing anisotropic conductive adhesive films for flip chip and TCP technologies. He has also worked on insulating adhesives for printed circuit boards and pressure-sensitive adhesives for surface-protective adhesive films. His research interests cover thermal, ultraviolet and electron-beam-curing technologies, and electrical and mechanical properties for connecting materials. He received his B.S. in Physical Chemistry from Kyusyu University and his M.S. in Polymer Science from Osaka University.

Chang Hoon Lee is a senior engineer with Samsung Display Devices, Co., Ltd., Kyungki-Do, Korea. Mr. Lee has been responsible for developing LCD panels over 6 years at Samsung's R&D center. He is also involved in developing IC packaging using FCOG technology for LCD application. A prototype ACA FCOG LCD developed by Mr. Lee was exhibited as a working display at the ISHM '93 and he recently code-veloped and applied for a patent on a new FCOG method with Zymet. Mr. Lee received a B.S. in Electronics from Chungnam National University, Taejon, Korea, in 1988.

Stacey Baba is a member of the technical staff and Project Leader, Acuson Corporation, Mountain View, California. Ms. Baba has been responsible for both process and product development. Prior to joining Acuson Corporation, she was employed by Raychem Corporation, Menlo Park, California. Her areas of interest have been materials development for high- and low-voltage electrical products, adhesive development for improved process development, and interconnect technology. She is currently working on transducer designs for medical application and novel interconnect schemes. Ms. Baba received a B.S. from the University of California, Berkeley, and an M.S. in Materials Engineering from San Jose State University. Her professional affiliations include IEEE and ASM.

William Carlomagno is vice president and general manager of Arrow-McBride Metal Products Company, East Palo Alto, California. Arrow-McBride manufactures products for the commercial electronics telecommunications and defense electronic industries. Prior to joining Arrow-McBride, Mr. Carlomagno was employed by Raychem Corporation, Menlo Park, California. Mr. Carlomagno holds several patents in technology areas including heat recoverable polymers, conductive polymers, and the design of high-performance multichip modules. Mr. Carlomagno received a B.S. degree from San Jose State University, San Jose, California.

Kazuhisa Tsunoi is a senior engineer of the Computer Circuit Engineering Department of Fujitsu Limited. He received a B.S. in Mechanical Engineering from Kanagawa University in 1982. He is specializing in packaging technology of small and middle scale computers.

Toshihiro Kusagaya is a manager of the Computer Circuit Engineering Department of Fujitsu Limited. He received a B.S. in Mechanical Engineering from Ibaragi University in 1975. He is specializing in packaging technology of small and middle scale computers.

Hidehiko Kira is in the PCA Production Engineering Department of Fujitsu Limited, Nagano Plant. He received a B.S. in Mechanical Engineering from Kansai University in 1988. He is specializing in printed circuit assembly products.

Larry L. Moresco is currently the Director of Research and Development at Fujitsu Computer Packaging Technologies, Inc. in San Jose, California. Prior to working at FCPT, he was a senior engineer on the Strategic Planning Staff at Hewlett-Packard Laboratories. He worked at HP from 1981 to 1991. At HP he worked on several packaging research and development projects related to computers and microwave instrumentation microelectronics packaging. He received his Ph.D. in Mechanical Engineering from the University of California at Santa Barbara. He is the author of technical papers and book chapters and has taught short courses on electronics packaging. In 1989 he was awarded Best Paper at the IEEE/IEMT Symposium in San Francisco. He is the editor of the *International Journal of Microelectronics Packaging* published by Gordon and Breach.

David Love is currently the manager of the Assembly and Quality Department at Fujitsu Computer Packaging Technologies, Inc. In this position he is responsible for research and development of interconnect technologies for Fujitsu computer products. Prior to working at FCPT, he was a manager in the Advanced Interconnect Group at Intel, in Santa Clara, California. At Intel, he was responsible for the transfer of C4 technology from IBM to Intel. He received his B.S. in Chemistry at the University of San Francisco in 1976. He is named on five patents, on four of which he is the lead inventor.

William T. Chou is currently a member of technical staff at Fujitsu Computer Packaging Technologies, Inc. His responsibilities include thin-film fabrication technology developments. Prior to working at FCPT, he was a principal engineer at Digital Equipment Corporation and a research scientist at Eastman Kodak Research Laboratories. He received his M.S. in Chemical Engineering from the University of California at Davis. He is named on two patents.

Ven R. Holalkere is currently a member of technical staff at Fujitsu Computer Packaging Technologies, Inc. His responsibilities include stress analysis and computer modeling in the area of electronic packaging. Prior to working at FCPT, he was a staff technologist at Amdahl in Sunnyvale, California. While at Amdahl, he was invited to participate in the development of MCM in Fujitsu at Kawasaki, Japan. He has authored several technical papers in the area of electronic-packaging-stress-related issues. He received his M.S.M.E. from West Virginia University at Morgantown.

Thomas DiStefano is Vice President of Marketing and Chief Technical Officer of Tessera, Inc. Prior to founding Tessera in 1990,

Dr. DiStefano spent 20 years with IBM Research where he was a senior manager in manufacturing research for packaging and semiconductors. He headed the Measurement Science and Technology Department which provided manufacturing technology assistance to 22 IBM sites. Dr. DiStefano holds a Ph.D. in Applied Physics from Stanford University. He has authored or coauthored more than 28 patents and 46 technical journal articles.

Joseph Fjelstad is a consultant and senior engineer with Tessera. Prior to joining Tessera he had been a full-time consultant in electronic interconnection manufacturing technology with more than 23 years of national and international experience in the industry including nearly a year at a research center near Moscow, Russia. In his earlier career he worked as analytical chemist, manufacturing chemist, process engineer, R&D manager, and as Educational Director of the institute for Interconnecting and Packaging Electronic Circuits (IPC). Mr. Fjelstad is inventor or coinventor of three patented inventions and currently has several other invention patents pending. He has written and presented numerous papers on electronics fabrication and is a frequent instructor at industry workshops. He has authored or coauthored chapters for three different books on electronics fabrication prior to this book and, in addition, is author of *Flexible Circuit Technology*. Currently, Mr. Fjelstad is serving as Vice Chairman of the IPC Design Committee as well as serving on the board of editorial advisors of *Electronic Packaging and Production* magazine.

Elke Zakel received her M.S. and Ph.D. degrees in Material Sciences from the Technical University of Berlin in 1989 and 1995, respectively. She joined the Microperipherics Technology Center of the Technical University Berlin in 1989, where she worked as a research scientist and project coordinator on several joint projects concerned with TAB, flip chip, green tape ceramic multilayers, and thin-film multilayers. Her main working area is the development of new chip interconnection technologies and the qualification of these techniques for industrial applications. In 1992 she became the head of the group Packaging and Reliability at the Microperipherics Center. In August 1993 she joined the Fraunhofer Einrichtung für Zuverlässigkeit und Mikrointegration Berlin (IZM) where she is the head of the Chip Interconnection and Reliability Department. She is author of several publications concerned with bonding techniques and reliability of flip chip, wire bond, and TAB interconnections. In 1991 and 1992 she obtained the best paper award from the International TAB Symposium (ITAB) in San Jose. She is a member of the IEEE, CHMT, and ISHM.

Herbert Reichl is the director of the Microperipherics Technology Center at the Technical University of Berlin and of Fraunhofer Einrichtung IZM Berlin. He received his M.S. and Ph.D. in Electrical Engineering from the Technical University of Munich, Germany. From 1971 to 1987 he was with the Fraunhofer Institute for Solid State Technology in Munich. Since 1987 he has been professor at the Technical University of Berlin, Germany, and director of the Microperipherics Technology Center. In 1993 he became the head of the newly established Fraunhofer Einrichtung IZM Berlin. He is author of several publications on packaging technologies for IC's and reliability of chip interconnections. He is member of IEEE, CHMT, and ISHM. Prof. Reichl is the general chairman of the SMT/ASIC/Hybrid Conference and MICROSYSTEM Technologies Conference in Germany.

Larry Gilg is a senior member of the technical staff at the Microelectronics and Computer Technology Corporation (MCC) where he has been the Known Good Die Program Manager for the past 3 years. Mr. Gilg received his B.S.E.E. from the University of Nebraska and his M.S. in Engineering from the University of Texas at El Paso. Mr. Gilg has over 23 years of engineering experience with 17 years devoted to the design and development of instrumentation and electronic systems. Since joining the technical staff at MCC in 1985 he has headed efforts to develop and evaluate test probes for advanced microelectronics packaging. Mr. Gilg is a registered professional engineer in Texas and California.

Glenn A. Rinne received a B.S. in Electrical Engineering in 1978 from Rochester Institute of Technology, Rochester, New York. From 1978 to 1990 he was employed at Data General Corporation, Portsmouth, New Hampshire. At Data General he was involved with test process development and failure analysis of printed circuit boards and integrated circuits. From 1988 to 1990 he was a member of Data General's Advanced Packaging group investigating multichip modules. In 1990 he joined the Microelectronics Center at MCNC as a member of the technical staff in MCNC's Interconnection and Packaging Technologies group. At MCNC he is developing test methods and strategies for multichip modules, flip chip, and known good die. He is currently the Principal Investigator for the Flip Chip Technology Center funded by ARPA.

Index